T0277067

VECTOR

VECTOR

A Surprising Story of Space, Time, and Mathematical Transformation

ROBYN ARIANRHOD

(THE UNIVERSITY OF CHICAGO PRESS • CHICAGO AND LONDON)

The University of Chicago Press, Chicago 60637
The University of Chicago Press, Ltd., London

Published 2024
Printed in the United States of America

33 32 31 30 29 28 27 26 25 24   2 3 4 5

ISBN-13: 978-0-226-82110-8 (cloth)
ISBN-13: 978-0-226-82111-5 (e-book)
DOI: https://doi.org/10.7208/chicago/9780226821115.001.0001

Library of Congress Cataloging-in-Publication Data

Names: Arianrhod, Robyn, author.
Title: Vector : a surprising story of space, time, and mathematical transformation / Robyn Arianrhod.
Description: Chicago ; London : The University of Chicago Press, 2024. | Includes bibliographical references and index.
Identifiers: LCCN 2023053637 | ISBN 9780226821108 (cloth) | ISBN 9780226821115 (e-book)
Subjects: LCSH: Vector analysis—Popular works. | Vector analysis—History. | Calculus of tensors—Popular works. | Calculus of tensors—History. | BISAC: MATHEMATICS / Vector Analysis | SCIENCE / History
Classification: LCC QA433.A75 2024 | DDC 515/.63—dc23/eng/20231201
LC record available at https://lccn.loc.gov/2023053637

♾ This paper meets the requirements of ANSI/NISO Z39.48-1992 (Permanence of Paper).

To Morgan

with all my love and heartfelt thanks.

CONTENTS

PROLOGUE

Every now and then, there's a spectacular breakthrough in our understanding of the world. The upending of the belief in our central place in the solar system, for example, and the relativistic revolution that changed the way we see time and space—and the way we see our place in the cosmos itself. The dramatic discovery of electromagnetic radio waves, which led to wireless technology and its wondrous transformation of our everyday lives—and the quantum revolution, with its seemingly magical new microtechnologies and its preternatural shattering of our notions of "reality." And then, of course, there's the digital revolution that is still changing the way we communicate with each other, especially now that AI has become so sophisticated. These breakthroughs have brought into being new technological and cultural eras, and much has been written about them. It's less well known, however, that these scientific and technological paradigm shifts went hand in hand with equally dramatic mathematical revolutions. This is a story about some of those unsung revolutions: the fascinating ideas behind them, and the people that made them possible.

At its core, the story I'll tell here is about the evolution of the way we humans record and make sense of all the data that swirl around us. In particular, I'll explore the dramatic mathematical transformations that gave us the remarkable concepts called "vectors" and "tensors," for they underlie so much of modern science—and much of our technology, too. They are languages that have helped us uncover mysteries of the universe as if we were gods.

One reason for this awesome power is that vectors and tensors made it possible to handle the dimensions of space in a new and transparent way—and this, in turn, made it possible to discover new laws of nature, and new technological applications of these laws. Anytime you want to utilize locations in space you need to handle these dimensions—rotating a robot arm, say, or designing a bridge or wind turbine; figuring out the effect of an electromagnetic force in a motor or generator, say, or predicting the path of an electromagnetic wave, a water wave, or even a gravitational wave; plotting the trajectory of a satellite or calibrating a guidance system such as GPS; or just about anything you need to analzye in space or space-time.

We'll see in more detail the physical power of vectors and tensors as the story unfolds. But these languages are not just about physical dimensions—they're about "dimensions" of information, too. You've likely read about "big data" and the information revolution, but it is vectors and tensors that help make data usable and comprehensible—the way the periodic table of chemical elements is both an organizational and theoretical tool in chemistry, except that the math in our story is so much more widely applicable.

Yet vectors and tensors themselves are remarkably simple—on the face of it, at least—for you can, indeed, begin by thinking of them simply as neat ways of representing information. For instance, you might remember from school that a vector can encode information about both the size *and* the direction of a physical quantity—say, a velocity or a force. So, you can represent it with an arrow pointing in the required direction, while the arrow's length gives the size or "magnitude." Tensors add in more layers of information, so they are like multidimensional arrays rather than arrows. But when mathematicians discovered the rules for how these arrows and arrays combine with each other, they realized they had found a brand-new language for thinking brand-new thoughts. And this is a rather wonderful idea.

A simple illustration of what I mean by this is that for thousands of years mathematicians worked only with numbers. The evolution of real number systems was remarkable enough, but these numbers express only one thing: quantity—the magnitude of a weight, height, distance, amount of money, number of apples, and so on. Vectors and tensors, on the other hand, encode several things at once, which is why they are such great ways of representing a lot of data. And this extra information means that vectors and tensors can offer a far richer picture of an industrial or IT problem, say, or a physical model, than a single number ever could.

The first major physicist to recognize the power of vector language was the gently eccentric nineteenth-century Scottish laird James Clerk Maxwell. We'll meet him properly later, but his theory of electromagnetism was the very first modern field theory, which he used to crack the long-standing riddle of the nature of light and to predict the existence of radio waves—all in one fell swoop. His initial theory was intuitively "vectorial," but once he learned that vectors were actually "a thing," with their own mathematical rules, he realized they were the right tools for expressing his discovery more succinctly and elegantly.

Not that many people took him seriously at first: as we'll see, his breakthrough application of "vector fields" to represent nature's electromagnetic fields was . . . well, just too *mathematical,* too "unphysical," for mainstream physicists. But if it was difficult enough for Maxwell to achieve recognition for his superb *application* of vectors, imagine the passion and self-belief the *creators* of this mathematical language must have had.

One of the many stars of this story is the Irish mathematician William Rowan Hamilton. He's the one who coined the term "vector" and first presented its mathematical theory—and he knew right away that he'd created something so new it broke a rule mathematicians had taken for granted for thousands of years. Yet when he glimpsed the possible applications of this new language, six years before Maxwell published his marvelous theory, he was over the moon. He wrote joyously to a colleague who was working with him on the subject, "*Could* anything be simpler or more satisfactory? Do you not *feel,* as well as think, that we are on a *right track,* and shall be *thanked* hereafter? Never mind *when.* . . ."[1] Hamilton was speaking not only about vectors here, but also his invention of *quaternions,* four-dimensional "numbers" that contain vectors. As we'll see, quaternions can do everything that vectors can do, but they are more efficient in programming certain tasks in spacecraft guidance and image processing, to mention just two modern applications.

Poor Hamilton never did receive sufficient thanks: he died in 1865, just a few months after Maxwell published his theory of electromagnetism but before he'd been able to recast it into full vector language.[2]

As for tensors, Maxwell died before they were developed, but I'm betting he would have recognized their power, too. He died the same year that Einstein was born, which is especially symbolic—not only because Einstein's theories were inspired by Maxwell's, but also because Einstein did for tensors what Maxwell had done for vectors: he was the first major physicist to show their practical power. They enabled him to create curved space-times, and the discipline of modern cosmology—and they enabled him to predict the existence of gravitational waves and lenses, and to accurately quantify the gravitational effect on time that is now used to make GPS directions so accurate.

It took experimental physicists a quarter of a century to verify in the lab Maxwell's prediction of radio waves, and it took a hundred years to detect Einstein's gravitational waves. That's an indication of how far ahead of the game these vector- and tensor-based theories were. This kind of ability to make predictions is one of the exciting things about mathematical language. It's as if the act of describing physical reality mathematically creates a magnifying glass, revealing, through mathematical patterns, underlying physical attributes that had long lain hidden. We'll see specific examples of this as we go, but here I'll just add that quantum theory also makes fine use of vectors and tensors—and, so far, none of its predictions have been disproved.

Since vectors and tensors are ways of storing and using information, they're useful far more widely than in physics alone, of course. As I intimated earlier, they are playing a fundamental role in a growing number of areas that need to handle a lot of data—from engineering and genetics to search engines and artificial intelligence, with much more in between.

Yet the development of the full power of these mathematical ideas was so astonishing, and so far-reaching, that I'm treating their discoveries as revolutions. It's helpful to think of tensors as a generalization of vectors, but that is hindsight: it took three hundred years to move from a fledgling form of vector language to a sophisticated language that incorporates vectors and tensors in a rigorous way. And to get to that first, nascent hint of the vector concept, it had already taken many centuries—millennia, in fact, if we go back to the oldest surviving mathematical records. For the history of vectors and tensors is linked with the history of the symbolic representation of data, and these ancient documents show that finding ways to represent information is at the heart of the story of mathematics itself.

So, I'll begin this tale with a brief tour back to the beginning of it all. Of course, my telling of this long history, here and as the story unfolds in later chapters, cannot be exhaustive. It's necessarily selective and subjective. One of my goals is simply to show just how long it takes—and how much intercultural cooperation is needed—for sophisticated mathematical ideas to develop. It was a long and winding road to the modern vector and tensor analyses that are so widely used today, and the story I want to tell is a journey of ideas—ideas that are often surprising, sometimes mundane, but which have always been, right from the beginning, rich with possibility.

Still, if at any point throughout the book you find you want to skim the details and get on with the story, that's fine, too!

BACK TO THE BEGINNING

It was some time about five thousand years ago that people living in the area around present-day Iraq began to write down information by scratching wedge-shaped signs into clay discs or sheets. These strange signs are known as "cuneiform" script, and the economic and administrative power of being able to record and control the exchange of tangible things such as goods and land must have seemed amazing. But it took another thousand years—and the help of computational tools such as fingers, pebbles, and eventually abaci and tables—for *abstract* number systems and the rules of arithmetic to develop.

The inventors of this cuneiform mathematics lived on the fertile plains between the Tigris and Euphrates Rivers—an area that the Greeks, a thousand years later, would call Mesopotamia (or "between two rivers"). It hosted a number of linked cultures, so the term "Mesopotamian" is still generally used to describe the ancient mathematical and other cultural innovations developed in this agriculturally and intellectually rich region. Of course, it was not the only place to move from simple counting to sophisticated numerical arithmetic. But we know so much more about Mesopotamian mathematics than that of other early cultures, simply because so many of those remarkable clay documents survived.

Some of the earliest of the more sophisticated numerical and mathematical tablets, dating from nearly four thousand years ago, contain multiplication tables. It might seem surprising that such things have been around for so long, but of course you need to be able to add and multiply to carry out basic economic tasks. Historians have gained insight into the nature of those early tasks, because these functions, too, were recorded on clay tablets. Tellingly, some of the oldest of these documents contain tables giving lists of the lengths of the sides of various square or rectangular fields, each matched with their areas—the kind of tabular layout that would later morph into mathematical matrices. Simple lists of information would morph into vectors—but more on all this mathematical morphing later, for it will show us how mathematicians went from simple accounting lists to modeling such complex things as electromagnetic waves, for example, or the qubits that underpin quantum computers. Meantime, these ancient tables were vital for working out potential grain harvests, seed requirements, the amount of labor needed to work the fields, and the wages and taxes to be paid—the kinds of things that any large society needs in order to create and distribute food and other necessities.[3]

In the earliest of these advanced Mesopotamian societies, the estimates of field sizes needed for the economy to run properly didn't need to be exact. Surveyors could mark out fields with pegs and ropes, and then measure their sides, but they didn't have to worry about making the land parcels *exactly* rectangular because the state owned it all anyway. From about 1900 BCE, however, things began to change, and soon ordinary folk could own land, too—and this meant that surveying needed to become more accurate, because land disputes were soon to begin their long history. (By contrast, many First Nations people practiced the old way right up until colonization disrupted the balance.) And so, hundreds of years after the earliest multiplication tables had helped accountants calculate the areas of approximate squares and rectangles, Mesopotamian surveyors worked out how to make perfect 90° corners—which suggests they may have discovered "Pythagoras's theorem" a millennium before Pythagoras.

Generations of schoolchildren have chanted this ancient rule: "the square of the hypotenuse equals the sum of the squares of the two adjacent sides." Pythagoras lived in the sixth century BCE, but cuneiform tablets dating from around 3,700 years ago—most famously those labeled Plimpton 322 and Si.427—contain compelling evidence that the rule is much older than its Greek namesake. The Plimpton 322 tablet (fig. 0.1) has been broken, but the surviving fragment lists fifteen pairs of numbers relating to the diagonal and the shorter side of a right-angled triangle. The choice of numbers used, and the headings in the table's columns, suggest that this was probably a list of implicit, sexagesimal "Pythagorean triples," as they are now called—a set of integer (whole number) triples that surveyors could choose from. For example, (3, 4, 5) is a Pythagorean triple, because $3^2 + 4^2 = 5^2$. The tablet Si.427 supports this interpretation, for it is a plan of the private subdivision of a land parcel into rectangular and triangular fields—each with perfectly rectangular corners and dimensions that fit Pythagoras's theorem.[4]

A thousand years later, ancient Greek-speaking mathematicians were interested in measuring a range of angles, not just 90°, because they wanted to survey not only the earth but the stars, too. Along with mathematics, astronomy is the oldest science. After all, a vast and dazzling night sky is a wondrous thing. Since they had no way of measuring the distance to the stars, these ancient Greek astronomers figured out that you could pinpoint them by measuring their angles—and so they also discovered two key things. First, trigonometry. Which is not to say that earlier cultures didn't have

FIGURE 0.1. Plimpton 322. This remarkable tablet is labeled 322 in the G. A. Plimpton Collection at Columbia University. Unfortunately, this tablet also represents colonial looting, because Plimpton bought it from an archaeologist-cum-antiquities dealer in 1922. Photographer unknown. Wikimedia Commons, public domain.

some form of angular tabulation and "trigonometric" calculation, too—historians are still debating and interpreting cuneiform texts. But the oldest extant, explicit trig table survives in Claudius Ptolemy's extraordinary 1,850-year-old compilation of Greek mathematical astronomy, *Almagest.* (If you've forgotten the basic idea of trigonometry, you can look ahead to fig. 3.4 in chap. 3.) Second, the Greeks developed the idea of representing positions in space with coordinates—a brilliant innovation that has a lot to do with the emerging idea of a vector.

Before I talk a little more about this, though, in the next couple of paragraphs I want to acknowledge the other ancient mathematical cultures that also connect with our story. For instance, Greek-speaking Ptolemy lived in Alexandria, Egypt, under Roman rule—a reminder of the brutal upheavals that tend to follow as empires expand and fall, and also of the fact that in math history "Greek" refers to the work of Greek-speaking mathematicians, no matter their race, ethnicity, or where they lived. A reminder, too, of the famous female Alexandrian mathematician Hypatia, who wrote a commentary on *Almagest* three centuries later. And while I'm emphasizing modern mathematics' multicultural history, I should note that *Almagest* is an Arabic word meaning "the greatest," a term that had long been the informal adjective used to describe Ptolemy's compilation, to distinguish it from others of the time. This truly "greatest" book survives today largely because of its medieval Arabic translators and annotators, so it is fitting that it has been known ever since by its Arabic name.

Like the Mesopotamians, ancient Egyptians had begun to represent numbers some five thousand years ago—in their case, in hieroglyphic symbols—and they, too, were interested in surveying the land. And of course the Greeks were not the first to survey the sky: the Mesopotamians, Egyptians, and many other ancient peoples had charted the paths of the moon and stars so that they could devise calendars, propitiate gods, navigate their way across land and sea, and notice important coincidences—such as the fact that each year the Nile would begin to flood when the bright star Sirius rose just before dawn.

It seems that Egyptian and Mesopotamian astronomical records were arithmetical rather than trigonometrical—they charted the number of days between celestial events, rather than the angular positions of stars and planets at different times. But these records were so accurate that later Greek-speaking astronomers were happy to make use of them. Most cultures from this time did not leave written records of their astronomical observations, but some still survive in stories—such as the Torres Strait Islander tradition of Baidam (or Beizam), the celestial shark. This "shark" is the constellation the Greeks called "the bear," aka Ursa Major or the Big Dipper, and its position in the night sky told the people when to plant crops—and when to stay out of the sea because the sharks were breeding! Then there's Stonehenge, New Grange, and other ancient astronomically aligned ceremonial structures, including the Wurdi Youang stone circle in Wathaurong country, Australia, which is likely thousands of years older than its Middle Eastern and European counterparts. The Chinese, Indians, Mayans, Incas, and many other people, too, developed an impressive knowledge of astronomy—and in 2017, the International Astronomical Union acknowledged the contributions of all these ancient cultures by including eighty-six indigenous star names in a new star naming system.[5]

As for ancient contributions to mathematics, we'll see a little more of those in the next two chapters. But now let me return to coordinates, and how they relate to vectors. Which brings me back to Ptolemy, who left us the earliest sophisticated record not just of mathematical astronomy, but also of the use of coordinates to represent positions in space. I cannot give him all the credit, for his astronomy or for his coordinates—it takes a long time, and contributions from many people, for brilliant ideas to come to fruition. But Ptolemy was very good at giving his forerunners due recognition—which is just as well for them, since most of their work was later lost to mold, pests, accidents, war, and the fact that *Almagest* was so successful that the earlier works were discarded.[6] Euclid of Alexandria is a notable exception, for the content of his famous *Elements* survives in full. Euclid composed this book around 300 BCE, some 450 years before Ptolemy's *Almagest,* and it gave us the rules we still use for understanding everyday (flat) "Euclidean" space.

As far as the surviving records go, though, it is Ptolemy who introduced the idea of latitude and longitude that we use today for locating places on Earth. In his book *Geography,* he lists thousands of geographical locations—and in *Almagest,* he has a star catalog of more than a thousand stars, located via celestial analogs of latitude and longitude. All these coordinates are angles—and to measure them, Ptolemy described how to find the sun's angle with a shadow stick and a graduated dial like a protractor, while to locate the other stars he used an armillary sphere. (This is a complex network of movable rings forming a skeletal globe that encloses a tiny Earth; the rings represent such things as the apparent yearly and daily orbits of the sun, the observer's horizon and meridian—which are related to the observer's latitude and longitude—as well as a ring marked with a degree scale and another with sighting holes to the stars.)

It took another twelve hundred years to get an inkling of the kind of coordinates on graph paper that we're used to in math classes today—particularly through the work of the fourteenth-century French mathematician Nicole Oresme (although Euclid's near-contemporary Apollonius of Perga had also glimpsed the idea). Three centuries later, another Frenchman, René Descartes, came closer still, and, following his lead, later mathematicians firmly established the familiar *x-y* "Cartesian" coordinates, whose axes meet in the middle, at the "origin" of the grid. And now we are homing in, in this short preliminary overview, on the idea of a vector.

So, I want to spend the rest of this introduction showing a little more about what modern vectors are, and what they can do. First, though, I want to reiterate the importance of vectors' long, multicultural prehistory that I sketched in the preceding pages, because most of this book will focus on the work of the modern European men who invented calculus, vectors, and tensors. Math today is a truly universal endeavor, but in the nineteenth century when much of our story takes place, the United Kingdom (which then included Ireland) was especially well placed to be a scientific powerhouse. It had long-established universities along with a growing number of Mechanics' Institutes and Working Men's Colleges for adult education. (Working women finally got their own college in 1874, and later we'll meet some of the women who studied at the first University Women's Colleges.) It also had a well-established banking system that enabled entrepreneurial technological investment, it had a growing railway network that facilitated communication as well as markets, and it had abundant raw materials—its own coal, water, and steel, as well as materials plundered from its colonies. These factors underpinned the Industrial Revolution of the eighteenth and nineteenth centuries, in which science and technology went hand in hand, each fueling the other—and not only in Britain, of course. Europe had similarly well-developed institutions and scientific communities—especially, for our story, France, Italy, and Germany. As we've seen, though, and as we'll see in the next two chapters, these modern scientific communities owed much to their ancient and medieval multicultural forebears.

A QUICK LOOK AHEAD: IMAGINING NEW DIMENSIONS, AND NEW WORLDS

You may remember from school that you can represent the direction from the origin O to a point P on a coordinate grid by drawing an arrow, as in figure 0.2 below. This is an example or model of a vector—it's a "position" vector, representing P's position in terms of its direction and distance from O. But the idea of something that encodes more than one thing—something beyond a single number—is actually quite sophisticated, and this story will take us through how it all came to be. And as we'll see, it took a very long time to create the idea of a single symbol having both magnitude and direction.

So, the next couple of pages are just a quick look ahead, and perhaps a reminder of two basic ideas you may have learned but didn't realize were so groundbreaking. We'll meet them in more detail later, but I mention them here to lay tracks for some terminology and concepts we'll need on the journey—and also to give a glimpse into why vectors are so important today. Tensors, vectors' multidimensional cousins, will come much later in the story, for they will build on the concepts and symbolism developed for vectors—and the journey to vectors is unexpectedly layered. But as we explore these layers, we'll also be building up ideas that underpin tensors, too.

The first idea is notation. Mathematical symbols can sometimes seem mysterious, so one of my aims in this book is to share with you some of the conceptual breakthroughs that made it necessary to invent new symbols for new ideas. Even those familiar x's and y's were a long time coming, as we'll see in the next chapter. We won't get to the inauguration of modern vector symbolism for several chapters—which goes to show just how long it took for these ideas to evolve. For now, though, I'll just lay tracks that connect this story with what you may already know. For example, position vectors are often denoted by $\mathbf{r}$, $\vec{r}$, or $\hat{r}$. Even if you're not *au fait* with vectors, you might remember seeing textbooks mixing up signs like this with ordinary letters and numbers. That's because boldface type—or "hats" or arrows on top of letters or squiggles underneath them—are common ways of denoting vectors today, to distinguish them from ordinary letters that represent numbers, magnitudes without direction. So, in this case r, with no bold type or hat or other marker, would represent just the magnitude of the distance, with no reference to the direction; quantities such as this are called "scalars," as opposed to "vectors." Another example

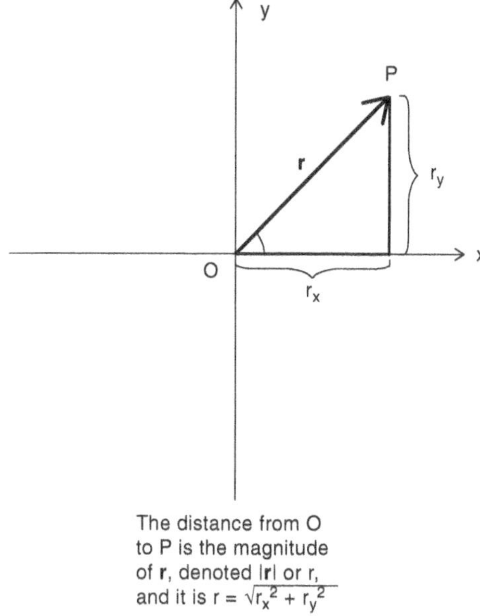
y
P
r
r_y
x
O
r_x
The distance from O
to P is the magnitude
of **r**, denoted |**r**| or r,
and it is $r = \sqrt{r_x^2 + r_y^2}$

FIGURE 0.2. *P*'s position vector ***r***, with components of magnitude r_x and r_y. The magnitude of ***r***—the length of the arrow, representing the distance from *O* to *P*—is denoted by $|\boldsymbol{r}|$ or *r*, and it is found from the components using Pythagoras's theorem.

of this notation is ***v*** for velocity, which includes direction as well as speed, and *v* for speed—the magnitude of the velocity.

The second key notion is that a vector has "components." In the two-dimensional space of a page in a book, say, or a piece of graph paper, you have two independent spatial directions—commonly represented as the *x* and *y* directions or axes—so your points have two coordinates and the arrow between them has two components. In other words, components are the values you get for your arrow when you measure it from each axis, as you can see in figure 0.2.

Similarly, in three-dimensional space, there are three separate directions and three components. Then there's space-time, where you have four coordinate axes, three spatial ones and a time axis, and your metaphorical arrows have four components—so your position vector would measure not only the distance traveled but also the time it took. It can be a bit trickier than simply measuring the three spatial components and reading off the time component from your wristwatch—because time and space coordinates change, in an intermingled way, if you're moving relative to what you want to measure. It's even more complicated when you're in a gravitational field—the province of general rather than special relativity. I'll talk more about relativity later, but here I just want to note that vectors in space-time have four components. (I should also note that this intermingling of space and time is why the spelling "spacetime" is perhaps more common than "space-time" in modern physics papers, because it shows that space and time are interwoven—they are not just two different things grafted on to each other. But both spellings are used in math and physics texts today, and the hyphenated version is the one used by the founder of the concept, Hermann Minkowski, and by Einstein. We'll meet Minkowski and Einstein later.)

One of the things that will prove extremely controversial in this story is the subtle difference between the arrow, so to speak, and its components. When the subtlety is finally grasped, the way will open for the creation not just of modern college-level vector math, but of tensors, too. We'll see that the confusion arose partly because you can also represent vectors simply by listing their components, and this is easier and more economical than drawing an arrow—especially when you want to move out of three dimensions. For example, suppose you're driving at 35 mph in a northerly direction (along a flat, two-dimensional street—mathematicians love to explore new ideas by starting with simplifying assumptions!), and you choose the north-south direction to be the y axis. Geometrically, your velocity vector would point upward along the y axis in figure 0.2, and it would be 35 units long. But you can also write this velocity as $\boldsymbol{v} = (0, 35)$, meaning you're moving 35 mph along the y axis and 0 mph along the x axis. So, the scalar, 35, gives your speed (in units of mph), and the vector, (0, 35), shows that the speed is in the northerly (y) direction. (Labeling (0, 35) with a vector symbol, as in $\boldsymbol{v} = (0, 35)$, distinguishes it from a single point on a Cartesian grid—so you can see the importance of notation.) You can represent any velocity like this, no matter the direction. For example, if you were travel-

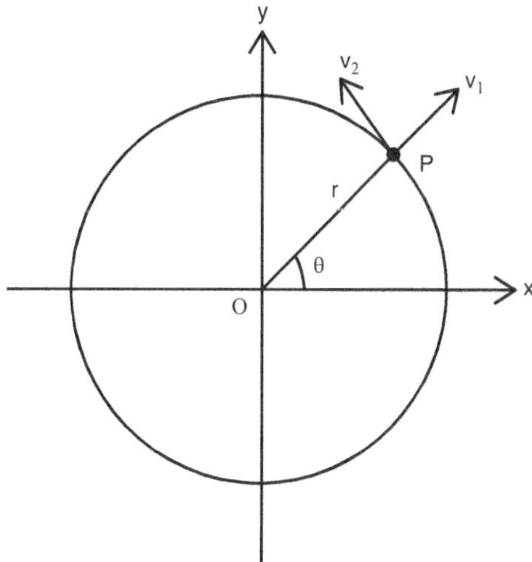

FIGURE 0.3. Polar coordinates of a point P on a circle of radius r; θ is measured anticlockwise. The velocity components are measured in the r direction and in the direction of increasing θ, as shown.

ing, say, *northeast* at 35 mph, your vector components would be $\left(\frac{35}{\sqrt{2}}, \frac{35}{\sqrt{2}}\right)$ (as I've shown in this endnote[7]).

More generally—and skipping blithely, for the moment, over the long road from numbers to symbolic algebra—you might represent velocity components in three-dimensional space by (v_x, v_y, v_z), indicating that you're going v_x mph in the x direction, v_y mph in the y direction, and v_z mph in the z direction. Even more generally, the components of velocity in 3-D space can be denoted by (v_1, v_2, v_3), because the Cartesian x, y, z coordinate system is not the only one available. For instance, in two dimensions polar coordinates (r, θ) are useful for locating points on the circumference of a circle, as you can see in figure 0.3; the vector (v_1, v_2) would mean you're traveling v_1 mph in the r direction, v_2 mph in the θ direction. To locate a point on a sphere, you can extend this to three dimensions using spherical coordinates (r, θ, φ), where φ is the angle between the vector and the z axis. But the details are not so important here: it's the story behind this symbolism that is so interesting—and, of course, the applications.

For example, in four-dimensional space-time, Einstein labeled his coordinates as (x_1, x_2, x_3, x_4), where the first three are the usual spatial coordinates and the fourth is the time coordinate. So, the components of a four-dimensional velocity vector (or "four-velocity"[8]) in this system would be labeled (v_1, v_2, v_3, v_4).

By analogy, it's possible to imagine any number of dimensions. This is the beauty of representing the information in a vector as a string of components: if you can write (v_1, v_2, v_3, v_4), what's to stop you writing $(v_1, v_2, v_3, v_4, v_5, \ldots)$ for as many (finite!) dimensions as you like? That's just the kind of thing string theorists have done when they suggest our universe might have not four but ten or eleven or even twenty-six dimensions—there are several different string theories. You can think of the extra dimensions as enabling space to accommodate the many ways in which tiny strings can vibrate, each way representing a different fundamental particle or force.

String theory is an example of the way the dimensions in a mathematical "vector space" can represent more exotic things than ordinary space and time. In quantum mechanics, to take another example, the axis of the magnetic orientation, or "spin," of an electron can be "up" or "down"—or a "superposition" of the two, as if the electron can't decide whether it wants its spin to be up or down. It's analogous to a two-dimensional velocity vector having one component in the direction x, say, and another in the direction y, so for electron spin you can let the two spin directions "up" and "down" be your axes.

Similarly, in computing you encode information with two binary digits ("bits"), 0 and 1, which are represented physically by, for example, turning an electric switch off and on. In quantum computing, the analog of a bit is the qubit (pronounced "cue-bit," short for "quantum bit"). Quantum 0's and 1's can be manifested physically as the two fundamental states of electron spin—0 might be represented by spin "up" and 1 by spin "down"—so qubits, too, can be represented mathematically as two-dimensional vectors.

Vectors are useful in modern business and tech applications, too. For instance, the axes or "dimensions" might represent different questions on a website questionnaire or political survey, or the different factors affecting house prices or other socioeconomic data, or the positions and colors of pixels in image processing. By the way, if you're familiar with image processing, you'll know that in some computing applications, the word "vector" can be used slightly differently from the usual mathematical meaning that I've been outlining. The "vector files" in image processing use equations to generate *shapes*, rather than vectors for storing *information* about such things as pixels. The shapes produced in these "vector files" do "carry" lines from point to point along the shape, and "carrier" is the original meaning of "vector." Looking at mathematical vectors as arrows, you can see how they, too, carry the line from the tail to the tip of the arrow. Still, the arrow, or more generally the string of its components, is the kind of vector used in most applications—in physics, business, *and* computing.

But William Rowan Hamilton didn't get so excited just because you can write down any number of components to represent various physical, digital, and economic systems or scenarios. Storing and representing data has been important ever since humans could count and write, but so has computation—and vectors and tensors enable you to do both the representing and the calculating at the same time. That's where the magic comes into it: despite the simplicity and utility of writing vectors as lists of components, what Hamilton was so thrilled about was that when you consider the *vector as a whole*, not just its individual components, the rules of vector arithmetic make vectors far more powerful than numbers alone.

For example, writing Maxwell's equations in whole vector form makes it relatively easy to deduce the equations for electromagnetic waves—a phenomenon whose very existence Maxwell had predicted. Magnetic field vectors and vector products also play a role in mathematically defining the spin of subatomic particles such as electrons (to take just one more example), thereby enabling spin to be applied in the real world—in magnetic resonance imaging (MRI), say, which is used to diagnose patients by "seeing" inside their bodies.

Hamilton couldn't have dreamed of such things. But once his vector arithmetic was out in the world, its patterns intrigued other mathematicians—especially those who like to see how far they can push the logical consequences of the rules and grammar of mathematics. And sometimes they push so far that they tumble into a new kind of reality, like Alice through the looking glass and down the rabbit hole.

BREAKING THE RULES

Not that Alice's creator, Lewis Carroll—aka Oxford mathematician Charles Dodgson—was one of those doing the creative pushing and tumbling in this case. It has even been conjectured that the Mad Hatter's Tea Party in *Alice's Adventures in Wonderland* is so delightfully absurd because it is a parody of Hamilton's rule-breaking vectors—although on the face of it, it's hard to tell whether Dodgson was delighted by them or thought them ridiculous.[9] Either way, Hamilton pulled the rug out from almost everyone when he discovered that, unlike ordinary arithmetic, there is more than one way to carry out multiplications in vector arithmetic. What's more, vector multiplication doesn't always follow the long-established rules of numerical multiplication, as we'll see.

Hamilton's seemingly absurd discovery about multiplication, and all the breakthrough vector and tensor developments and applications that followed, required the use of algebraic symbolism. I spoke of "vector arithmetic," but more generally this is "vector algebra." Speaking loosely, arithmetic works with numbers, and algebra works with letters that symbolize numbers or quantifiable quantities, such as temperature or speed. This symbolism enables mathematicians to generalize. For example, writing the basic rules of ordinary arithmetic symbolically—such as the commutative law, $a + b = b + a$ and $a \times b = b \times a$—enables you to see patterns that hold for all numbers, so you don't have to list each example separately the way the earliest mathematicians had to do. Hamilton had developed his rules of vector algebra by following, as best he could, these basic arithmetic patterns, so he was as shocked as anyone when he realized he had to broaden the rules for multiplication.

But symbolic algebraic thinking had been a long time coming—long after the Egyptians and Mesopotamians, the ancient Greeks and Chinese, the medieval Arabs and Indians, and all the other mathematical cultures before the seventeenth century. Even in the nineteenth century many mathematicians mistrusted purely abstract symbolic algebra, so unmoored from everyday experience did it seem. So, to set the scene for the challenges that Hamilton, Maxwell, and the other vector pioneers would face, the next chapter moves back from this overview to a story about how people began to represent and calculate with letters in the first place—and how this seemingly simple step has enabled mathematicians to think more creatively.

First, though, we'll walk in Hamilton's shoes for a mile or two, reliving the moment he had his wild, transgressive realization that the ancient rules of mathematics could, after all, be broken.

(1)

THE LIBERATION OF ALGEBRA

Every year on October 16, mathematically inclined Dubliners take a communal walk across the fields from Dunsink Observatory and down to the Royal Canal. The canal has many crossings, but they're heading for one in particular: Broome Bridge. They are celebrating Hamilton Day—that's William Rowan Hamilton, Ireland's greatest mathematician—and they're reenacting one of the most famous walks in the history of mathematics. They are commemorating October 16, 1843, when Hamilton—Ireland's royal astronomer—was on his way to preside over a meeting of the Royal Irish Academy.

Hamilton's wife, Helen, had joined him on his walk, but lovely as the setting was and much as he was drawn to Romanticism, this was not a romantic walk in any traditional sense of the word: he was much too preoccupied. He'd been wrestling for years with a seemingly intractable problem that simply would not leave him alone: how to represent *algebraically* geometrical operations such as rotations, in *three*-dimensional space—the kind of rotations you need to manipulate a robot, a realistic computer image, or a spacecraft, say, not that Hamilton was thinking about such high-tech possibilities. He was interested simply in solving the *mathematical* problem of 3-D rotations, so he was still trying to invent the math that would eventually help make these applications possible. And then, as he passed by Broome Bridge, it hit him. You could only do 3-D rotations with 4-D math.[1]

This was an intriguing enough insight, but there was more. To make his 4-D math work, Hamilton had to change the fundamental rules of multiplication. No longer could you assume the commutative law $a \times b = b \times a$, which is so obviously true when you multiply ordinary numbers. You might remember wrestling with the "right-hand rule" for vector multiplication, but we'll meet it properly in chapter 4, for it gives a neat way of showing that vector products aren't commutative. School and college vector analysis is hindsight, however. Back in 1843, this breaking of the so-called commutative law was astoundingly daring, even presumptuous: In what universe did it make sense to say that some analog of 2×3 does not equal 3×2? Perhaps in Lewis Carroll's Wonderland, where the Mad Hatter tells Alice that "say what you mean" is *not* the same as "mean what you say," for "you might just as well say that 'I see what I eat' is the same thing as 'I eat what I see'!"[2] But it certainly did not make sense in the intuitive mental world mathematicians had inhabited for several thousand years. Hamilton was so excited that he took out his penknife and scratched his magic formula right then and there on the stone bridge.

The site of this legendary piece of mathematical graffiti is the destination of those modern walkers on Hamilton Day. The original carving has long been lost to the wind and rain, but we know it existed from Hamilton's letters—and for the past half century, a plaque has marked the spot where he experienced the inspiration that came to him, as he later recalled, as if an electric circuit closed and a spark flashed forth. A 2019 commemorative artwork in the pavement by the bridge built on this electrical metaphor, illuminating with electric light the substance of the famous graffiti,

$$i^2 = j^2 = k^2 = ijk = -1.$$

If only Hamilton could have seen it![3]

You might have already spotted that the i, j, and k in Hamilton's formulae are "imaginary" numbers, for no "real" number squared can equal a negative number. (When you're taking the square of a real number, you're always multiplying two numbers of the same sign, so you always get a positive square. Which means that if you get a *negative* square, as in $i^2 = -1$, then such a number i is not a real number; rather, it's a so-called imaginary one.)

I'll explore the meaning of Hamilton's equations later, but for the moment, I want to emphasize the reason pilgrims from around the world continue to visit Broome Bridge. This seemingly simple line of equations contained the key to a new kind of four-dimensional construct, which opened up a whole new mathematical language that has become indispensable in all sorts of fields today. Hamilton gave the name "quaternions" to his 4-D creations; they contained two parts, a 1-D real number, which he called a "scalar," and a 3-D (three-component) quantity having magnitude and direction, which he called a "vector." If you're familiar with modern vectors, then for practical purposes you can think of Hamilton's vector in just the same way. (We'll see the subtle conceptual difference between them later. It will play a controversial role in our story.)

Two decades after that storied walk, James Clerk Maxwell created his vector field theory of electromagnetism. The more complicated quaternions had to wait nearly 150 years to find the real-world applications I began to list earlier—robotics, CGI, molecular dynamics, the rotations of our mobile phone screens, spacecraft control, and much more. Moonwalker Neil Armstrong, for one, knew all about the role quaternions now play in airplane and spacecraft navigation. He was an aeronautical engineer as well as an astronaut, and toward the end of his life he paid his own personal tribute to Hamilton during a visit to Dublin.[4]

Before we take a closer look at how Hamilton succeeded in his quest—and at what you can do with quaternions and vectors—I want to tell you how it came about that he could envisage such things as imaginary numbers and higher-dimensional math. In the prologue, I skipped through the move from three dimensions to four or more, but this transition in thinking was far from straightforward. That's because for thousands of years there had essentially been only two kinds of mathematics, arithmetic and geometry. Arithmetic was about counting concrete quantities, such as money, weight, and distance—and geometry, too, was visualized in a palpable way, through points, lines, planes, and shapes that you could draw on a two-dimensional page or represent in three-dimensional space. Three dimensions are as far as we can go if we want to physically model the kinds of things we can readily imagine with our everyday experience of physical reality.

So, what of Hamilton's 4-D math? How can you do such math—and how can it represent something as concrete as an object rotating in ordinary space? It's quite a long tale, because it took a very long time for mathematicians to learn to think abstractly—to free their imaginations from the tangible. And to do this, first they had to learn to think symbolically—to move from concrete arithmetic and geometry into the world of abstract, algebraic symbolism. So, if you ever struggled with algebra, take heart: this long history shows that generations of famous mathematicians have done so, too.

LEARNING TO THINK SYMBOLICALLY

Algebra has been part of mathematics since records began nearly four thousand years ago, but not always in the symbolic form we learn today. In fact, for most of those four millennia it was written entirely in words and numerals—although works such as Euclid's famous 300 BCE textbook *Elements* also included geometric diagrams, to help prove such things as Pythagoras's theorem, and to show how to expand squares that we would write today as $(a + b)^2$. So "algebra" was communicated in cumbersome word problems or increasingly complicated diagrams—although geometry did have its advantages. For instance, it's the easiest way to prove Pythagoras's theorem. In figure 1.1, I've given an algebraic adaptation of such a proof, although the ancients simply rearranged the diagram to show visually that the shaded area is equal to the sum of the areas of the squares on the adjacent sides of the triangle—a pretty clever approach![5]

It took a long time for algebra to emerge from arithmetic and geometry as a separate subject. It didn't even get its name until medieval

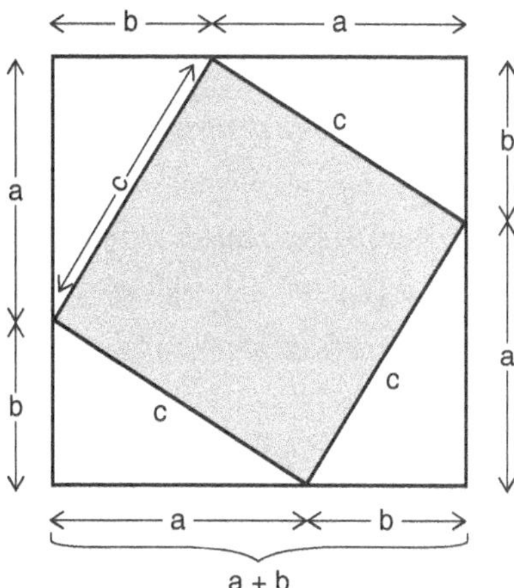

FIGURE 1.1. Proving Pythagoras's theorem. Four copies of the required right-angled triangle make up the four corners of the larger square. The hypotenuse of the triangle is c, so you find the area, c^2, of the shaded square by subtracting the total area of the four triangles (that is, 4 times $\frac{ab}{2}$) from the area of the larger square (whose sides are of length $a + b$): $c^2 = (a + b)^2 - \frac{4ab}{2} = a^2 + b^2$.

times, and that was thanks to the ninth-century Persian mathematician Mohammed ibn-Mūsā (al-)Khwārizmī. (The Arabic prefix al- was not part of Khwārizmī's Persian name, but since he worked in Baghdad, al-Khwārizmī is the name by which he became known.) He studied at Caliph al-Ma'mūn's pioneering Baghdad-based university, or "House of Wisdom," when the great Arabic translation movement was at its height: Greek, Indian, and other ancient manuscripts were being collected from all corners of the burgeoning Islamic empire and translated into Arabic. Imperialism is rarely ethical and often violent, but it can ultimately lead to cultural cross-fertilization, and in this case the visionary translation movement was so important that by the twelfth century, Europeans were learning Arabic in order to translate these manuscripts into Latin—including Ptolemy's *Almagest* and Euclid's *Elements*, along with new Arabic works such as those of al-Khwārizmī. The name "algebra" famously comes from the first word in the title of his book *Al-Jabr wa'l muqābalah*—which means something like "The Compendious Book on Calculation by Completion and Balancing."[6]

Judging from the problems al-Khwārizmī included, an example of what he meant by "Completion" is "completing the square," the method you might have learned in school to solve quadratic equations. Such equations play a part in the story of Hamilton's imaginary numbers. For example, consider $x^2 + 1 = 0$. Today's students can write down the solution immediately: $x = \pm i$, where i stands for the "imaginary" number $\sqrt{-1}$.[7] Frenchman René Descartes introduced the term "imaginary" to mathematics in the seventeenth century. He noted that "for any equation we can imagine" it is possible to find solutions, and yet, "in many cases no quantity exists that corresponds to what one imagines." In other words, he was saying that you can "imagine" solutions with square roots of negative numbers, but they don't really exist—at least, not as numbers in the traditional sense.[8]

If Descartes thought such numbers didn't exist, then it's not surprising that eight hundred years earlier, in al-Khwārizmī's day, the square root of −1 was considered "impossible," and the equations that gave rise to it were deemed unsolvable. So, like most of the ancients, al-Khwārizmī focused only on quadratic equations with positive solutions, for even negative numbers seemed to have no practical meaning. (The seventh-century Indian mathematician Brahmagupta was way ahead of his time, for he considered both positive and negative solutions.)

Al-Khwārizmī didn't write equations in the symbolic form we use today, either. In fact, to modern eyes his book is more arithmetical than algebraic, and one of its important impacts in Europe, when it was translated into Latin, was the popularization of the Hindu-Arabic decimal system of numeration that eventually evolved into our modern one. Yet Al-Khwārizmī is often called the "father of algebra." He may have used words rather than symbols, and the problems he included may have been simple—his purpose, he tells us, was to teach students how to solve basic problems in "cases of inheritance, legacies, partitions, lawsuits and trade, and in all their dealings with one another, or where the measuring of lands, the digging of canals, geometrical computation, and other objects of various sorts and kinds are concerned." But he systematically set out word-form linear and quadratic equations, with algorithmic methods for solving them—that is, for finding the "unknown numbers," our modern *x*'s and *y*'s. In fact, the English word "algorithm"—meaning a set of rules for performing a calculation or other operation—comes from "algorismi," an early Latinized attempt at Al-Khwārizmī.[9]

Such an epithet as "father of algebra" makes me want to ask, was there a "mother," too? Al-Khwārizmī didn't arise in a vacuum—not even Newton or Maxwell or Einstein did. Most of Al-Khwārizmī's predecessors have been lost to history, but perhaps there were women involved somewhere along the line. As it is, the mysterious Hypatia may be the closest we have to a known "mother," although the extent of her originality is not known, for only fragments of her work remain. But letters from the time show that she did write a learned commentary on another contender for the title of "father of algebra," the third-century Greek-speaking Alexandrian Diophantus, who took a significant step in the process of developing symbols when he used word abbreviations in setting out algebraic word problems.

In modern times, of course, we've had some fine female algebraists, including the University of Western Australia's Emeritus Professor Cheryl Praeger, who has been a contemporary mathematical role model for many young women. Moving back in time a little, there's the groundbreaking Emmy Noether, a colleague of Einstein. She has been dubbed the "mother of modern algebra," for her work on modern algebraic concepts that go way beyond the school-level algebra I'm talking about here. Half a century earlier—to take just one more pioneering mathematical woman—in 1872 Mary Somerville had been studying Hamilton's quaternions the day before she died, at the age of nearly ninety-two. She had been a famous expert on the latest developments in mathematical astronomy, and was known to her contemporaries as the "Queen of Science." She has become celebrated again in recent times, including—by popular vote, and beating fellow Scot Maxwell for the honor—as the face of the new polymer ten-pound note issued by Royal Bank of Scotland in 2017.

As far as the more distant past goes, my favorite "father of algebra" is the enigmatic Elizabethan mathematician Thomas Harriot. I'm in good company: in an 1883 letter to Arthur Cayley—an early admirer of Hamilton's quaternions—the British algebraist James Joseph Sylvester described Harriot as "the father of current algebra." Harriot, too, was indebted to earlier mathematicians, but his posthumous *Artis Analyticae Praxis* (*Practice of the Analytical Art*) is the first algebra textbook where the equations are written entirely symbolically, using essentially modern symbols. It was published in 1631, some eight centuries after al-Khwārizmī and fourteen centuries after Diophantus. That's an indication of how long it took for mathematicians to learn to think symbolically.

Harriot himself didn't publish any of his mathematical and experimental work: with the colorful and controversial Sir Walter Ralegh as his first patron, he was too busy sailing the high seas, dodging heretic hunters, and, above all, developing new mathematical applications to aid Ralegh's navigational enterprises. (In Elizabethan times, spelling was not yet fixed, and the American town named after Sir Walter is spelled Raleigh. The man himself used this spelling, too, but most often he used Ralegh, so this is the spelling many scholars prefer today.)

Although Harriot never got around to publishing his discoveries, he left behind thousands of pages of handwritten calculations and observations—his thoughts in progress as he searched for hidden patterns and generalities in the vast amounts of data and equations he'd collected. And that is, indeed, a key point of algebra: pattern and generality. The beauty of symbolic equations is that it's much easier to see these general patterns when you can see a problem at a glance. Compare this:

> Take the square of the unknown number, then add the unknown number to itself and take the sum away from the square; now let the total be eight.

with this:

$$x^2 - 2x = 8.$$

And there's more: the earliest mathematicians solved each equation separately, but it's easier if you can see that whatever method works for the equation $x^2 - 2x = 8$ will also work for any equation of the same form, $x^2 - ax = b$. Eventually, ancient mathematicians did begin to recognize this, but progress was relatively slow because they had to keep all these patterns in their heads, or in long, convoluted sentences, and it was easy to lose track.

The first to publish *any* equation in a transparent, recognizably modern symbolic form were Harriot's executors in 1631, and then Descartes in an appendix to his 1637 *Discourse on Method*.[10] (There were a few earlier attempts, but the symbolism—more properly called abbreviation—was tortured and idiosyncratic.)[11] Even the +, −, =, and × signs we take for granted only came into widespread use in the seventeenth century. Which means that the earlier algebraists we know of—the ancient Mesopotamians, Egyptians, Chinese, and Greeks, the medieval Indians, Persians, and Arabs, as well as the early modern Europeans—all had expressed their equations mostly in words or pictorial word images.

• • •

It is a singular skill to think symbolically, as this long history shows. Take the word problem I gave above: it is an example of algorithmic thinking. But symbolic thinking is algorithmic and more, for its symbols sometimes contain the seeds of a new kind of creativity—a new kind of far-reaching yet economical thought.

A classic case is Albert Einstein's $E = mc^2$. Einstein did not set out to find the connection between energy and matter. Rather, he simply wanted to calculate the kinetic energy of a moving electron according to his new theory of relativity, so that his theoretical prediction could be tested experimentally. A few months later, however, twenty-six-year-old Einstein began to realize the significance of his equation. He wrote it up in his fifth groundbreaking paper of 1905, his annus mirabilis, but it would take him two more years to tease out the full, dramatic implications of this symbolic relationship. To realize that this wasn't just a calculation about a particular form of energy and a particular type of matter, it was *general*: if a body gains (or loses) energy, it also gains (or loses) mass. This bizarre idea is alien to all our commonsense experience—but there it was, hidden in the symbols of his equation. It took experimental physicists decades to experimentally confirm this astonishing mathematical prediction.[12]

A much simpler and earlier example is the sequence of powers x, x^2, x^3, and so on. The first "power" is 1, so x is really x^1, where the 1 was traditionally linked geometrically to a 1-D line. The next two, x^2 and x^3, are pronounced "x squared" and "x cubed" by analogy with the area of a square and the volume of a cube. These names highlight the way that early mathematicians thought geometrically rather than algebraically, because of the tangible nature of geometry. By contrast, symbolic algebra is abstract: you have to give it meaning, even if it is simply the display of an interesting pattern such as x, x^2, x^3, x^4, . . . But this flexibility is algebra's great strength. You can write down as many (finite) higher powers as you like, *without* having to visualize them as physical objects.

This may sound obvious today, but it took three and a half *thousand* years for mathematicians to move from solving quadratic equations—"quadratic" derives from the Latin for "square," so quadratic equations are those whose highest power is x^2 (the unknown multiplied by itself, as the ancients put it)—to solving "cubic" and higher equations. These higher-degree equations are much more difficult, of course; but part of the reason solutions didn't come easily was that algebra was tied to words and concrete images for such a very long time.

For instance, I mentioned Al-Khwārizmī's "completing the square" in order to solve a quadratic equation. It's actually a four-thousand-year-old problem, dating back (as far as the historical record shows) to cuneiform tablets made by mathematicians living, like al-Khwārizmī, in the region of modern-day Iraq. These ancient Mesopotamians solved quadratic equations by literally completing a square. Here is a typical teaching problem of the time: "Add 20 of my length to the area of my square, [to get] 21. How square is my square?"[13] This type of problem, and the algorithm for solving it, is similar to those taught today—except that four millennia ago, the method was worked out entirely geometrically. First, draw a square

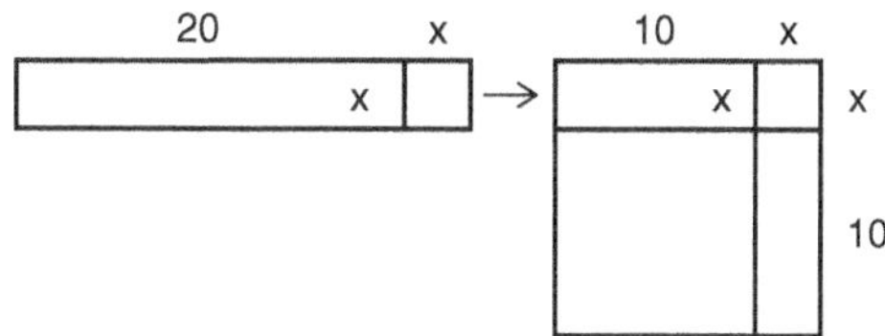

FIGURE 1.2. In modern algebraic symbolism, this ancient problem is the word form of the equation $x^2 + 20x = 21$. Like its ancient geometric counterpart, the algebraic algorithm for completing the square involves memorizing the step of adding "(half the coefficient of x) squared" to both sides, so that the equation becomes $(x + 10)^2 = 121$. Then the required side of the original square is $x = \sqrt{121} - 10 = 1$.

of arbitrary side x (in modern notation); then add to it a rectangle of dimensions $20 \times x$. Now split this additional rectangle into two equal smaller ones and arrange them beside and below the original square. Finally, complete this new, larger square, as in figure 1.2.

The Mesopotamians had practical problems in mind when they developed this method, at least initially. Living in a land where water was at a premium, their tablets contain many problems relating to canal and reservoir excavations, the capacity of cisterns, the construction and repair of dams and levees, and administrative accounts relating to these tasks—and to solve these problems, these ancient mathematicians had to solve equations relating to areas and volumes.[14] Nearly three thousand years later, al-Khwārizmī, too, focused on similar practical problems, and he used a similar geometrical method of completing the square—and so did other mathematicians right up to the seventeenth century.

The Islamic mathematician Sharaf al- Dīn al-Ṭūsī was one of the earliest to make progress in the search for solutions of *cubic* equations, in about 1200 CE, but the first to publish correct *general* cubic algorithms was the Italian mathematician Girolamo Cardano, in his 1545 book *Ars Magna* (*The Great Art*).[15] Like everyone before him, he still wrote his solutions in words (or abbreviations of words) rather than symbols, and he still devised his method geometrically—literally completing a cube in a stunning feat of visualization.

A MATHEMATICAL DUEL, A PESKY EQUATION, AND AN IMAGINARY NUMBER

As well as being a talented mathematician, Cardano was a physician, an astrologer, a gambler, something of a philosopher, and a mystic who believed that his best ideas came from a spirit who visited him at night. In the case of cubic equations, however, he received his inspiration from his countryman Niccolò Tartaglia rather than his faithful ethereal advisor. Cardano had heard that Tartaglia had cracked the problem, and he was so intrigued that he badgered him to reveal his method—he even offered to use his connections to put the impecunious Tartaglia in touch with influential people who might pay him for his work on such useful topics as ballistics. Tartaglia finally relented, on condition that Cardano keep the method secret—Tartaglia naturally wanted to publish it himself or, better still, offer it to a future patron.

Some years later, while Tartaglia was still holding onto his secret, Cardano discovered that Scipione del Ferro had also found the solution, before Tartaglia. So, Cardano felt he could break his promise and publish—he always had his eye on a publicity opportunity. But he fully acknowledged both men, and he went beyond them in solving a wider, more general range of equations. Still, Tartaglia was furious—so much so that he challenged Cardano to a public duel, not with swords but with a problem-solving competition. Cardano prudently refused: reputations (and jobs) were easily won and lost in these fiercely competitive Renaissance spectacles. Besides, Tartaglia had already taken on del Ferro's student Antonio Fior, who knew of his teacher's cubic method—and Tartaglia had won that match.

In his book, Cardano explained his general algorithm in a page of ingenious geometrical analogy and then gave specific illustrative examples. This is how he explained his method for solving $x^3 = 6x + 40$ (to use modern notation, which I'll also use to make Cardano's algorithm a little easier to follow; bear with me, even if you just skim through it, because the form of the expression in the last line has surprising relevance to the story of imaginary numbers and vectors): "Raise 2, one-third the coefficient of x, to the cube, which makes 8; subtract this from 400, the square of 20, [which is] one-half of the constant, making 392; the square root of this added to 20 makes $20 + \sqrt{392}$, and subtracted from 20 makes $20 - \sqrt{392}$; and the sum of the cube roots of these, $\sqrt[3]{20+\sqrt{392}} + \sqrt[3]{20-\sqrt{392}}$, is the value of x." Phew! You've got to admire his patience in coming up with something so convoluted.[16]

The interesting thing, from the point of view of the story of vectors—and of the development of mathematics in general—is what happens when the number under the square root sign in such a solution is negative. That is, when you have an imaginary number such as $\sqrt{-121}$.

The Mesopotamians had ignored negative and imaginary solutions of quadratic equations because they had no relevance to the practical problems they were trying to solve—you can't have negative or imaginary dimensions of fields and canals. Similarly for the Greeks, through to al-Khwārizmī and al-Ṭūsī, and right up until Cardano was forced to wrestle with these "impossible" numbers. He was studying the mathematics of equations simply for its own sake, for the intellectual challenge of it—and he was flummoxed by the fact that if he took the same method he'd used for $x^3 = 6x + 40$ and applied it to $x^3 = 15x + 4$, then the value of x turned out to be

$$\sqrt[3]{2+\sqrt{-121}} + \sqrt[3]{2-\sqrt{-121}}.$$

Cardano concluded that such a solution was "sophistic," and "as subtle as it is useless"—because aside from the unwelcome $\sqrt{-121}$, he already knew that in fact $x = 4$. He knew this because he would have begun to understand the problem by guessing the solution—something mathematicians have always done. It is especially useful when there isn't a known algorithm for solving a problem, so it is the way ancient algebra began. For Cardano's equation $x^3 = 15x + 4$, you can see the idea by trying a simple possible value such as $x = 3$; comparing each side you see that this is too small, so try $x = 4$. In this case it works straightaway, but sometimes you have to try intermediate values. This is still the way mathematicians solve difficult problems "numerically," although they have algorithms (and now computers) to choose their guesses efficiently and exhaustively.

Fifteen years later, around 1560, yet another excellent early modern

Italian algebraist, Rafael Bombelli, took another look at Cardano's conundrum. The question was, what did $x = \sqrt[3]{2+\sqrt{-121}} + \sqrt[3]{2-\sqrt{-121}}$ have to do with it, given that the solution was $x = 4$? After a great deal of thinking Bombelli suddenly had what he called "a wild thought": what if you could factor $\sqrt{-121}$ like this: $\sqrt{121} \times \sqrt{-1}$, to get $11\sqrt{-1}$? And could you then find what is nowadays called a "complex" number—a mix of real and imaginary numbers—whose cube is $2 + 11\sqrt{-1}$? Amazingly, with trial, error, and a lot of patience he found that $2+\sqrt{-1}$ is a solution of $\sqrt[3]{2+11\sqrt{-1}}$, as you can see if you multiply out $(2+\sqrt{-1})^3$. Similarly, he found that $2-\sqrt{-1}$ is a solution of $\sqrt[3]{2-11\sqrt{-1}}$. Adding these together as in Cardano's solution, you get

$$x = 2+\sqrt{-1}+2-\sqrt{-1},$$

which, seemingly miraculously, gives $x = 4$. Mystery solved!

It was solved only for this special case, though, and when Bombelli knew beforehand that $x = 4$—he'd had a brilliant insight about manipulating imaginary numbers, but he had no general algorithm. He didn't write his equations in the transparent modern symbolic form I've given here, either—and like Cardano, he disparaged $\sqrt{-1}$ as "sophistic." But he did put this strange number more firmly on the mathematical radar when his book *Algebra* was published in the 1570s. Little did he or anyone else know back then just how useful it would become—in Hamilton's quaternions and vectors, and in fields such as engineering, computing, and quantum theory.[17]

As for cubic equations, it was Harriot who first found general, *symbolic* algebraic solutions, sometime around 1600—and with no reference to geometry for his proofs. John Wallis—perhaps the best British mathematician between Harriot's time and Newton's—was one of the few near-contemporaries to recognize Harriot's achievement in liberating algebra from geometry, treating, as Wallis put it, "algebra purely by itself, and from its own principles, without dependence on geometry, or any connexion therewith."[18] This sets some mathematical context for where I'll be going with Hamilton, for he, too, wanted to find a way to represent purely *algebraically* a *geometrical* operation—in his case, not cubes but rotations in space, the problem whose solution will lead him to invent vectors.

Using algebra to envisage geometry expands not just algebra but geometry, too, and we'll see that these two kinds of math went hand in hand, each influencing the other, as vectors and tensors emerged. But the first step had been to see, as Harriot and Wallis did, that algebra was a subject in its own right, just as geometry was.

Harriot had taken his lead from the versatile early modern Frenchman François Viète, who had begun to use uppercase letters for unknowns, and whose treatise on cubic equations Harriot studied assiduously. Harriot used lowercase letters as we do today, and he used symbols so completely that he became a master of symbolic thinking. One of his insights was to show that polynomial equations can be generated by multiplying their factors—for example, two linear factors generate a quadratic, three give a cubic, four a quartic, and so on. This "factor theorem" may seem obvious now—you may have learned it in a senior high school algebra class—but no one before Harriot had written symbolic equations such as

$$(x - l)(x - m)(x - n) = 0.$$

Actually, Harriot didn't use separate round brackets for products, but wrote the factors one on top of the other with a square bracket around the group. And he tended to use a rather than x for the unknown, and aa instead of a^2. We owe the notation x, x^2, x^3, x^4, . . . to Descartes, who published it in 1637, although he still sometimes used xx and even aa, like Harriot. Either way, what this equation hints at is that a *cubic* equation has to have *three* solutions, $x = l$, $x = m$, $x = n$, whether they are positive or negative, real or imaginary. By contrast, Cardano's algorithm had spoken of "the" solution, as if there were only one—which is what you'd expect if you were imagining it in terms of a material cube.[19]

To see the advantage of Harriot's symbolism, which was not too different from the modern version I'm using here, and just as transparent, consider solving that pesky Cardano equation from Bombelli's starting point of knowing that $x = 4$. Harriot's method suggests that first you write $x^3 = 15x + 4$ as $x^3 - 15x - 4 = 0$. This is just what you would have done in high school, and you'd then divide $x^3 - 15x - 4$ by $x - 4$ to get $x^3 - 15x - 4 =$

$(x-4)(x^2+4x+1)$. This equals zero when $x = 4$ *or* when $x^2 + 4x + 1 = 0$. You can complete the square to solve the quadratic, to find two *additional* solutions, $x = -2+\sqrt{3}$ and $x = -2-\sqrt{3}$, making a total of three solutions. In this case, all the solutions turn out to be real, and Cardano's complicated expression $x = \sqrt[3]{2+\sqrt{-121}} + \sqrt[3]{2-\sqrt{-121}}$ doesn't come into it. Or so it seems. . . . Later mathematicians, however, would connect the historical dots by discovering that in fact complex numbers *themselves* each have three cube roots. So, the three real solutions of Cardano's pesky equation *can* be recovered from his algorithm! I've shown how in the endnote.[20]

THE POWER OF THINKING SYMBOLICALLY

The factor approach is elementary today, but it was a huge breakthrough four hundred years ago. Harriot didn't always use it, and its full, more general implication (the fundamental theorem of algebra) would not be proved rigorously for another two centuries. So, following Viète and Cardano, he also devised a whole list of algorithms for solving various types of quadratic, cubic, and quartic equations. But he was clear about the value of algebraic symbolism. "What need is there for verbose precepts," he said (for even Viète was wordy), "when our kind of reduction exhibits all the roots [that is, all the solutions] directly, not only for this type of equation, but for any other case you like."[21]

What he was getting at is that generalization is far easier in symbols than in words. And when you can generalize—when you can see common patterns that apply to an unexpectedly wide range of problems—you can make extraordinary progress in science and technology as well as mathematics. For instance, Maxwell was able to show the electromagnetic wave nature of light, and to predict the existence of radio waves, because his mathematical analysis of electromagnetism turned up the very same kind of equation that had been used to describe the wave pattern you see when you pluck a guitar or violin string. And Noether brilliantly generalized the relationship between mathematical patterns of symmetry and the conservation of physical quantities such as energy and momentum.

More on these examples later; meantime, Harriot scholar Muriel Seltman sums up neatly both Harriot's importance and the power of algebraic symbolism:

> There is a reciprocal relation between symbolism and mathematical thought-processes, and it would be hard to overestimate the effect of Harriot's techniques and clarity of thought expressed in a symbolism that directs what you do *visually* and therefore makes mathematics accessible in a totally new way. . . . The visualizability is obvious but profoundly important. It is now possible to *manipulate the symbol* as if it were the *non-visualizable concept* of which the symbol is only the embodiment.[22]

Which brings me to where I'm ultimately heading. Hamilton's four-dimensional quaternions don't have traditional concrete geometrical analogs, and it's impossible for most of us to visualize Einstein's four-dimensional space-time (let alone such things as ten-dimensional string theory); but the mathematical symbolic equations that describe them have their own kind of tangibility. They can be manipulated as if four dimensions "really" do exist, because algebraically speaking x^4 is just as valid as the "square" x^2 and the "cube" x^3, and—jumping ahead to coordinates and vectors—(a, b, c, d) is the same kind of thing as (a, b, c).

With the rise of algebraic symbolism, a new kind of abstract thinking arose, which opened the way for the development of calculus—and ultimately to vector and tensor calculus. This marvelous language has enabled mathematicians to solve a vast range of new, more complicated problems—problems whose solutions have given us new technologies as well as new ways of seeing reality. So the next chapter will offer a brief tour through the extraordinary story of calculus.

(2)

THE ARRIVAL OF CALCULUS

A decade before his "electric" insight at Broome Bridge, Hamilton had used differential calculus to make his first major discovery: the theoretical existence of a new optical phenomenon—a special type of refraction in certain crystals, which no one had ever observed before.

Optics, the study of how light behaves, was an exciting topic for both physicists and mathematicians, even though no one knew what light actually *was*. Yet a growing number of physicists were confident that it wasn't a particle, as Newton had believed, but a *wave* of some sort. Maxwell will take this to its ultimate conclusion—highlighting the importance of vector calculus in the process—but the tide had begun to turn in 1801, with a remarkable experiment that is now a classic in the story of physics: Thomas Young's "double slit" experiment.

The idea was to observe what happens when two beams of light interact with each other, and to see if the resulting interaction patterns were those of waves or particles. You can see the pattern that waves produce if you drop two pebbles into a pond. The impact of the stones as they hit the water produces ripples that spread out in concentric circles centered around each of the stones—and you can watch the places where the ripples from one stone jump over those around the other. And when they do, the water level rises, so the two ripples combine to produce a bigger wave. If you could do the same experiment with light—and if it did travel as a wave—you'd see this effect as an increase in brightness as the two light waves reinforced each other. It's harder to see that at some places on the water surface the ripples cancel each other out—the peak of one wave has met up with the trough of the other—and so the water surface is flat. The optical analog of this effect would be a dark patch—there'd be no light because the waves canceled each other. But *was* light a wave? That is what Young set out to explore.

He created two beams of light by shining light through two tiny pinholes, and he manifested the resulting "interference pattern" by letting the two beams shine onto a screen. If light were made of particles, each beam would just go straight ahead, so you'd expect to see just two bright patches in line with the two pinholes—analogous to two piles of letters dropped into a post-box with two openings, or two streams of ping-pong balls fired at a screen. If light traveled as waves, however, then it would bend (or "diffract") around the pinholes to produce circular waves like the ripples on the pond—and the crisscrossing ripples would produce an interference pattern of light and dark patches across the screen. And sure enough, this is just what Young found. It was a beautiful experiment. A century later, however, Einstein would introduce the idea of a light quantum, later called a photon. Photons are not Newtonian-style material particles, but in their interactions with atoms they can behave in a particle-like way. To add to this bizarre finding, quantum physicists have adapted Young's experiment to show that both light photons *and* subatomic material particles such as electrons exhibit wave interference patterns. In other words, both light and matter can manifest wave-like or particle-like behavior, depending on the situation. But this is hindsight, so let's return to Young's wave theory of light, and what Hamilton did with it.

Young was a friend of Mary Somerville, and she recalled that few in England took his light-wave deduction seriously at first, because no one wanted to diminish the great Newton. (In fact, at the time he'd had good reasons for rejecting a wave theory.) But it wasn't just Newton's reputation that got in the way of the reception of Young's experiment: there was also the fact that no one knew what a light wave could possibly be made of, or how it managed to travel through empty space in its journey from the sun to Earth. Still, this hadn't stopped early nineteenth-century mathematicians such as Hamilton from exploring light waves theoretically—something researchers had been doing since Christian Huygens in Newton's time.[1]

In his new analysis of the way light waves should behave when they pass through crystals of various shapes, Hamilton found that when these waves passed into a certain type of crystal (called "biaxial") in a certain direction, they should be refracted into a *cone* of rays rather than the familiar band you see with an ordinary prism. Today, the "conical refraction" of laser beams is finding various practical uses, including as optical tweezers for manipulating chemical molecules or blood cells, and in free-space optical communication.[2]

Back in early 1832, however, no one had ever seen such a phenomenon as conical refraction (let alone lasers). So Hamilton asked his experimentalist friend Humphrey Lloyd to see if he could find this new kind of refraction—and when he did, Hamilton became a sensation, for it was perhaps the first time that mathematics had been used to predict the very existence of a physical phenomenon, rather than explaining what was already known. Many more such breakthroughs were to come: the prediction of the existence of Neptune would follow a decade later, and then a host of predictions, from radio waves and radiation pressure to photons and $E = mc^2$ to the Higgs boson, gravitational waves, and much, much more.

• • •

Hamilton had presented his first mathematical paper on optics when he was still a teenager, and he was still only twenty-seven when his mathematical prediction of conical refraction was verified. And among the tools he'd needed to make his remarkable prediction was calculus.

Calculus has two branches, differential and integral. Differential calculus is essentially about rates of change—as in speed, the rate of change of distance with time. In his work on refraction, Hamilton used derivatives to show how certain quantities (related to optical path length and wave speed) changed at the faces of the crystal, but all manner of smoothly changing phenomena can be modeled by differential calculus: heating and cooling, growth and decay both biological and atomic, waves of various kinds, ecological systems, thermal and topographical gradients, the gradients of mathematical functions (which can be used also to solve optimization problems), financial trends, and so on.

As for integral calculus, it developed as a calculation tool to solve just the kinds of problems that ancient mathematicians had grappled with, when they wanted to calculate such things as the areas of fields and floor plans or the volumes of cisterns and canals. In today's high-tech society there are many more such applications—calculating the area of the lifting surface of an airplane's wing, for example; determining the square footage of material needed for manufacturing products, from computers and car bodies to bridges and skyscrapers; finding the weight (via the density and volume) of materials used in construction, to ensure the structure is sufficiently load-bearing; monitoring the volume of a patient's blood during surgery . . . I could go on, but you can see the scope.

You likely first learned integral calculus by approximating the area under a curve $y = f(x)$, where $f(x)$ represents the function of x that describes the curve you're interested in; for instance, for a parabola centered and based at the origin, you'd have $f(x) = x^2$. You'd approximate this area with a lot of skinny rectangles, each of area ydx, as in figure 2.1; the more rectangles you add, the thinner each one becomes and the better the approximation. It's an intuitive geometrical approach, but with algebraic formulae for the algorithms of integral calculus, you can "add up" an *infinite* number of these infinitesimal rectangles to get the exact area. You do something similar to find volumes, adding up an infinite number of cylindrical slices or shells, each of infinitesimal volume.

Saying "adding up an infinite number" trips easily off the tongue today, but it's really quite a marvel to be able to manipulate infinity. Ancient mathematicians had shied away from the concept altogether—just as they'd done with those imaginary numbers we met earlier.

Today, for instance, if you wanted to derive rigorously the formula for the area (πr^2) or circumference ($2\pi r$) of a circle, you'd integrate its

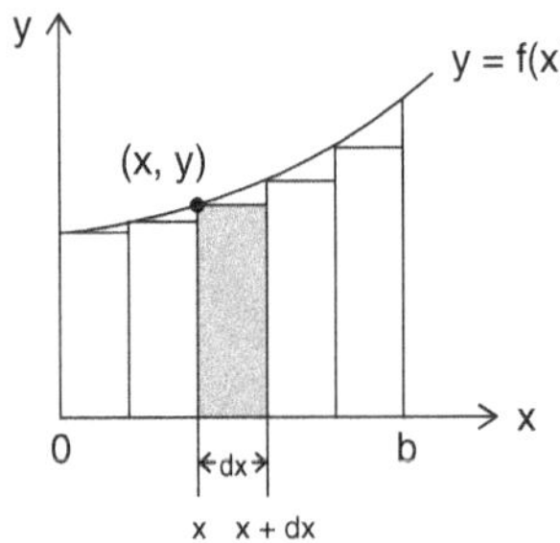

FIGURE 2.1. Area under a curve. The shaded rectangle corresponds to the general point (x, y) on the graph, and its area (length by width) is ydx. When you "integrate" the function $y = f(x)$ between $x = 0$ and $x = b$, you're effectively finding the sum of all these rectangles when dx is infinitesimally small. This integral is written as $\int_0^b ydx$, or more specifically for this function, $\int_0^b f(x)dx$, and it equals the area between the x-axis and the curve $f(x)$ for x values between 0 and b.

algebraic equation, as we'll see in figure 2.3. Three and a half thousand years ago, by contrast, the Egyptian scribe Ahmes recorded—on a papyrus scroll that miraculously still survives—a simple *geometric* construction that approximates a circle with an octagon. The resulting area computation gives a surprisingly good approximation for the constant we now call π, as you can see in figure 2.2. To find the circumference of the approximating polygon, Ahmes just had to add up the sides of the octagon—using Pythagoras's theorem to find the lengths of the corners.[3]

A thousand years after Ahmes recorded the math of his day, Greek-speaking mathematicians developed "the method of exhaustion" (as it was later described), in which circles were successively approximated by polygons with an increasing number of sides, up to a desired accuracy. It was a brilliant attempt at rigor—many ancient mathematicians were interested in the idea of proof as well as in computation—and Archimedes of Syracuse, for one, made fine use of it. To find the circumference of a circle, he started with an inscribed regular hexagon, and then doubled the number of sides, comparing the perimeters of the two successive polygons. He kept on doubling and comparing, and, in this way, he found an algorithm for adding and comparing the segments of polygons with up to *ninety-six* sides. It was certainly an impressive feat of computation, and if you imagine trying to draw it, you'll see that a ninety-six-sided polygon is a pretty fine approximation to a circle—good enough that Archimedes was able to estimate the value of (what we call) π as somewhere between $3\frac{10}{71}$ and $3\frac{10}{70}$. The upper limit is the value that many of us learned in school: $3\frac{1}{7}$ or 22/7 (which is 3.14286 compared with the modern value of 3.14159, both rounded to five decimal places).

Still, adding up ninety-six sides is not the same as adding up the infinite number of them required to turn the polygon into a true circle. With calculus, by contrast, lengths of curves are found by integrating the algebraic formula for the length of a minute segment of the curve, denoted by *ds*. It's an infinitesimal version of one of Archimedes's ninety-six sides, and its length is found by analogy with Pythagoras's theorem. For the example in figure 2.3b, you can see there's an ancestral similarity even with Ahmes's approach.

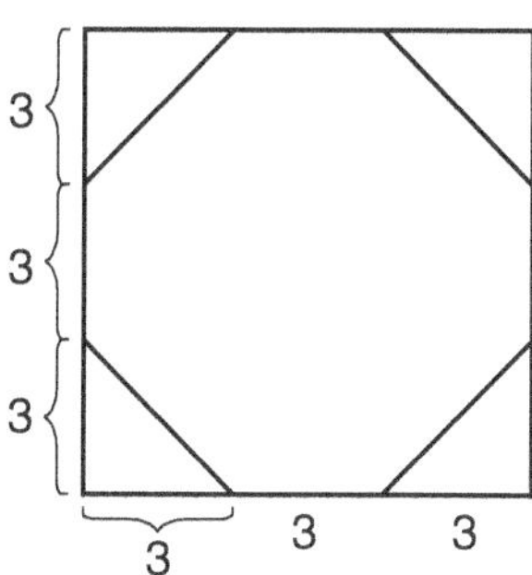

FIGURE 2.2. Ancient Egyptian estimate of the area and circumference of a circle. The square has sides of length 9 units, so its area is 81 square units. The area of the inscribed octagon = area of square minus area of the four corner triangles $= 81 - 4\left(\frac{3\times3}{2}\right) = 63$ square units. This seemed pretty close to 64 square units, and $\sqrt{64}$ was a lot easier to contemplate than $\sqrt{63}$—so Ahmes recorded the area of the circle as equivalent to the area of a square of side 8 units. Since the octagon is taken as an approximation of the circle inscribed in the original square, the circle has a radius of $\frac{9}{2}$ units. So their area formula was (in modern terms) $A = Nr^2$, where N stands for their version of our π: $N = \frac{A}{r^2} = \frac{64}{\left(\frac{9}{2}\right)^2} = 64 \times \frac{4}{81} = \frac{256}{81}$, which is about 3.16; the modern value is a little over 3.14.

Adding up the sides of the octagon (using Pythagoras's theorem for the corners) and comparing this with the modern formula for the circumference of a circle of radius $\frac{9}{2}$, you get the Egyptian approximation of π to be 3.2—still in the same ballpark as their 3.16.

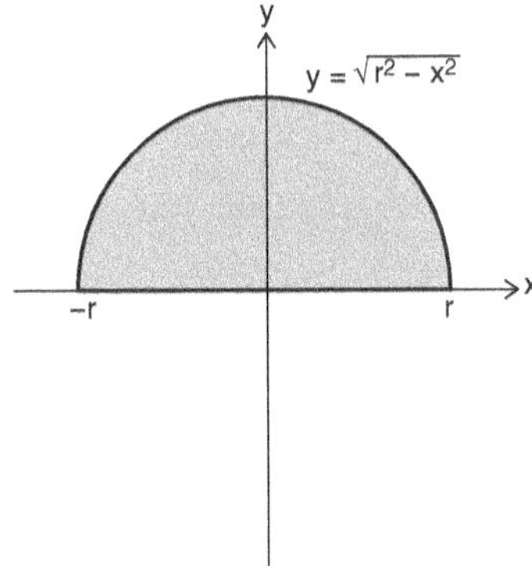

FIGURE 2.3A. Finding the area of a circle using integral calculus. The idea is to find the area under the semicircle $y = \sqrt{r^2 - x^2}$ and multiply by 2.

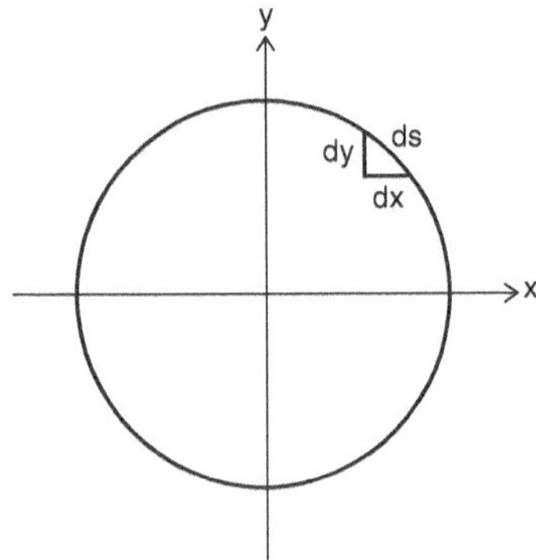

FIGURE 2.3B. Finding the circumference of a circle using integral calculus. Here the idea is to add up (integrate!) all the little segments *ds*. (I'll show some of the calculations in the related box, but here as elsewhere in the book, if the details are not your thing, I hope you don't stop reading but just skip over them and move on with the story.)

CALCULATIONS FOR FIGURE 2.3A

Choosing polar coordinates, let $x = r\cos\theta$, so for a fixed radius r, $dx = -r\sin\theta\, d\theta$; also let $y = r\sin\theta$. (Fig. 3.4 in the next chapter shows how $\sin\theta$ and $\cos\theta$ are defined, although the radius there is 1.) Then, using the double angle formula

$$\cos 2\theta = (\cos\theta)^2 - (\sin\theta)^2 = 1 - 2(\sin\theta)^2,$$

and changing the order of the limits of integration in the second integral to take in the minus sign in dx, the area of a circle of radius r is:

$$A = 2\int_{-r}^{r} \sqrt{r^2 - x^2}\, dx = 2\int_0^{\pi} (r \sin\theta) r \sin\theta \, d\theta = 2r^2 \int_0^{\pi} (\sin\theta)^2 \, d\theta =$$
$$r^2 \int_0^{\pi} (1 - \cos 2\theta)\, d\theta = r^2[\theta - (\sin 2\theta)/2]\big|_0^{\pi} = \pi r^2.$$

You can also do this, rather more readily, with a double integral (which we'll meet briefly in chap. 6).

CALCULATIONS FOR FIGURE 2.3B

Again let $x = r\cos\theta$, and $y = r\sin\theta$, so that for a fixed radius r, $dx = r\sin\theta\, d\theta$ and $dy = r\cos\theta\, d\theta$; then the length of ds, the infinitesimal segment of the circumference, is:

$$ds = \sqrt{(dx)^2 + (dy)^2} = \sqrt{r^2(\cos\theta\, d\theta)^2 + r^2(\sin\theta\, d\theta)^2} = r\, d\theta.$$

Integrating this around the whole circle from $\theta = 0$ to $\theta = 2\pi$, we find that the circumference of a circle of radius r is $2\pi r$.

Archimedes might have come even closer to inventing calculus if he hadn't been killed during the Roman siege of Syracuse in 212 BCE. He was about seventy-five years old, and the soldier who killed him apparently had no respect for elders or for mathematics—Archimedes was reputedly so immersed in his calculations that he hadn't heard the soldier's orders. Two thousand years later, during the dark days of another violent upheaval, the French Revolution, a remarkable young girl read this story. She was so excited by Archimedes's thrall that she decided to learn the evidently thrilling subject of mathematics—but she had to teach herself because most girls were denied a proper education at that time. And she had to do it in secret, because her parents believed, like almost everyone else, that because women were generally physically smaller and weaker than men, serious study would damage their health, stopping them from having babies or even driving them mad. She was Sophie Germain, and she became one of the best female mathematicians before the twentieth century.

She used differential calculus in her most important work, pioneering the mathematical theory of vibrating surfaces.

Even despite Archimedes's tragic death, calculus might have arrived sooner if his most important manuscript on the subject hadn't been partly washed off and written over by a medieval scribe. It was recovered only in 1906, and it showed that Archimedes had found some truly extraordinary results. Working just with concrete geometry and mechanical analogies, he'd managed to find the volumes of spheres and curved-shaped containers, using methods that prefigure the volumes of "solids of revolution" taught today in undergrad integral calculus classes. But this work of genius, this treasure of information, was lost for so long that early modern mathematicians had had to rediscover its ideas for themselves.

NEWTON, LEIBNIZ, AND ZENO: SEARCHING FOR THE INFINITESIMAL

And so we come to that early modern incarnation of genius, Isaac Newton. According to the *Oxford English Dictionary,* the modern definition of "genius" is "an exceptional intellectual or creative power or other natural ability or tendency." It's a word that tends to be overused in everyday speech, but it certainly fits Newton, and the other exceptionally creative mathematicians and mathematical physicists we'll meet in this story—the ones whose ideas were so innovative they played foundational roles in the mathematical revolutions that gave us vector and tensor calculus. Of course, even these standout thinkers needed inspiration and help from their colleagues, because the "lone genius" stereotype rarely applies in real life. For two hundred years, though, Newton was widely deified as singularly brilliant—a perception that shaped and limited the idea of what it meant to be a mathematician and a theoretical physicist until late in the nineteenth century. After all, we can't all be geniuses, and many fine mathematical thinkers throughout history have made important contributions. We'll meet some of them throughout this story, including the colleagues who helped the feted geniuses to make their breakthroughs.

One of the hallmarks of Newton's genius is that he was an outstanding contributor to so many branches of mathematics and physics: in modern terms, he was an experimental physicist, especially on the subject of light; he was a theoretical physicist, especially with his theories of motion and gravity; he was an applied mathematician—a category that in his case, and for several others in this story as well as for many researchers today, intersects with theoretical physics when mathematics is applied to natural physical processes—and he was a pure mathematician, one who invents new mathematical methods, concepts, and proofs, with a focus on rigor and beauty. These "pure" techniques often underpin the later real-world applications of the applied mathematicians and theoretical physicists—and calculus is a classic example.

Newton was the first to find general, symbolically represented algorithms for the two types of calculus, differential and integral, clearly showing the connection between them. That is, he showed that integration is the inverse operation of differentiation. Gottfried Leibniz got the idea, too, for calculus was "in the air," thanks to forerunners such as Harriot and Galileo, Pierre de Fermat (of Fermat's last theorem fame), Isaac Barrow (Newton's mentor at Cambridge), and John Wallis. It is through their works—and those they built on, from ancients such as Archimedes and Euclid to medieval scholars such as Abu Ali Ibn al-Haytham and Nicole Oresme—that Leibniz and Newton were able, each independently, to put calculus into its first definitive form.

These two founders could not have been more different. Newton was rather secretive about his discoveries because he couldn't stand the thought of being criticized. He was rather weird about a lot of things, although I do have some sympathy for him. He made his initial breakthroughs with calculus, optics, *and* the theory of gravity when he was still only in his early twenties—tucked away at his mother's farm, isolating during the plague years 1665–67. It's hard not to be awed by such an achievement. And it's hard not to feel for the frightened child inside. Before he was born, his father had died in the Civil War between Oliver Cromwell's Parliamentarians and King Charles I's Royalists. It must have been a terrifying time for his mother, Hannah, with her husband dead and the fighting so fierce—it came as close to her farm as the nearby village. And so she married Barnabas Smith, a well-off rector, and moved to another town;

unfortunately—and apparently at Smith's insistence—she left young Isaac behind with her mother. He famously wished he could have burned Smith's house down with his mother and Smith inside, but it seems the only outwardly destructive behavior he engaged in was the odd bit of artistic graffiti or the occasional nighttime pranks with exploding kites that frightened the wits out of superstitious villagers. Only when Smith died did Hannah return to Isaac, then aged about ten, with three unwelcome half-siblings in tow.[4]

Leibniz was a much sunnier person than Newton. A true Renaissance man, he was "a lawyer, scientist, inventor, diplomat, poet, philologist, logician, and a philosopher who religiously defended the cultivation of reason as the radiant hope of human progress," to quote Leibniz scholar Philip Wiener.[5]

When Leibniz was six, his father, a philosophy professor, died. His mother's father had also been a professor—of law, in his case, the field in which young Gottfried later earned not just a degree but also a doctorate. By contrast, Newton's father had been an illiterate farmer, and his mother had not understood his intellectual passion and talent: she took him out of school when he was seventeen, until he made such a mess of running the farm that she finally accepted the advice of both the schoolmaster and her Cambridge-educated brother, and allowed Isaac to prepare for university. But he had to pay his own way through Cambridge, working for a wealthy student as a "sizar" or servant. Fortunately for him and for us, he soon became a professor there, and was free to make his amazing discoveries.

Leibniz had no such worries as a student, and after graduating he led a rather glamorous life. Traveling through Europe on diplomatic missions for his princely patrons and conversing with scientists and philosophers along the way, he built up a huge network of some six hundred correspondents—he was a prolific letter writer, with big ideas about universal peace, philosophy, and science. Given all his wide-ranging interests and his professional duties, he showed an extraordinary talent for mathematics. Not surprisingly, the academic Newton was the better mathematician and experimental physicist, but the eclectic Leibniz made significant discoveries nonetheless—and he was the better symbol-maker.

In the previous chapter, I highlighted algebra's slow progress from words to symbols, but in the early days of modern calculus—half a century after Viète, Harriot, and Descartes introduced true algebraic symbolism—the situation was reversed: calculus symbols stood for ideas that no one at the time could adequately explain in words. This is the special power of symbolic thinking: it can take you into new places that are, at first, beyond ordinary understanding.

For instance, although I've been talking easily about "infinity" and "infinitesimal" quantities, they are concepts that are very difficult to define meaningfully. A modern dictionary definition of "infinitesimal" is "extremely small," but that's not much help to mathematicians. They need something more precise, although in this case it took two hundred years after the birth of Newtonian-Leibnizian calculus to find such a definition (via the theory of limits, continuity, and functions). You can see the difficulty when you consider that to find a moving object's speed at any given time, you have to compare its position at that time with its position an instant later, and then divide the difference by the instant of time. But what do you mean by an "instant"? And how do you define the incremental change in position? Obviously, they are both very small quantities, but how small?

Newton called these infinitesimal changes "moments," denoting them by the symbol o (a "not quite zero" represented by the Greek letter omicron). Leibniz called them "differences" or "differentials," and denoted them dt, dx, and so on. For the rate of change of a distance x with respect to time, Newton used a dot, as in $\dot{x}$. Leibniz, on the other hand, wrote his "ratios"—his rates of change of x with respect to t, for instance, and of y with respect to x—as $\frac{dx}{dt}, \frac{dy}{dx}$, and so on. Actually, Leibniz mostly used the notation $dy{:}dx$, where the colon symbolized a ratio; it was his followers—especially Johann Bernoulli, whose first name is commonly written in the French way, Jean, and occasionally Anglicized as John—who popularized this ratio as $\frac{dy}{dx}$. You can see whose symbols eventually won the day—although the dot notation is still used for differentiation with respect to time, which is important in the study of motion, waves, fields, and various other quantities that change over time.

Both Newton and Leibniz were aware of the conceptual problem underlying the word "infinitesimal." Leibniz had a stab at explaining it by refining the ancient method of exhaustion, and Newton made a reasonable fist of an early definition of what we now call limits. (You can see their attempts, and by comparison a modern definition, in the endnote,[6] and I've sketched the limit idea in fig. 2.4 below.) In their calculations, though, Leibniz and Newton often did what any canny student would do: ignore these tiny increments and instants as too small to worry about—a little like when you buy an item for $124.99 and don't worry about the change out of $125.00. It works well in practice, except that like a cent, an "instant" of time is not actually zero. If you act as though it *is* zero, you run into theoretical problems when you want to divide by an "instant": to find an instantaneous speed, for example, you'd be doing the impossible and dividing by zero. Worse, the tiny change in distance is approximately zero, too, so you'd be trying to calculate 0/0.

It was hard enough trying to add infinitesimal quantities, as Archimedes found, let alone divide them to find derivatives—and as long ago as 450 BCE, the legendary Zeno had highlighted the difficulty in his famous paradoxes. For instance, he said, to run from A to B, first you have to run halfway to B; but before you reach halfway, you have to run a quarter of the way, and so on, indefinitely. Which means you can never really get started. Of course, everyone knows that you *can* run from A to B. What Zeno seemed to be showing here—from a modern perspective, at least—was just how difficult it is to define the idea of motion, with its infinitesimal increments in distance and its tiny "instants" of time. Of course, Zeno himself may have had a mystical rather than a mathematical purpose, so it's possible that his paradoxes—like the teachings of his Buddhist contemporaries—were intended to suggest that time and space are illusions.

SAILING THE HIGH SEAS AND BREAKING WARTIME CODES: THE RISE OF ALGEBRAIC CALCULUS

It took mathematicians almost two and a half thousand years to sort out the theory of limits, which is needed to give a satisfactory resolution of

Zeno's paradoxes as well as of calculus. One of the first steps involved the idea of "converging infinite series," where it does indeed turn out to be possible to add up an infinite number of increasingly small terms and to mathematically prove that the result is finite. Math buffs might recognize that Zeno's A to B problem is the geometric series $\frac{1}{2}+\frac{1}{4}+\frac{1}{8}+\ldots$, which equals 1—so you can, indeed, traverse the whole distance between A and B! Not all infinite series "converge" (or add up to a finite number), though, so limit theory is essential in order to tell the difference. Integral calculus, too, requires adding up an infinite number of quantities, so you can see that its history is intimately linked to the early work on infinite series.

Although the ideas of limits and convergence weren't proven rigorously until modern times, ancient and medieval mathematicians had used ingenious geometric constructions to find the sums of a limited number of series. They wrote both their series and their algorithms in words—Harriot seems to have been the first to rewrite series in terms of number and letter symbols as we do today, which he did sometime around 1600. In particular, in a long and virtuoso calculation, he used an infinite geometric series and an intuitive limit argument to find an algebraic expression for the length of the spiral traced over the curved Earth by a ship following a fixed compass bearing. (His patron, Sir Walter Ralegh, had wanted the best scientific expertise to help make navigation safer and more efficient, for he was planning voyages from England to America across an uncharted ocean.) Harriot began by breaking up the spiral into small segments like Archimedes's circle, except he assumed an infinite number of them, of ever-decreasing size. He took dozens of pages and many months to conclude this work—the first extant algebraic derivation of the length of a curved line, other than those early estimates of the circumference of a circle. Today, integral calculus is used to find such lengths, as figure 2.3b showed. Still, Harriot would have been chuffed to know that his symbolic approach to algebra directly influenced the English calculus pioneer John Wallis, and Wallis influenced the young Newton.[7]

Figure 2.3b and the related box also showed that to use calculus to find the length of a curved line, you need a formula for your curve—and this is the province of "analytic geometry." Harriot had made a start at this, too, but it was Descartes who really put "analytic" or algebraic geometry on the map, and we call the rectangular x, y, z coordinate system "Cartesian" in his honor. Even so, Descartes barely glimpsed the remarkable power of being able to represent geometric curves by algebraic equations written in terms of these coordinates (such as $x^2 + y^2 = r^2$ for the circle in fig. 2.3 above). His main purpose had been to show that algebra and geometry together make problem-solving easier: algebraic symbolism frees the mind from having to visualize complicated geometric constructions (such as Cardano's completed cube), while geometry gives concrete meaning to algebraic operations. Twenty years later, Wallis was less evenhanded about the matter, definitely favoring algebra over geometry.

Wallis also favored the English over the French, because he'd been miffed by the French response—in the persons of Fermat and several others—to his own early work. He also swore, and he wasn't *entirely* alone in this, that Descartes had cribbed from Harriot's book *Praxis*.[8] Anyway, Wallis took a significant step toward an algebraic calculus in his 1655 book *Arithmetica Infinitorum*. He'd even created a symbol for infinity—the same one we use today, ∞. This symbol had been used occasionally by the Romans to represent 1,000, and it also cropped up every now and then in various contexts, including a version of the Egyptian image of a snake eating its own tail and symbolizing the endless cycle of birth and death. No one knows why Wallis used it for infinity; perhaps he was impressed with its ancient mystical connotations, although as a Calvinist he more likely chose it simply because it represents a never-ending curve.

Wallis's story gives a glimpse into the mathematical and political context at that time, just before Newton burst onto the scene. He was the son of a village rector in Kent, and fortunately his family recognized and supported his academic bent. He eventually went to Cambridge, although he didn't learn much math there. He'd only realized the subject existed when he came home for the Christmas break and noticed his younger brother studying arithmetic in preparation for a trade. Intrigued, he'd asked for a lesson. As he put it later,

> Mathematicks were not, at that time, looked upon as Academical Learning, but the business of Traders, Merchants, Sea-men, Carpenters, land-measurers, or the like. . . . Of more than 200 at that time in our College [Emmanuel, Cambridge], I do not know of any two that had more Mathematicks than myself, which was but very little; having never made it my serious studie (otherwise than as a pleasant diversion) till some little time before I was designed for a Professor in it.[9]

They do say that the best way to learn something is to teach it! And Wallis must have learned it well, for he became one of the best mathematicians in the generation before Newton. It had been quite a journey, though. First, he graduated with a master of arts in divinity. Then he took up a fellowship at Queen's College, Cambridge, but it was short-lived: fellows had to remain unmarried in those days, so in 1645, the newly wedded Wallis became a minister instead of an academic. And then, quite out of the blue, he discovered he had a knack for code breaking. The Civil War was raging, and a fellow clergyman had shown him an intercepted message in cipher, asking half-jokingly if he could make sense of it. To his own amazement, Wallis was able to decipher the message within a few hours. It was a turning point in his life, for when the king was executed in 1649, it was very handy if you'd proved your loyalty to the victors. Wallis had broken a number of codes for the Parliamentarians, showing his mathematical skills into the bargain—and when Oxford's Savilian Professor of Geometry was sacked for having been a Royalist, Wallis was appointed instead. (Although he supported the Parliamentarians, Wallis had spoken out against the execution of the king, and had signed, rather bravely, a document protesting it. So when the monarchy was later restored, he remained in favor. Besides, he preferred mathematics to politics, and as he told his Royalist friends, most of the messages he'd unlocked hadn't been of all that much use to the Parliamentarians anyway.)[10]

It was just six years after his appointment at Oxford when Wallis published his algebraic *Arithmetica Infinitorum*. At the time, most mathematicians still preferred using tangible geometric ideas to develop ways of handling infinite sums and infinitesimal increments—and Thomas Hobbes was particularly vociferous on the subject. His fame today rests on his political philosophy—and his dour conclusion that life is "nasty, brutish, and short" without the protection of a state. But he also dabbled in mathematics, and he expressed this pro-geometrical sentiment by pronouncing Wallis's *Arithmetica Infinitorum* a "scurvy book" and "a scab of symbols." In fact, he denounced "the whole herd of them who apply their algebra to geometry." He evidently was not a sunny person.

Despite Hobbes and his fellow geometers, a decade later it was Wallis's book that first set young Newton on his path to calculus. Two decades after that, Newton wrote his magnificent *Principia*, whose full title, in English, is *Mathematical Principles of Natural Philosophy*; like most European scholars at the time, Newton wrote in Latin. I should add here that the term "physicist" is a nineteenth-century one; before then, theoretical physics was called "natural philosophy." The difference, in hindsight, is that natural philosophy focused on logical arguments, while theoretical physics offers explanations and predictions that can be experimentally tested—a change of viewpoint that Newton spearheaded. For example, traditional natural philosophers invoked the "logical" hypothesis that something mechanical, such as gargantuan ethereal whirlpools, carried the planets in their orbits. There was no actual evidence these whirlpools existed, aside from the planetary motion they purported to "explain." By contrast, Newton had worked out an explanation for planetary motion based on observable facts, from which he derived a quantitative formula for the gravitational attraction between any two bodies, including a planet and the sun: it is proportional to the product of their masses and inversely proportional to the distance between them. In symbolic algebraic form this is, of course, the famous inverse square law $F = \frac{GmM}{r^2}$, where the masses are m and M, say, the distance is r, and G is the constant of proportionality. It was also in *Principia* that Newton first published his general calculus algorithms, which he used in his science of motion. These laws of motion, together with the law of gravity, were so general that they explained everything from the fall of a leaf to the pull of the tides, from the path of a ball thrown in the air to the motion of the planets. It was an astonishing achievement.

Einstein explained in detail the importance of this achievement: "Before Newton," he said, "there existed no self-contained system of physical causality [about] the deeper features of the empirical world." For instance, he added, Kepler's observation-based laws of planetary motion, including the elliptical shape of planetary trajectories, established *how* the planets move, but not *why*. That's because "these laws are concerned with the movement as a whole, and not with the question *how the state of motion of a system gives rise to that which immediately follows it in time*; they are, as we should say now, integral and not differential laws."[11] We'll see Newton's differential second law of motion shortly, and later we'll see that this same preference for differential over integral laws is key to James Clerk Maxwell's success in uncovering the secret of light.

DRY DRUDGES AND PLAGIARISM: THE RECEPTION OF *PRINCIPIA*

Such glowing recognition didn't come easily at the time. As soon as *Principia* was published in 1687 it had its detractors—including Leibniz. Like most of his contemporaries he was dazzled by Newton's mathematical genius, but as a philosopher first and foremost, he had a different idea from Newton (and later Einstein) about what a theory should do. Newton had explained that the cause of planetary motion was gravity, but Leibniz thought that a theory of gravity should also explain gravity itself. In particular, he thought it should suggest a *concrete mechanism*—such as those ethereal whirlpools—for how gravity physically effected the moment-by-moment changes in planetary motion that Newton had described *mathematically*. Since this is how natural philosophers had long defined the hallmark of scientific theory, many other critics agreed with Leibniz. But Newton inaugurated the modern view: it was obvious that gravity *exists*, and he'd quantified its *effects*—notably on planetary motion—but he didn't want to include untestable speculation about what it *was*. He certainly would have liked to explain gravity itself, but he knew that what he *had* achieved was unprecedented in its conception and scope, and that it would be up to later researchers to find a deeper understanding of the nature of gravity. (That's where Einstein, and his groundbreaking use of tensors, will come into our story.)

Even more distressing for Newton than these criticisms was Robert Hooke's accusation of plagiarism. Hooke hasn't had a good rap among scientists, because he tended to big-note himself—and he seemed to take special delight in criticizing Newton publicly. Recently, though, historians have taken a closer look at his work, and it seems he deserves credit for some significant original thinking about a gravitational explanation of planetary motion—including the idea that all planets have gravity, and that it weakens with distance "probably" in inverse square proportion, because that is the way light's intensity weakens. Although Newton had been thinking about the problem for years beforehand, it was Hooke's letters to him, in 1679, that catalyzed his thinking on the subject—as he acknowledged in the first edition of *Principia*. Hooke suggested the idea that orbital motion was made up of two different "motions," one tangential to the orbit and one directed to the center of the motion. It was an intuitive idea of motion being made of vector-like *components*, although it wasn't quite correct: Newton would show that the *force* is directed inward but the *motion* (velocity) is tangential.

Hooke didn't extend gravity to material bodies other than planets, and he wasn't able to prove the inverse-square law by direct experiment or calculation. Rather, he made some ingenious analogies based on careful observation of the motion of mechanical models.[12] And unlike Newton, he didn't generalize planetary motion to mathematical laws for all types of motion—and consequently he didn't deduce, as Newton did, the conservation of what we now call kinetic and potential energy. (Leibniz, too, pioneered the notion of conservation of energy.) Nor did Hooke use his idea of gravity to estimate the oblate shape of the earth or to explain the tides, the precession of the equinoxes, or any of the other mysteries that Newton explained. So, when Hooke accused him of plagiarism, Newton was furious: while Hooke's mathematical ability was greater than has long been assumed, it was simply no match for Newton's. No wonder Newton fumed to Edmond Halley that if Hooke could claim to be the inventor of the theory of gravity, then the mathematicians that "find out, settle and do all the business must content themselves with being nothing but dry calculators and drudges." He had a point. It is interesting that in copyright law today, it is forms of expression rather than ideas that are protected. And although recently there has been an important move among math educators to emphasize the role of calculation in the history and application of mathematics, many modern mathematicians are in tune with Newton, seeing beauty and creativity in their work, not just computations and applications.[13]

Not that Newton was averse to computation. Much of his research was devoted to it, for it is not always possible to find exact algebraic solutions of equations, let alone to solve more complex problems. So you have to come at it numerically, plugging in numbers until you get a suitably accurate answer—somewhat reminiscent of the ancient method of exhaustion. Cardano did this to get to his solution $x = 4$, but Newton developed not one but several systematic numerical approaches.

As for *Principia*, fifty years after its first publication there were still some who thought Newton's approach was too mathematical—that an equation couldn't provide a causal explanation the way a mechanical model could. Among Newton's early champions in France were Émilie du Châtelet and her partner, the provocative playwright Voltaire. In the 1730s, they collaborated on a popular book aimed at bringing Newton's theories of gravity and light to a wider audience. (Aside from believing that light was a material particle, Newton did many meticulous experiments, including those showing that ordinary sunlight is made up of all the colors of the rainbow. Du Châtelet expanded on this by suggesting that the different colors are associated with different amounts of heat—a cutting-edge idea at the time, which she tested by dyeing a bedsheet in rainbow colors and timing how long it took each color to dry: violet was first, and then the other colors in order, with red taking the longest. It was a beautiful DIY experiment, suggesting that violet is the coolest color and red the warmest, although it would take later scientists to confirm and explain it.)[14] Du Châtelet also wrote a book comparing the ideas of Leibniz and

Newton. But her most enduring legacy is her translation of *Principia* from Latin to French—the first such translation into an everyday language other than English. It was so good that it is still the definitive French version. By contrast, the original English translation has long since been surpassed. (The modern version is by I. Bernard Cohen and Ann Whitman, published in 1999. They found Du Châtelet's translation particularly helpful in clarifying some obscure passages using seventeenth-century technical terms that have since been modernized.)

Du Châtelet hadn't come to mathematics until she was about thirty—as a girl, she'd been denied formal education (just like Sophie Germain and Mary Somerville a century later) and had been married off at eighteen to a suitable aristocrat, the Marquis du Châtelet. The story of her intertwined love affairs with intellectual discovery and with Voltaire is wonderfully romantic, but her development as a mathematician in her own right is truly inspiring. Setting up house—and a state-of-the-art library and scientific laboratory—in her château in Champagne, she and Voltaire attracted writers and scholars from far and wide. (Actually, it was her husband's château, but he was a decent man who had come to accept Voltaire's place in his wife's heart. After all, his had been an arranged marriage, not a love match, and he was often away fighting wars, as aristocrats were expected to do, and dallying with his own lovers. Besides, Voltaire paid for much of the necessary renovation of the old castle.) Visitors reveled in the lively conversation and the little plays written especially for them by Voltaire, as well as the peaceful space for research. It was one such visitor who inspired Du Châtelet to translate *Principia*, for she had become a recognized authority on the latest mathematical physics.

Still, it was an incredible undertaking for a mostly self-taught woman at a time when women had no intellectual standing. Initially she had had to teach herself math—although later she employed as tutors two of France's best mathematicians, Pierre-Louis Moreau Maupertuis and Alexis-Claude Clairaut. Voltaire had joined some of the lessons, but he soon deferred to her mathematical superiority, playfully dubbing her "Madame Newton du Châtelet."

In addition to translating Newton's 500-odd pages, Du Châtelet provided a 180-page appendix. This included a 110-page outline of developments in gravitational theory since Newton—including those by her friends Maupertuis and Clairaut—and 70 pages of additional mathematics, designed to clarify Newton's working. He had famously chosen to present the proofs of most of his theorems using innovative geometrical constructions rather than symbolic calculus calculations—even when he was using calculus.[15] Perhaps Hobbes's "scab of symbols" still resonated. Trouble was, Newton's proofs were so clever and idiosyncratic that they couldn't readily be generalized—which meant that his ideas of motion couldn't readily be extended to new applications and insights. Generations of scholars have been frustrated with him: He invented calculus, so why didn't he use it more transparently? Evidently, he thought geometry was the most rigorous *and* intuitive form of mathematics at that time.

For example, it is certainly very helpful to imagine a derivative, such as speed, geometrically—as the gradient of the hypotenuse of a triangle whose sides represent the infinitesimal changes in distance and time. Figure 2.4 shows the idea, and Leibnizian algebraic symbolism—the familiar $\frac{dy}{dx}$—reflects this geometrical, intuitive interpretation.

In the modern conception of a derivative, however, Leibnizian notation is a little misleading, because $\frac{dy}{dx}$ is not a ratio or fraction at all, in the sense of a number dy divided by a number dx. Rather, $\frac{d}{dx}$ is an *operator* acting on a *function* $y(x)$—to use modern language; it just means that the symbol $\frac{d}{dx}$ directs you to "operate" on $y(x)$ by taking its derivative. Nevertheless, the Leibnizian symbols for derivatives can be manipulated as if they really are ordinary fractions. A classic case is the "chain rule," where you can write equations such as $\frac{dy}{dx}\frac{dx}{dt} = \frac{dy}{dt}$, which does look for all the world as if you simply canceled the dx's.

Newton was more cautious about rigor, and he didn't have such a

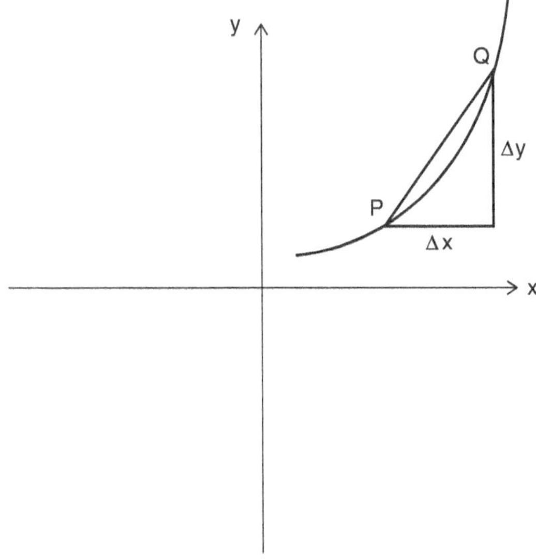

FIGURE 2.4. The slope of the line PQ is $\frac{\Delta y}{\Delta x}$; it approximates the gradient of the graph between P and Q. To be more technical—and modern—about it, we need limits: To get the exact gradient at P—and thereby find the derivative at P—you move the point Q closer and closer to P until Δx and Δy are infinitesimally small. Then the derivative of the graph at P is the limit, as $\Delta x \to 0$, of $\frac{\Delta y}{\Delta x}$. When the limit is taken, Leibnizian notation $\frac{dy}{dx}$ replaces $\frac{\Delta y}{\Delta x}$. If you had a graph of distance x versus time t (instead of y versus x as shown here), the gradient, which would now be $\frac{dx}{dt}$, would represent the speed v—the distance per time.

transparent symbolism. For instance, with Leibnizian symbols you can write $vdt = \frac{dx}{dt}dt = dx$, which readily shows that an object moving at speed v travels a tiny distance dx in the instant dt. By contrast, in Newton's notation this would be $vo = \dot{x}o$, so a physical interpretation is not so obvious from the symbolism. At any rate, he seemed to think that *Principia* would be more accessible if he stuck to literal geometric constructions: knowing the theory of gravity would be controversial, he didn't want to make too much of the new calculus with its disappearing differentials and with Zeno's paradoxes still unresolved. When he needed calculus, he tended to write it in terms of areas under graphs and geometric ratios like that in figure 2.4, although he did sometimes use algebra to explain his algorithms and diagrams (as I illustrated in the previous endnote).

Half a century later, thanks initially to the formidable Swiss mathematician Johann (or Jean) Bernoulli, calculus symbolism had become more acceptable—on the continent, at least. Bernoulli had been Leibniz's most avid defender during the infamous priority dispute with Newton, and later he'd corresponded with Du Châtelet's circle. In the appendix to *Principes Mathématiques* (her French *Principia*), Du Châtelet reworked dozens of Newton's proofs in terms of algebraic calculus. And, in what the British no doubt saw as an unwelcome irony, she did it using Leibniz's symbolism, not Newton's.[16]

Yet, by translating Newton's geometrical calculus into Leibnizian symbols, Du Châtelet and Bernoulli and his students helped bring Newton's vision into full view. It is thanks to them, as well as to Newton, that calculus has proved so important in physics and technology.

• • •

It is *vector* and *tensor* calculus, however, that we'll explore in the rest of this book, along with the ideas and applications of vectors and tensors themselves. I spoke just now about speed as a derivative, but as I mentioned in the prologue, "velocity" is a more precise quantity: it gives the direction of the motion as well as the speed, so it contains information about two different things. It is what Hamilton called a "vector." Speed alone, like any "scalar" or number, represents just one attribute.

When you put it like this, the concept of a vector seems very simple. As the preceding sections have suggested, though, its usefulness to mathematicians, physicists, and engineers will also hinge on its symbolic representation. So, how to represent, and calculate with, a vector? That's a topic we'll now begin to explore—along with answers to various intriguing questions. For instance, if vectors really are so simple, what was so important about Hamilton's epiphany at Broome Bridge?

It is time to dive more deeply into the story.

(3)

IDEAS FOR VECTORS

Since mathematicians are not known for their graffiti, William Rowan Hamilton's scratchings on Broome Bridge have attracted a lot of attention. But the kind of lightning-bolt inspiration that led to this ecstatic act of minor vandalism rarely comes without a long history of deep, creative thinking. In the case of vectors, it took hundreds of years, and many mathematical thinkers, for all the ideas to fall into place so that Hamilton could score his own deep conclusions in stone. So let me begin this chapter of the story by revisiting the redoubtable Newton.

He doesn't always come off well these days, but recently I was asked if I'd like to invite him to dinner. I've mentioned already my admiration of his genius, but I also like the little human touches—dog-earing his books, for example, and becoming so engrossed in his work that he forgot to eat. Then there's the fact that the legendary apple tree is still growing in the garden of his family home and has descendants around the world—including at my own university. More seriously, though, I took this invitation to sit with Newton as a chance to dip back into *Principia*, and for the first time I was struck not just by its scope and its pioneering applications of calculus, but by its clear exposition of the fundamental nature of "vectorial" quantities.

I'd never paid this any attention before—I'd simply taken vectors for granted, so I hadn't realized how significant Newton's treatment was. I hadn't realized how telling it is that, in creating the blueprint for evidence-based, testable physical theories of nature, he saw the need to define fundamental physical quantities, such as force, velocity, and momentum, so that they have *both* direction and magnitude. Today's high school math students learn to express this dual nature as an arrow, as we've seen. But the vector itself was a surprisingly long time in coming, and not even Newton got the full concept.

His key insight in this context was his definition of force in terms of its effect on material bodies. The effective force, he said in his famous second law of motion, is proportional to the resulting "alteration of motion," and this "alteration" *is in the same direction* as the force. Using word-form equations, Newton had already given a careful definition of "quantity of motion" (which he just called "motion" in the second law) as the "quantity of matter" (mass) times the velocity—so in symbols the quantity of motion is mv, which we call "momentum." In modern language, then, Newton's "alteration of motion" is the rate of change of the body's momentum, and in terms of vector calculus, the second law is:

$$\boldsymbol{F} = \frac{d(m\boldsymbol{v})}{dt}.$$

If the body's mass doesn't change measurably during the motion—which it doesn't unless it is moving extremely fast relative to the measurer—then this is the same as $\boldsymbol{F} = m\frac{d\boldsymbol{v}}{dt}$, or mass times acceleration, $\boldsymbol{F} = m\boldsymbol{a}$. I've written the $\boldsymbol{F}$, $\boldsymbol{a}$, and $\boldsymbol{v}$ in bold type to indicate that they are vectors, but this is hindsight. Although it was clear that Newton's force and momentum had both magnitude and direction, they were not yet full mathematical vectors, for there was no general concept of how to add or multiply them. You might think that worrying over adding and multiplying is an arcane distinction, given that the most obvious characteristic of a vector is its ability to represent both magnitude and direction at the same time. As we'll see, though, multiplying vectors opens up a whole new range of physical applications. And it is the *algebra* of vectors, treated as mathematical quantities rather than simply as physical analogs, that makes vectors applicable

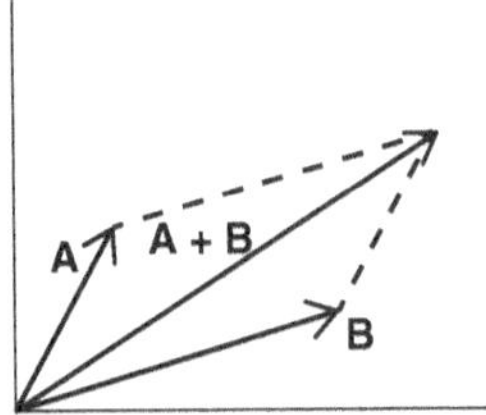

FIGURE 3.1. Adding two vectors *A* and *B*. Newton and the other early pioneers who used the parallelogram probably did not think in terms of adding vectors or "lines" here, but this is how we think of it today: *A* and *B* are components of the resulting vector *A* + *B*. A key feature of the parallelogram rule is that the two components in a vector sum such as this act independently, in the sense that each behaves as if the other weren't there. In other words, when you add the two components using the parallelogram rule, you don't change their original sizes or directions, but simply translate them to form the opposite edges of the parallelogram.

not only in theoretical physics, but also in fields such as numerical modeling, data analysis, engineering, AI, and robotics.

Meantime, in Newton's day there *was* a practical notion of simple addition, where two vectorial quantities were added geometrically via the "parallelogram rule" shown in figure 3.1. Actually, the term "addition" here is probably anachronistic before the formal development of vectors—"composition" is a more accurate term. In the first two corollaries to his laws of motion, Newton gave a clear explanation of this rule, and showed explicitly how it enables a force in the diagonal direction to be composed from two other "component" forces (and conversely, it can be "decomposed" into its components). The parallelogram rule applies to velocities, too, of course, and Newton went on to make spectacular use of it in his theories of planetary and everyday motion.

It seems such a simple concept, "motion," but, as Zeno showed, it's not so easy to come up with a useful definition—especially one that allows you to predict in advance how a body will move under various forces. Newton's rival Leibniz spent many wordy pages trying to figure it out, in his 1695 *Essay on Dynamics*—but he didn't mention *Principia*. Newton later returned the favor by removing his acknowledgment of Leibniz's calculus from *Principia*'s third edition—the unfortunate calculus priority dispute was in full swing. Leibniz's neglect of Newton in his *Essay* was more than churlishness, though: it shows how difficult it can be to understand new ideas. Leibniz was perhaps the greatest philosopher of his generation, and an innovative scientific thinker, yet even he didn't seem to realize that Newton had made a great step forward in the analysis of motion when he defined force in terms of the resulting change in momentum. (Later physicists did recognize this achievement, of course, and ultimately named the SI unit[1] of force the "newton.")

It is intriguing to realize, though, that not even Newton developed the idea of vector algebra, let alone vector calculus. It highlights just how hard it is to come up with new mathematical concepts—and it suggests that the seeming simplicity of vectors is deceptive.

• • •

In fact, when Newton was writing *Principia*, it had already been a long road, peopled with some of the best thinkers in the world, just to arrive at the parallelogram rule. Like Newton and Leibniz, these thinkers had been trying to work out how to understand how things move and how forces act. You can get a sense of the difficulty they faced by imagining how you would go about defining "force" from scratch. The everyday definition given in the *Oxford English Dictionary* is "power; exerted strength or impetus; intense effort." That gives you the general idea, but it's not going to help you launch a communications satellite or design a wind turbine. You need a quantitative, vector calculus definition for that kind of application.

Of course, it's obvious that if you push twice as hard on an object it will move twice as far—a quantifying step that ancient philosophers had taken long ago. In this case, though, both pushes are in the same direction as the resulting motion, so it isn't apparent that direction as well as strength should be part of the definition of force itself. But the ancients did glimpse the idea of composing a trajectory from two different "components" in different directions—such as in Ptolemy's mechanical model of planetary motion, where the planets whirled about as if they were powered by gears that also rolled slowly along their orbits.

The first hint of combining different motions via the parallelogram rule seems to be in *Questiones Mechanicae*, which was apparently written by a member of Aristotle's school at the end of the fourth century BCE, nearly five hundred years before Ptolemy. It was one of those long-lost manuscripts that were only rediscovered in Europe during the Renaissance, in the wake of the great medieval Arabic translation movement—and its treatment of mechanics proved inspirational to early modern physicists such as Tartaglia and Galileo. They were trying to work out how to compose the paths of moving objects from component motions, such as the combination of an oblique upward motion and the downward effect of gravity when a tennis ball is lobbed into the air.[2]

Actually, Tartaglia and Galileo were interested not so much in tennis balls as in arrows and cannonballs; such were the times, with their seemingly endless religious and imperial wars. Guns and cannon were becoming more sophisticated—for example, new cast-iron techniques meant cannonballs were now lighter but more powerful—and it was increasingly clear that knowing a missile's trajectory in advance would enable these new weapons to be deployed more efficiently. Gunners' manuals were being published, too, setting down records of the observed ranges and angles of various shots, and mathematicians began to use these data to try to describe the path of a projectile mathematically.

As a first step, in the early 1530s—a few years before his acrimonious quarrel with Cardano over cubics—Tartaglia managed to work out that if you want a cannonball to go as far as possible, the cannon should be pointed at an angle of 45° to the ground. If you just pointed it straight ahead at a distant target, gravity would likely catch you short. And if you pointed it high and hoped for the best—which had long been the usual practice—then you'd often miss your target and destroy something else instead. But Tartaglia suffered a crisis of conscience, for he knew he was dallying with people's lives while he worked on weapons of war. So he burned his papers, fearing God's wrath for engaging in such "vituperative and cruel" work.[3]

These were fearful times, however. As a child growing up in northern Italy, Tartaglia had suffered a terrible wound from a French soldier's saber. He was twelve or thirteen and had been hiding with his mother and sister in a church as the French stormed their town, but the slashing sword caught young Niccolò across the mouth and jaw. The wound's legacy lives on in his name, Tartaglia—his actual surname seems to have been Fontana, but after the injury to his palate he acquired the nickname "Tartaglia," the Stammerer. Despite all this, and despite his lack of education, he became one of the best mathematicians of the early sixteenth century. As for his work on ballistics, he soon decided that as a good Christian he'd better talk about it to his patron, for the Islamic Ottoman Empire was expanding ever further into Christian Europe. It already extended through much of Greece and Turkey, and as far north as Belgrade and Sofia, but by the 1530s Emperor Süleyman's troops had invaded Hungary and were skirmishing with Austria. The Austrian Habsburgs were extending their own influence, too, and the French and the Ottomans formed an alliance against them—and so it went: war seemed to be everywhere.

So Tartaglia kept working on the subject. He followed up his advice about angling cannon at 45° by designing a kind of quadrant for measuring such angles. He also tried to compose a theoretical ballistic trajectory. It was intuitively evident that if you ignored air resistance—a realistic enough assumption for a cannonball although not for a feather—such a path was built from just two different components: one in the direction the shot was fired, and one in the direction of gravity. But Tartaglia's attempt wasn't successful. First, his components were not clearly defined forces or velocities but just intuitive "motions"—after all, he was writing 150 years before Newton's careful, groundbreaking laws of motion. What's more, Tartaglia never added (or "composed") these "motions" vectorially. In his commentary on the rediscovered *Questiones Mechanicae,* he'd simply ignored the parallelogram rule—like most his contemporaries, he didn't understand its significance.[4]

Second, he didn't have the idea of *independent* components, each one acting as if the other weren't there. This is key to the parallelogram rule, as you can see in figure 3.1. An example of what this means is that if you took away gravity in this analysis so that only one of the components acted—the one in the direction the cannon was fired—you should deduce that the ball would keep moving at the same speed along that initial direction. This is

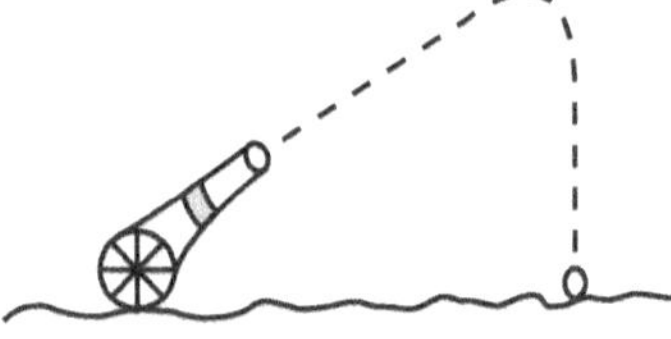

FIGURE 3.2. Sketch of how early theoreticians such as Tartaglia conceptualized the path of a projectile.

Newton's first law of motion, although others, notably Galileo and Harriot, had intuited it before him.

Tartaglia, however—like almost everyone since Aristotle—believed it was not possible for a body to have two independent components of motion acting *at the same time*. So he didn't have the idea of *combining* the two components, via the parallelogram rule, to give the resulting observed motion. As you can see in figure 3.2, he composed his trajectory from an initial oblique component *followed by* a downward one. He didn't have the idea that both were acting together from the outset, so that the ball's path actually starts out curved, not straight, because gravity is already pulling it down. Instead, in the middle of his trajectory, he has a curved section representing the cannonball's changing course when gravity "suddenly" overcomes the initial outward force, and then the ball drops straight down.

While Tartaglia's was the first significant foray into the study of ballistics, the flaw in his analysis suggests that no one really knew how gravity operated, let alone understood vectors. Enter Galileo, and Harriot, half a century later. War was still raging, but now the tension between Protestants and Catholics was bringing new unrest. Yet there was an intellectual and aesthetic fascination with "the art of war," too. War games were an entertaining pastime for people such as Harriot's second patron, the Earl of Northumberland; he'd served briefly in the military and was so fascinated by military strategy that he spent countless hours playing an elaborate board game with 140 brass soldiers, each with a wire pike, and 320 lead soldiers carrying tiny muskets. War was a subject of art, too: continental Renaissance artists such as Georgio Vasari and Albrecht Altdorfer had been painting bird's-eye battle scenes, and now, in England, William Shakespeare and Christopher Marlowe were evoking huge battles through powerful language and clever staging, in plays such as *Henry V* and *Tamburlaine*. *Tamburlaine* dramatizes the terrifying conquests of the fourteenth-century Tartar warrior Tamerlane. It was a reminder, perhaps, of the more recent exploits of Süleyman, although Marlowe wrote it in the shadow of a more immediate danger, the Spanish Armada, which for various long-simmering religious, geopolitical, and economic reasons attacked England in 1588. There is some evidence that Marlowe knew Harriot—and it has been conjectured that they discussed the "art of war," Marlowe as a poet and playwright, and Harriot as a mathematician working on ballistics.[5]

There's no evidence that Galileo and Harriot knew each other, but they shared the cutting-edge scientific interests of their time—and not just in ballistics, but in all motion that is affected by gravity. As a result, they independently codiscovered the general law of free fall—that in everyday situations, if you take away air resistance, all bodies, no matter their weight and size, fall at the same rate. For Tartaglia, though, a cannonball fell faster than a tennis ball, so they'd each have differently shaped projectile trajectories. But Harriot and Galileo showed that in a vacuum, all projectiles trace the same shaped path: a parabola. You can see this shape easily if you tilt a hose with a jet of water coming out, because each droplet, one after the other, traces out the shape of a projectile. *Proving* it's a parabola, with a distinctive equation of the form $y = -ax^2 + bx + c$, is another matter. But if you've studied math or physics at high school or college level (or if you peek ahead at fig. 8.1), you'll know that it can be done in a dozen lines using Newton's laws, vectors, and the rules of calculus. Without these techniques, it took Galileo and Harriot many pages and many arduous months as they searched for the right equation to fit their data.

By contrast, you can see in figure 3.2 that Tartaglia's estimated trajectory was definitely not a parabola! Galileo and Harriot succeeded because they realized that component motions do act independently—which is also how they came to intuit Newton's first law.[6] Galileo's work on projectiles influenced Newton, who cited it in *Principia*. But Galileo never quite got to the full parallelogram rule. He used it to calculate the magnitudes of his resultant forces, but not the directions—and the same was true for most other contemporary pioneers. René Descartes did use the rule to calculate magnitudes *and* directions, but he thought of them as separate quantities, not as separate aspects of a *single* quantity, such as velocity or force. So, you can see why there is more to vectors than we, with our hindsight and habit, might realize![7]

Harriot came close to the idea, though, in his 1619 analysis of the mechanics of collisions. As well as war and war games, the game of billiards was in the air—at least for wealthy aristocrats such as Harriot's patron Northumberland. Unfortunately, the earl soon landed in the Tower of London, a prisoner of King James I—both he and Harriot were arrested in 1605 when the Gunpowder Plot was foiled, because they had unknowingly dined with one of the Plotters just the day before. (The Plotters were disaffected Catholics chafing under the rule of the Protestant king and his parliament, and one of them happened to be Northumberland's cousin.) Harriot was released after a few terrifying weeks, but the innocent earl was not so lucky: he spent the next sixteen years in prison. He managed to while away some of the time playing war games—and by having a bowling alley constructed in the Tower's grounds. But he also spent his time thinking about science, and to build on his interest in bowls and billiards he asked Harriot to explain the mechanics of collisions.

Harriot's paper was the first detailed analysis of the topic—and one of the most sophisticated vectorial analyses before Newton. I've reproduced Harriot's vectorial diagram in figure 3.3. The masses of the balls *a* and *A* are indicated by the size of his circles, and they move toward each other from the left, collide in the middle (at *b* and *B*), and bounce off to the positions denoted by the dotted circles *c* and *C*. I've explained his calculations in the endnote,[8] but you can see just by looking at Harriot's diagram that he used parallelograms to analyze the two types of collision he's shown here. And although his terminology is old-fashioned—he speaks of motion and impetus instead of velocity and force—he has the essence of the modern parallelogram rule: both magnitudes and directions are specified, and he

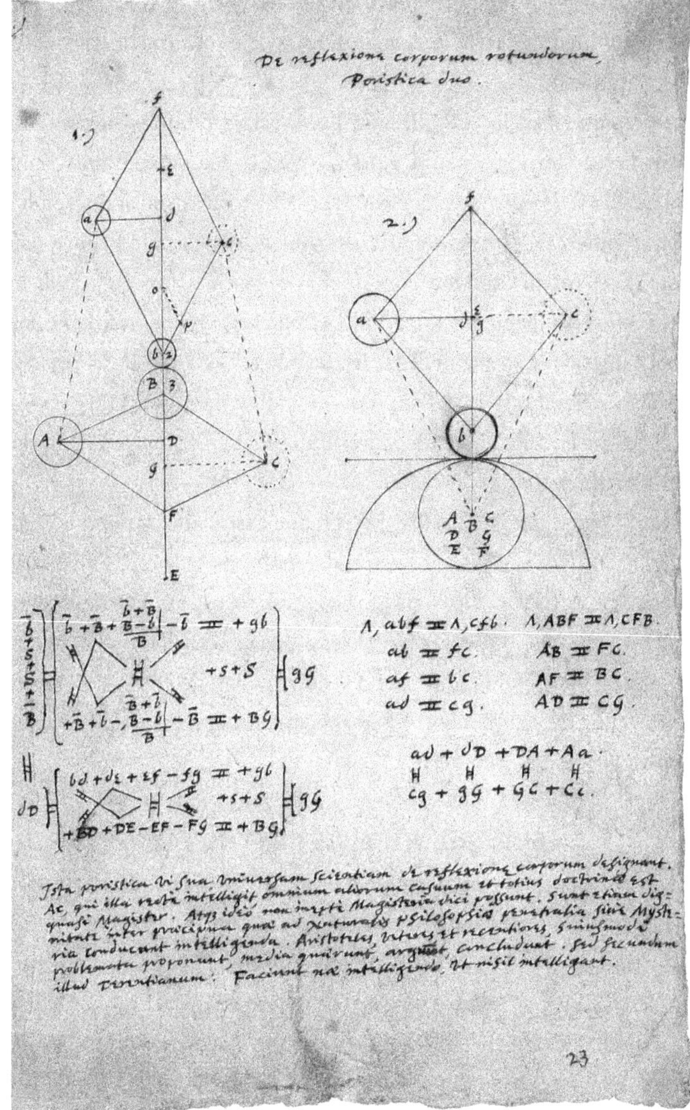

FIGURE 3.3. Harriot's use of the parallelogram rule to analyze the mechanics of collisions. The manuscript: West Sussex Record Office, PHA HMC 241 Vol VIa, f23r. By courtesy of the Rt. Hon. Lord Egremont, and with acknowledgments to the County Archivist, West Sussex Record Office.

has independent components, each acting "as if" the other weren't there, which combine to give the resultant motion or force. The only trouble is, he didn't publish it.[9]

His countryman John Wallis may have seen it, though—he'd certainly seen some of Harriot's manuscripts, as well as his posthumous algebra textbook *Praxis*. Wallis was inspired especially by Harriot's fully symbolic algebraic notation, as I've mentioned, but he also tackled the topic of collisions—Newton cited his work on this as another important influence on *Principia*. Wallis clearly stated the parallelogram rule in a 1671 paper on mechanics, and, in the decades leading up to 1687 when *Principia* was published, several others did so, too—including the remarkably versatile Fermat. (Like Harriot's algebraic forerunner Viète, Fermat's day job was in law, not mathematics.)[10]

You might think that, with the parallelogram rule coming of age in *Principia*, and with Newton's vectorial definitions of force, acceleration, and velocity, it was he who sparked the search for a complete, mathematical vector analysis—but it wasn't, at least not directly. His work did contribute later, by showing how useful vectors would be in physics, but that was after the first vector algebra had appeared from a surprising and completely different direction.

A SURPRISING TWIST IN THE TALE

You can perhaps guess this surprising direction from Hamilton's graffiti. It contains the imaginary number i, which comes up when you try to solve an equation such as $x^2 + 1 = 0$. More complicated equations give an even stranger quantity, a mixture of a real and an imaginary number, such as Rafael Bombelli's $2 + 11\sqrt{-1}$, which we also met in chapter 1. I mentioned, too, that Descartes had coined the term "imaginary number," about a century after Bombelli—and then, in 1777, more than a century after Descartes, the Swiss mathematician Leonhard Euler created the modern symbol i to denote $\sqrt{-1}$.

Euler also gave us the symbol e for the exponential (or Euler) number—which I've defined below—and he popularized William Jones's symbol π. But just as with Jones's π, Euler's *i* didn't really take off until another standout mathematician used it several decades later—the German Carl Friedrich Gauss. Gauss also coined the term "complex number," for the mix of real and imaginary numbers we write today as $a + ib$ or $a + bi$—the order in which you write the *i* doesn't matter. But the term "complex" didn't catch on straightaway, either. All of which goes to show the slow evolution of innovative ideas—and how crucial it is to publish if you want to get some belated credit!

Incidentally, Gauss was also the unknowing mentor of the self-taught female trailblazer Sophie Germain. Like most men of his time, Gauss believed women were incapable of higher mathematics, and Germain had prudently written to him—about an idea for proving Fermat's last theorem—using the pseudonym Monsieur Le Blanc. When she finally revealed that she was a woman, he was amazed, and very impressed.

But what has *i* to do with vectors? It hinges on the question of *how to represent information*—and the related question of what kind of numerical constructs qualify as objects of mathematical study, by obeying clear mathematical rules.

In the case of *i*, if mathematicians were going to take this number seriously, how were they to think about it? It had taken long enough to extend the number line to the left of zero to represent negative numbers—a seemingly simple step that made it possible to think of negative numbers simply as counterparts to positive ones, and just as real. It was Wallis who took this momentous step—although he also thought that negative numbers must be larger than infinity, which shows how difficult it was for even the best mathematicians to get a handle on negative numbers (let alone infinity). It was also Wallis who first tried to do something similar for complex numbers, but he wasn't successful.[11]

It would take another century for this representation problem to be solved, and to inadvertently open the way for the discovery of vectors. Meantime, mathematicians needed to find out more about the mathematical nature of complex numbers. Sure, they came up as solutions of quadratic and cubic equations, but did they play a broader role in mathematics—a role that would make it imperative to find a way to incorporate them into an expanded system of numbers?

One of the first to find answers to this question was Euler. He was one of the most prolific mathematicians in history, and he had such a phenomenal memory and such extraordinary dedication that not even blindness stopped him. He lost sight first in his right eye, in 1735 when he was only twenty-eight, and became almost blind at fifty-nine, but with the help of a secretary and his children and grandchildren, he continued researching until he died suddenly at seventy-six. He had been a student of the formidable Johann (or Jean) Bernoulli—the fierce defender of Leibniz and scourge of Newton—and he was one of the first to rewrite Newton's laws explicitly in terms of symbolic calculus. He began working on complex numbers with Bernoulli in 1727, but his work on this topic really began to come together in the 1740s, when he connected i with the "circular" functions $\sin\theta$ and $\cos\theta$. (Bear with me here, because circles are related to rotations, and coming to grips with the algebra of rotations is what will lead Hamilton to create vectors. But first he will need to understand the connection between rotations and i.)

If you represent $\sin\theta$ and $\cos\theta$ on a circle with radius 1, as in figure 3.4, then, using Pythagoras's theorem, you get the helpful identity

$$(\cos\theta)^2 + (\sin\theta)^2 = 1.$$

Euler realized that you could factorize this equation as:

$$(\cos\theta + i\sin\theta)(\cos\theta - i\sin\theta) = 1.$$

From there, using a fledgling definition of e (which you can see in the endnote),[12] he came up with what is now known as Euler's formula:

$$e^{i\theta} = \cos\theta + i\sin\theta.$$

(It follows easily if you know, as Euler did, the Taylor series for $\sin\theta$, $\cos\theta$, and e^{θ}, but these details are not important for our story.)

You can do a lot with Euler's remarkable formula. It enables you to do pure mathematical things such as finding the square-, cube-, and higher

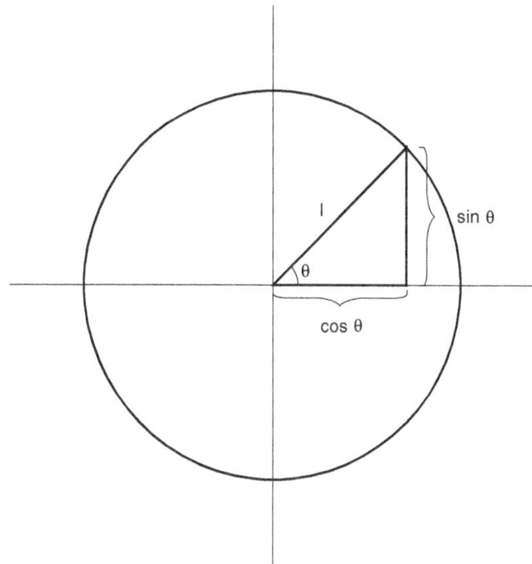

FIGURE 3.4. Using Pythagoras's theorem and the definitions of sin θ and cos θ shown in the diagram, you get $(\cos\theta)^2 + (\sin\theta)^2 = 1$.

roots of complex numbers mentioned in chapter 1. And it turns out to be incredibly useful in practical applications, too, such as modeling electric circuits and other wave-like phenomena—including in quantum mechanics, as in Schrödinger's equation (which governs the "wave function" representing the possible state of a moving subatomic particle such as an electron). That's because mathematically, waves are periodic cycles, such as you get from cos θ and sin θ when you turn θ through a whole circle (as in fig. 3.5).

Rotations around a circle will also turn out to be the key to understanding the relationship between real and imaginary numbers. In fact, since π plays such an important role in the formulae for the circumference and area of a circle, you can already see there's some sort of connection between circles and imaginary numbers simply by choosing $\theta = \pi$ in Euler's formula. You get what is often voted the most beautiful equation in mathematics:

$$e^{i\pi} + 1 = 0.$$

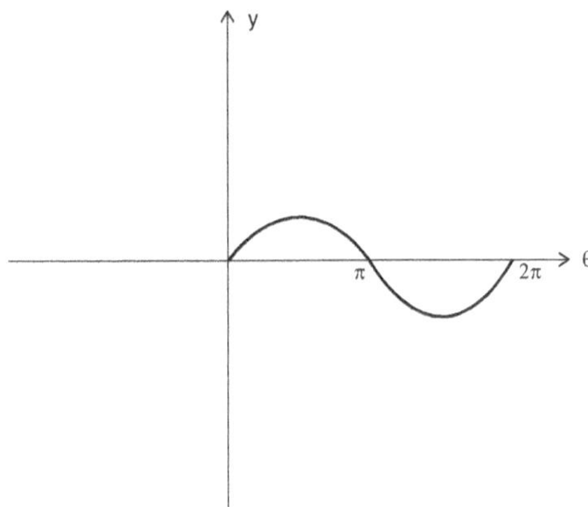

FIGURE 3.5. A sketch of the graph of $y = \sin\theta$.

It's beautiful because it is elegantly simple, yet profound. It has just seven symbols, and they are interconnected in an entirely unexpected way: aside from the basic operation signs + and = there is *e*, the base of the natural logarithms; π, the mysterious number at the heart of all circles; 0 and 1, the most important integers; and the imaginary number *i*. Who'd have thought that all these different types of number were related? It is a marvel of an equation, although Euler never actually wrote it down. But since it is such an obvious step from Euler's formula, and since there is no record of who was the first to take it, today it is justifiably called Euler's identity.[13]

Euler was still confused about the nature of complex numbers, though, and he didn't make any progress on representing them on some sort of "number line," either. So, no one really knew *how* all these numbers were connected.

Still, Euler did go on to use complex numbers algebraically, in what might at first seem a surprising way: trying to prove Fermat's last theorem. Or, should I say, trying to prove a special case of it—that aside from 0 and 1 there are no integers x, y, z such that $x^3 + y^3 = z^3$. Since this is a cubic equation, it is possible to try to "solve" it with Cardano's algorithm, which has a habit of turning up complex numbers. The idea is that when you find a solution, you then prove that it contradicts the assumptions inherent in the problem—such as the fact that x, y, z have to be integers! This process is called proof by contradiction, and it is very powerful in pure mathematics. Euler took an approach similar to that of Bombelli, by exploring cubes and cube roots of numbers of the form $a + b\sqrt{-3}$ (that is, $a + ib\sqrt{3}$).

Everyone was intrigued by Fermat's last theorem, which says that if n is an integer larger than 2, there are no nontrivial x, y, z such that $x^n + y^n = z^n$. It was intriguing because there was no problem in finding solutions to $x^2 + y^2 = z^2$—the ancient Plimpton 322 tablet has a whole list of them. In the 1630s, Fermat had famously scribbled a note indicating that he had found a truly marvelous general proof of his conjecture—but if he did, no one has been able to find it. He did give a proof when $n = 4$, and Euler's proof for $n = 3$ was the next step, albeit a century later.[14]

Over yet two more centuries, many mathematicians—including Germain—contributed proofs of special cases, until Andrew Wiles proved the whole thing in the 1990s, with the help of new ideas and massive computer power. But Euler was the trendsetter on the road to a proof, and this meant that his use of algebraic numbers of the form $a + ib$ also got noticed, although others had already done something similar—including Frenchman Jean Le Rond d'Alembert, who was miffed that Euler hadn't acknowledged him. Euler was terrific at developing others' fledgling ideas and putting them in their definitive form, the way he did with Newton's laws—but he was rather careless about acknowledging his sources. Perhaps he read so widely and took ideas into such deeper territory that he simply lost track of where he'd read what.[15]

Euler didn't go any further with complex numbers, but what he achieved with these two projects was significant—linking i with the circular functions, and treating $a + ib$ as an algebraic quantity that behaved like an ordinary number if you added and subtracted the real and imaginary parts separately. But it was this last bit that William Rowan Hamilton had trouble with, half a century later, for he thought it was like adding apples and oranges.[16]

Of course, a lot had happened since Euler's time. In particular, mathematicians had finally found a way to *represent* these strange, complex numbers on some sort of number line—or rather, as it turned out, a plane. It was geometric rather than algebraic, but it was brilliantly intuitive, and it went a long way toward helping mathematicians feel secure in treating these hybrid beasts as numbers. More than a century after Wallis's attempt, it was a solution whose time had come, for several mathematicians independently came up with ideas similar to the modern representation in figure 3.6. The most notable of them are the Norwegian surveyor Caspar Wessel in 1799, although his pioneering effort lay largely buried for the next century; Gauss around the same time, although he didn't publish till 1831; Argand (whose biography is uncertain but who was probably the Swiss-born Parisian bookstore manager Jean Robert Argand), in 1806—and whoever he was, we often call the "complex plane" in figure 3.6 the "Argand plane" in his honor—and Englishman John Warren in 1828.

And finally, some of the disparate ideas I've been talking about began to come together: components, circles, imaginary numbers, and a glimpse of vectors.

First, if you take the idea of the unit circle from figure 3.4 and superimpose it on the complex plane in figure 3.6, you get figure 3.7—a unit circle centered at the origin of the real and imaginary number lines. The

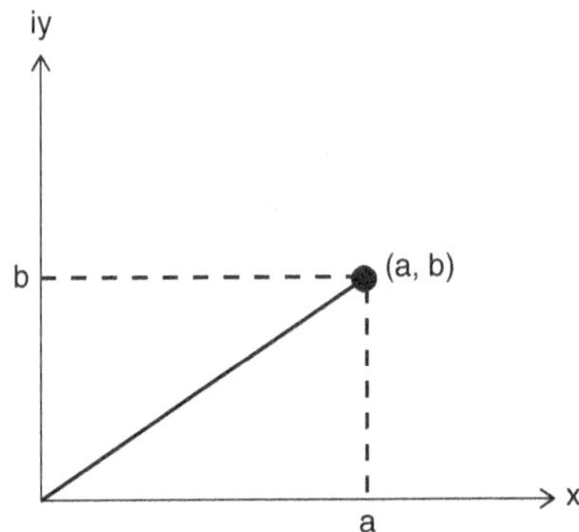

FIGURE 3.6. The Argand or complex plane has the real number line as its horizontal axis and the imaginary number line as its vertical axis. Points (x, y) in this plane represent complex numbers $x + iy$. The diagram shows the particular complex number $a + ib$, represented as the point (a, b).

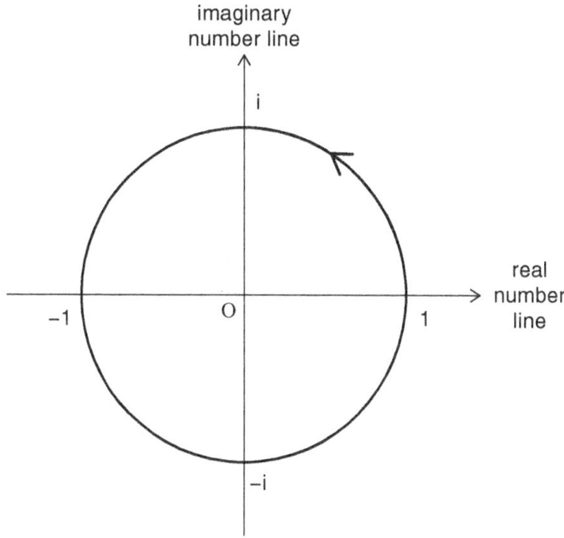

FIGURE 3.7. Multiplication by imaginary numbers, envisaged as rotations in the complex plane.

number +1 corresponds to $\theta = 0$; i corresponds to $\theta = \frac{\pi}{2}$ radians or 90°; −1 corresponds to $\theta = \pi$ or 180°, and $-i$ corresponds to $\theta = \frac{3\pi}{2}$ or 270°. (Or you can go clockwise so that $-i$ corresponds to $\theta = -\frac{\pi}{2}$ or −90°, and so on.) But here's the fascinating part. If you mentally move along the circle from the point 1 to the point i, you've rotated through 90°. In other words, a rotation of 90° takes the number 1 to the number i—just as when you write $1 \times i = i$. It's as if the act of rotating the real number 1 through 90° is equivalent to multiplying it by i. Now rotate the point at i by 90°: you get −1, just as you do when you say $i^2 = -1$. Multiplication by an imaginary number does indeed seem to be nothing more than a simple rotation! So, the wonderful thing about the complex plane is that it shows just *how* imaginary numbers can be related to real ones: via geometric rotations.

What's more, a *complex* number, made of real and imaginary parts and expressed as $a + ib$ (or $a + bi$, whichever you prefer), can be represented geometrically as a *point* (a, b) in this complex plane (as in fig. 3.6). And if you imagine, with hindsight, an arrow from the origin to this point, you see that it has a real and an imaginary component. It obviously has a direction, too, and a magnitude calculated from the components via Pythagoras's theorem. (Actually, a complex number has a "modulus" rather than a magnitude: the modulus of the complex number $a + bi$ is defined to be $\sqrt{a^2 + b^2}$, analogous to the magnitude of a vector, which came later.) So, you can see that the inventors of this geometrical way of representing complex numbers—Argand, Gauss, Wessel, and Warren—were closing in on the idea of a vector, just as Newton had done. Newton had approached the problem from the perspective of physics, emphasizing the magnitude and direction of physical quantities such as force, while these early nineteenth-century mathematicians were coming at it from a number theory point of view. It would take Hamilton to define vectors explicitly, and to bring the two approaches together.

HAMILTON STEPS IN

In the 1830s, it must be said that Hamilton still had quite a ways to go. Initially, he knew only of Warren's work on the complex plane, not that of Gauss or the others. But Warren's approach inspired Hamilton to imagine a way of thinking about complex numbers and their geometrical rotations so that they could be handled algebraically, the way you handled ordinary numbers when you wanted to solve equations—and with no overt mixing of apples and oranges.

He got around the difficulty in a rather clever way. If you start by taking any two of these hybrid numbers and adding and multiplying them as if they were ordinary binomial expressions, except that $i^2 = -1$, you get this:

$$(a + ib) + (c + id) = (a + c) + i(b + d),$$
$$(a + ib)(c + id) = (ac - bd) + i(ad + bc).$$

We do it all the time in math classes today, and Euler, Gauss, and others had done it, too, although back then it still seemed somehow like fudging, because no one was yet comfortable with this kind of arithmetic—or, should I say, with this kind of number. As Gauss told the astronomer Peter Hansen, "The true meaning of $\sqrt{-1}$ reveals itself vividly before my soul, but it will be very difficult to express it in words, which can give only an image suspended in air." And Hamilton's friend Augustus De Morgan claimed to have "shown the symbol $\sqrt{-1}$ to be void of meaning, or rather self-contradictory and absurd."[17]

So, Hamilton said, why not define complex numbers without the distracting i, the stray orange among the apples, *yet still keep the same rules of addition and multiplication*. Why not think of complex numbers entirely in terms of "couples" (or what modern mathematicians call "ordered pairs") of *real* numbers—(a, b), (c, d), and so on—and then *define* addition and multiplication so that they followed the same pattern:

$$(a, b) + (c, d) = (a + c, b + d),$$
$$(a, b) \times (c, d) = (ac - bd, ad + bc).$$

Earlier mathematicians, especially Gauss, had come close to this idea, but Hamilton spelled it out clearly. Not that it met with much success at first. For thousands of years multiplication had been just one number multiplied by another, so Hamilton's multiplication of couples, with its jumbled-up mixture of multiplications, additions, and subtractions, was something very different.

But its time was near, for his English colleagues De Morgan and George Peacock had already begun exploring the idea of a completely symbolic algebra—moving on from Harriot and Wallis to an idea of algebra divorced not just from geometry but from arithmetic, too. In other words, algebra defined in terms of symbols that didn't have to mean anything concrete, but which did have to obey clearly established rules—just as Hamilton's "couples" do. Peacock and De Morgan had begun by expressing the rules of ordinary algebra formally, even though they'd been used intuitively for hundreds, perhaps thousands of years. After all, it doesn't matter if you add three apples to two or two apples to three, so you'd think it hardly needed stating as the "commutative law for addition." Unless, of course, you wanted to move algebra beyond numerical quantities to a science of purely abstract symbols and operations, as Peacock and his former student De Morgan did.

Peacock's interest in symbolism had initially taken shape in the context of calculus. In chapter 2, I mentioned the power of the suggestive Leibnizian differentials *dx*, *dy*, and so on, compared with Newton's dot notation, and Peacock had been part of a group of young Cambridge academics—including Charles Babbage of prototype-computer fame, and John Herschel, son of the famous astronomer who discovered Uranus, and a significant astronomer in his own right—who'd agitated for a curriculum reform to replace Newton's calculus with Leibniz's. As Babbage impishly put it—punning on both formal religion and the awe in which Cambridge still held its former student Newton—these young mavericks supported "the principles of pure *d*-ism as opposed to the *dot*-age of the university."[18]

As for purely symbolic algebra, Hamilton himself initially straddled both camps. He agreed that geometrical representations on the complex plane were illustrative rather than fundamentally mathematical. But he wasn't a pure symbolist, because he felt that algebra should be about something more tangible than a set of rules. He'd even wondered if the two camps were linked like space and time. The German philosopher Immanuel Kant had said that our existence depends on *preexisting* space and time, and that geometry is the fundamental language of spatial representation—so, thought Hamilton, perhaps algebra was the science of time? After all, he said, the way we mark time and space is the means "by which thoughts become things, and spirit puts on body, and the act and passion of mind are clothed with an outward existence, and we behold ourselves from afar."[19]

Poetry aside—Hamilton had been writing poetry since his youth—his algebra of time sounds odd today, when we are so used to algebraically manipulating arbitrary *x*'s and *y*'s or *a*'s and *b*'s or whatever symbols we choose. His symbolist colleagues didn't like it either. But as Leibniz and others had pointed out long ago, our notion of time (and of space, too) is relational—*this* happened before *that*, this is happening *now*, and so on. Einstein would take this idea to its logical conclusion, turning up in the process some bizarre new ideas about time slowing down and "now" itself being relative; in the meantime, Hamilton realized that these relations between moments of time could be expressed in terms of algebraic expressions. His focus was pure mathematics rather than physics—otherwise he might have anticipated Einstein; as it is, he cited Newton, who had appealed to "moments" of increase and the "flux" of time when describing how things change, in his concept of what we now call a derivative but which he had called a "fluxion." Hamilton took the time analogy into more basic territory by suggesting that if you think of numbers themselves as representing the mental "steps" between "moments" on a directed line segment, you can "explain" the laws of arithmetic as a science of time.

These "steps" were rather esoteric, although later he moved on from moments in time to points in space, and he identified his "steps" as vectors; either way, you can get his drift in an example: the confusing but seemingly inescapable fact that the product of two negative numbers is positive, so that $(-3) \times (-1)$, say, is equal to 3. As Hamilton put it, in his philosophy you could explain this bizarre result by saying that "two successive reversals restore the direction of a step." The take-home message here is the importance of direction in his conception of a vector: reverse the direction of an arrow twice and you're back where you started. It's analogous to saying "I will not not go"—two successive "nots" cancel each other out, just as two minus signs cancel or "restore" us to the positive direction. It certainly helps make sense of things when you're multiplying two numbers that are each less than zero and yet you end up with a positive![20]

Hamilton did eventually move closer to the purely symbolic approach that seems evident, today, in his definition of the arithmetic of "couples." It had been almost evident to De Morgan, too. He was "inclined to think" that if you left time out of it and focused simply on real numbers, Hamilton "may finally" have provided a way to think about complex numbers through symbolic algebra, not just intuitive geometric analogy.[21]

Today, it's easy to wonder why anyone would worry about the peculiarities of negative and complex numbers and the rules of algebra—we are so used to just getting on with it and using these things when we need to, without worrying about underlying conceptual problems. But, of course, it is because some of the greatest mathematical minds did worry about such things that we have both modern math and much of our technology. When Hamilton introduced his paper on complex couples, which he published in 1837, he said right up front that his focus here was purely mathematical. Little did he know that today NASA would be using the ultimate fruits of his research to help drive its spacecraft! So even if his initial focus on time seems misguided, it was this kind of out-of-the-box thinking—this concern with the very foundations of number and arithmetic—that would ultimately lead Hamilton to the invention of vectors and quaternions, which have so many practical applications today.

It is telling that Hamilton's sources of inspiration included not just mathematicians but also philosophers such as Kant—and writers and poets, too: William Wordsworth was a good friend, and so were the famous novelist and educational theorist Maria Edgeworth along with the Romantics Francis Beaufort Edgeworth (Maria's younger half-brother) and Samuel Taylor Coleridge. Wordsworth said Hamilton and Coleridge were the two most wonderful and gifted men he had ever met. And when Maria Edgeworth first met Hamilton, who was then only nineteen, she thought he had "both the simplicity and the candour which make a true genius."[22]

Such eclectic influences were not as unusual in the nineteenth century as they are today, now that mathematics, like science, has become so sophisticated and specialized. But Hamilton was no ordinary man, even in those days.

A DAZZLING PRODIGY

He was born in Dublin at midnight, as August 3 turned into August 4, 1805. His father, Archibald, was an attorney, and his mother, Sarah, was considered intellectually gifted. But William—the only son among five surviving children—was *so* gifted that his parents entrusted his education to his university-educated uncle, James Hamilton, an Anglican clergyman, classicist, linguist, and diocesan schoolmaster. From the age of three, William lived mostly with his uncle, who nurtured his precocious talent so well that by the time he was thirteen, he reputedly had a good knowledge of Greek, Latin, Hebrew, French, German, and Italian, and was studying Sanskrit, Persian, and Arabic.

Uncle James wasn't quite so able when it came to William's equally prodigious talent for calculating, but he did his best to find suitable textbooks for him. Young William was thereby able to use his French skills to teach himself calculus from a French textbook—so he learned Leibnizian calculus rather than Newton's form—and at sixteen he began to study Pierre-Simon Laplace's *Traité de Mécanique Celeste* (*Treatise on Celestial Mechanics*). This was a monumental update of Newton's *Principia*, incorporating all the advances in mathematical astronomy that had taken place in the intervening century. Laplace was a great mathematician, but the teenaged Hamilton spotted a flaw in his reasoning—about, as it happens, the parallelogram rule for composing forces. Evidently, he was already on the road to pioneering vectors.

This was in 1821. At the same time, a self-taught Scottish woman was also privately studying *Mécanique Celeste*, and ten years later she would astonish the British mathematical community with her textbook *Mechanism of the Heavens*. She was Mary Somerville—we met her earlier, when she was studying Hamilton's quaternions in her nineties—and her book was an expanded, explicated 610-page English version of the first two volumes of Laplace's five-volume tome, plus a 70-page "preliminary dissertation" giving an overview for the nonspecialist. Maria Edgeworth read the overview and was entranced: she found that the simplicity of her friend Somerville's writing superbly suited "the scientific sublime." As for the academics, George Peacock was one of the Cambridge mathematicians who were impressed with the whole book—so much so that they used it as a text for their advanced astronomy students. (True, its use of Leibnizian calculus also helped further Peacock's curriculum reform—but the book was so good that it remained the standard text for the next century.) Somerville was thrilled—especially when reviewers commented on how extraordinary it was that a self-taught woman had written a book that few men could understand. But she was furious, too, because women were still denied a decent education—and in the late 1860s she would be the first to sign John Stuart Mill's (ultimately unsuccessful) petition for votes for women.[23]

Hamilton, by contrast, did go to university—to Trinity College Dublin, where the teaching was excellent, in both classics and mathematics. He did so brilliantly that he was made professor of astronomy and royal astronomer of Ireland when he was not quite twenty-two. A few years later he became famous for his mathematical prediction of conical refraction, as I've mentioned. Then, still in his twenties, he formulated a new interpretation of dynamics that would become significant in many areas of physics—in the mechanics of systems of moving objects, for instance, and in Schrödinger's equation describing the (quantum) mechanics of an electron. He was still only twenty-eight when he presented his initial ideas on complex number "couples" to the Royal Irish Academy in 1833, and he was knighted at thirty for his services to mathematics.

At the end of 1837, thirty-two-year-old Hamilton was elected president of the Royal Irish Academy, and one of the first things he did was seek advice from Maria Edgeworth, who had just turned seventy. The Academy was designed to foster science and the humanities—Edgeworth's father had been a founder—but Hamilton felt that literature wasn't getting enough of a look-in. "It is known to all the world," he wrote to her, "that you are not only a lover of literature but a successful pursuer and powerful promoter of it, and that on any point connected therewith, your opinion must be most valuable." Edgeworth's works may be largely forgotten today—although in 2009 two of them made the *Guardian*'s list of "1000 novels everyone must read"—but in the early nineteenth century she was Ireland's most famous novelist, and it wasn't just Hamilton who admired her but the likes of Walter Scott and Jane Austen, too. She gave Hamilton some sound advice—such as offering medals for essay competitions, and setting subscription fees at an affordable rate for ordinary literature-lovers—and he gratefully drew on most of it in his inaugural presidential address. Unfortunately, he wasn't brave enough to tackle her advice to allow women to attend special Academy discussion nights. She had long been promoting women's right to education and their right to engage with science, and she wasn't impressed with the excuses he made in his reply to her. Still, he would preside over her election as an honorary member of the Academy in 1842—the first local woman to be so honored. Three foreign women had already achieved this distinction—including Mary Somerville and Caroline Herschel (John Herschel's aunt, and an astronomer like her

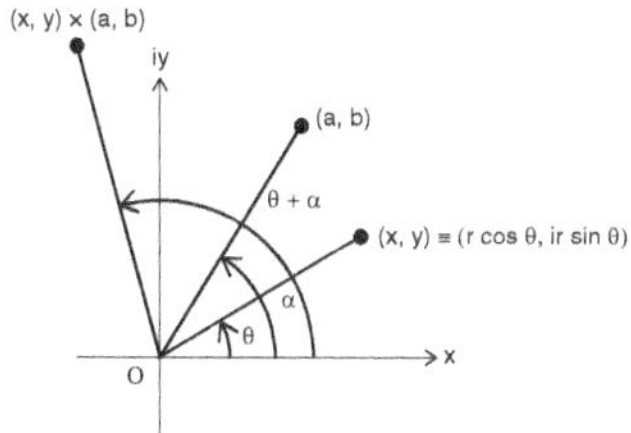

FIGURE 3.8A. Multiplication of complex numbers represented by rotations.

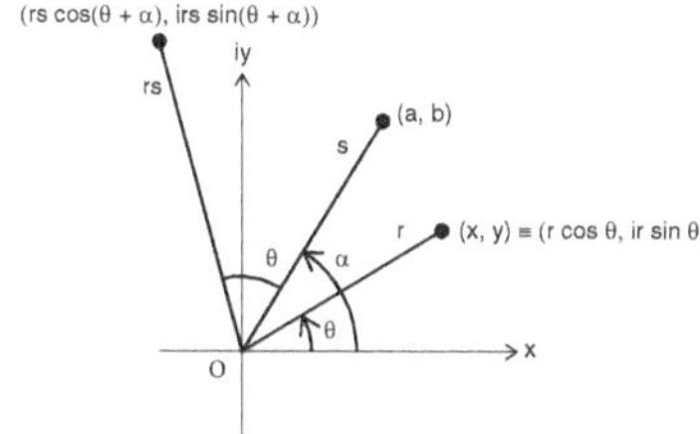

FIGURE 3.8B. How to calculate multiplications via rotations: Use polar coordinates and Euler's formula to write $x + iy = r\cos\theta + ir\sin\theta = re^{i\theta}$, and $a + ib = s\cos\alpha + is\sin\alpha = se^{i\alpha}$; then, using the index laws, you see that the product is $rse^{i(\theta+\alpha)}$. So, multiplication of $x + iy$ by $a + ib$ has extended the magnitude of the original vector and rotated it through an angle of α radians. Simply adding one complex number to another can be viewed as a translation in the plane.

nephew and her brother William)—but, living abroad, they were unlikely to ruffle male feathers by wanting to attend meetings. The first full female members would not be admitted until 1949.[24]

Disappointing as it is that Hamilton was not as ahead of his time on the so-called woman question as he was in mathematics, it is what he did next that is important for the story of vectors. First, while Gauss and others had seen that a complex number could be represented as a *point* in the complex plane, Hamilton saw it as a "directed line segment"—in other words, an *arrow*.[25] Wessel had something like this idea, too, but neither Hamilton nor anyone else in the mainstream knew about it. Argand also considered "directed line segments," but it was Hamilton's next step that would be truly foundational for vectors. If you multiply two "couples"—two "arrows" (x, y) and (a, b) in the Argand plane—you get a rotation in a two-dimensional plane, as in figure 3.8. So complex number multiplication is linked to the geometrical operation of rotation—just as it is for multiplication by *i*. Was it possible, Hamilton wondered, to represent rotations in *three*-dimensional space via the multiplication of "triples"?

He began working on the problem in earnest in 1841, when he read a paper in which his friend De Morgan concluded that it simply didn't seem possible to represent three-dimensional geometry algebraically. The imaginary number *i* had performed its rotational magic in two dimensions, but De Morgan argued that there were no more such symbols available—no new algebraic ideas that could move geometry from the two-dimensional page to all of space.

It was just the challenge Hamilton needed.

(4)

UNDERSTANDING SPACE (AND STORAGE)

Hamilton mulled over his algebra of triples for years, and by the fall of 1843 even his children were following the saga. Although they were very young, Hamilton had tried to explain his basic idea to eight-year-old Archibald and nine-year-old William Edwin, and now every morning at breakfast they would ask, "Well, Papa, can you *multiply* triplets?" And every morning Hamilton would shake his head sadly, saying, "No, I can only *add* and subtract them."[1]

Several of Hamilton's letters have touching references to his explaining math to "my boys," and in the case of triples it *had* seemed, at first, as if the answer would be child's play—for a prodigy such as Hamilton, at least. After all, multiplying by a complex "couple" of real numbers (x, y) was equivalent to rotating the directed line $OP = x + iy$ in the two-dimensional plane, as in figures 3.6–3.8—so surely in 3-D space you'd have "triples" (x, y, z), and you'd just be rotating $OP = x + iy + jz$. You'd have to invent j, of course, another imaginary number like i, but that seemed straightforward enough: if you extended the Argand plane to three dimensions, with

Sir William Rowan Hamilton with one of his sons (circa 1845); by permission of the Royal Irish Academy © RIA. In those days photographs required long exposure times, which likely explains why Hamilton is never smiling; but by all accounts, he had a great sense of humor.

j in the direction of the third axis, you'd have a new imaginary number, different from i because of its direction. And if you defined multiplication by j via analogy with i—that is, as a rotation in the x-z plane—you'd have $j^2 = -1$. (In fact, today engineers use j to denote $\sqrt{-1}$, because they use i to denote electric current.) But then, what was the *algebraic* difference between this new number j and the original i?

The first thing Hamilton did was to check in with the laws of algebra, which De Morgan and Peacock had begun to formalize. It seemed obvious that if you added numbers like $x + iy + jz$ and $a + ib + jc$, it wouldn't matter in which order you added them, and you could be pretty sure that the associative law would hold, too. (I've given a reminder of these laws in the endnote.)[2]

Then there's the idea of algebraic "closure," as modern mathematicians put it. If you add two integers, you get another integer; if you add two real numbers, you get another real number; and if you add two complex numbers, you get another complex number. It's obvious that closure holds if you add two "triples," too. So that was all nice and easy. It was multiplication that was going to be tricky.

Hamilton had found the rules for multiplying his couples simply by expanding $(x + iy)(a + ib)$. But if you multiply out $(x + iy + jz)$ and $(a + ib + jc)$, term by term, you can see that you're not only going to need $i^2 = j^2 = -1$, you're also going to have to figure out what ij and ji mean. On the face of it, these extra terms suggest that multiplication of triples is *not* closed—and to a pure mathematician, that means triples do not form a consistent multiplicative algebra. (If you never know what type of number you'll get in a product, you can't devise consistent algebraic rules that work for every possible number in your initial set—and if there are no rules, you can't solve equations.)

In good mathematical research style, Hamilton started out by trying to simplify matters, assuming $ij = 0$. That certainly would make multiplication closed, but it was a stretch—in ordinary arithmetic and algebra, you can't multiply two nonzero quantities and get zero. (You *can* get zero from the product of two nonzero *matrices* if the determinant of one or both is zero, such as when you have the matrix equation $AX = 0$—but matrix algebra came *after* quaternions.) And there were other laws to check, too—Hamilton was worried especially about what he called the "law of moduli," which I've explained in the endnote.[3] To make this work for triples, however, Hamilton found that ij could *not* equal 0. As he told John Graves—his old friend from university, who had also been thinking about extending the complex plane, and whose brother Robert would later write a sympathetic biography of Hamilton—he then tried $ij = -ji$ in desperation, even though it was as contrary to ordinary arithmetic as $ij = 0$. He even tried $ij = j$ and $ji = i$. But while all these bizarre possibilities did simplify some aspects of multiplication, his new numbers still didn't satisfy the law of moduli.[4]

Hamilton must have been sorely tempted to give up. Perhaps it really was impossible, just as his friend De Morgan had said.

A MATHEMATICAL FRIENDSHIP, TWO MARRIAGES, AND A QUEST FINALLY FULFILLED

De Morgan was an interesting person, quite aside from his pioneering work on the foundations of abstract algebra. Vivacious and liberal-minded, he hadn't taken to formal religion—even at school he'd preferred pricking out equations on the pews to listening to the sermons. He considered himself an "unattached" Christian and was passionate about religious and intellectual freedom—so much so that although he'd studied at Newton's old college, Trinity, he refused to take the Anglican oaths that were required for fellowships at Cambridge and Oxford. Newton had likewise refused the oath, not because he was in favor of religious tolerance—far from it—but because he was a secret Arian, which meant he didn't believe in the divinity of Jesus. But he was so brilliant that the king had given him a special dispensation and he got his Cambridge fellowship anyway. De Morgan would soon publish some astute essays on Newton. He admired immensely both Newton's genius and his essential morality. But his own sense of justice compelled him to speak out about the great mathematician's flaws, too, particularly in connection with the priority dispute with Leibniz. This was something very few had dared to do, so godlike did Newton's mind seem and so worshipped had he been for two centuries.[5]

Like Hamilton, De Morgan was not quite twenty-two when he'd first taken up a professorship—in his case, at England's first nonsectarian university, the newly established London University, soon renamed University College, London. The two men were also born within a year of each other, although as a baby De Morgan had become blind in one eye. He'd been bullied at school for his ocular disability, but like Euler, he didn't let it prevent him from becoming a first-rate mathematician.

De Morgan was lucky enough to find a kindred soul in his wife, the well-educated social activist, polemicist, and spiritualist Sophia Frend. They were both so passionately committed to freedom that they'd taken the unusual step of getting married in a registry office, as a way of declaring their religious independence. By contrast, Hamilton had what one of his disciples would call "a deeply reverential nature," and he was a committed Anglican, while his wife Helen was reputedly particularly religious.[6]

Hamilton was also an emotional man and couldn't work easily without Helen nearby. But sometimes—especially when she was pregnant or nursing—she felt anxious in the dark, remote grounds of Dunsink Observatory, five miles from Dublin, where Hamilton had lived since his appointment as Ireland's royal astronomer. She began spending time away from home, staying with relatives nearby, and Hamilton often stayed with her there. But then, in 1841—a year or so after their daughter Helen Eliza was born—she left her husband and children to stay with her sister in England. She was away for about a year, and all sorts of rumors began to fly around, about the supposedly disastrous marriage and how Hamilton had apparently taken to drink. It didn't help that he'd famously lost his first love, Catherine Disney, through the kind of misunderstandings engendered by the strict romantic etiquette of the day, which Jane Austen had highlighted in her novels just a decade or two earlier. It didn't help, either, that attitudes to social drinking dramatically tightened in the 1830s and 1840s, as the temperance movement gathered pace. So, with each passing decade the stories became more and more exaggerated, unfairly tainting Hamilton's reputation during his lifetime and continuing right up until today.[7]

While recent scholarship suggests that the Hamiltons had a much happier marriage than the rumors suggest, the De Morgans seem to have had a charmed domestic life. Sophia's parents were friends of Lady Byron, the poet's estranged wife, and for a couple of years between 1840 and 1842, while the Hamiltons were dealing with Helen's anxiety and ill health, De Morgan was tutoring the Byrons' remarkable daughter, Ada Lovelace. (Her first mathematical mentor had been Mary Somerville, who had since moved to Italy for the healthier climate.) A year later, Lovelace wrote up her mathematical development of Charles Babbage's design for an advanced calculating machine—and her commentary included what is often considered the first computer program. It was published in the very same month that Hamilton discovered quaternions—a prophetic coincidence of timing, as it turns out, for we'll see shortly that it is computer programming that has brought to light the true computational economy of quaternions.[8]

Meantime, Hamilton was still struggling in his quest for quaternions, and when De Morgan claimed the impossibility of a three-dimensional complex algebra, Hamilton had taken it as a serious challenge. After all, both he and De Morgan were searching for ways to make sense of algebra—to give it a logical foundation, the way geometry had had since the time of Euclid two thousand years earlier. It was this Euclidean rigor that had led Newton to justify the theorems of *Principia* geometrically, rather than via more compact but still methodologically shaky algebraic calculus. As for more fundamental algebraic problems, I've already indicated that it was the lack of a concrete explanation of negative and imaginary numbers that had led earlier mathematicians to dismiss them—but as De Morgan wryly noted in this context, "Nothing could make a more easy pillow for the mind, than the rejection of all which could give any trouble."[9]

Hamilton was definitely not one to use an easy pillow. So he kept plodding away, trying first this and then that, and never giving up on the idea that if you could write the algebra of ordinary, two-dimensional complex numbers in terms of real numbers (x, y), then you should be able to extend it to (x, y, z)—and even, as he'd told De Morgan in 1841, to any number of dimensions, as in $a = (a_1, a_2, \ldots, a_n)$, where the $a_1, a_2, \ldots, a_n$ are real numbers. The single symbol a, he said presciently, is "indicative of one (complex) thought." He went on to elaborate this in terms of his mathematical philosophy of time, suggesting to De Morgan that these n numbers—where you can choose any value of n, depending on the problem—represent a chronological ordering of events, and therefore a "notion of cause and effect." But while Hamilton would ultimately move on from his fascination with time, he had glimpsed something important here: the idea that a single symbol can encode many pieces of information, or "one complex thought." As I've mentioned, this is part of the power of vectors and tensors, and it will also help explain the computational economy of quaternions.[10]

Hamilton worked so hard in his quest that his study was a mess of papers, scattered over his desk and piled high on the floor, and he often skipped meals, surviving on snacks brought to his room. But then one day his persistence paid off, and suddenly he had a startling idea. As he explained to John Graves, "there dawned on me the notion that we must admit, in some sense, a *fourth dimension* of space." What he meant was that if he introduced a *third* imaginary number, k—which immediately suggests the existence of a new "4-D" complex number $a + ib + jc + kd$—and if he defined k to be ij, then the law of moduli for triples would work at last. You can see why in the following endnote, but it was this marvelous insight that struck him, as if "an *electric* circuit *closed* and a spark flashed forth," as he and Helen were walking by Broome Bridge on that autumn Dublin day in 1843.[11]

When it came to naming his new 4-D numbers, I'm not sure why he didn't follow his "couples" and "triples" or "triplets" with "quadruples" or "quadruplets"—but "quaternion" does have a ring to it, and Hamilton had an ear for poetry. As De Morgan would later recall, "Hamilton himself said, 'I *live* by mathematics, but I *am* a poet.' Such an aphorism may surprise our readers, but they should remember that the moving power of mathematical *invention* is not reasoning, but imagination." Hamilton's friend Wordsworth, however, felt that Hamilton's imagination was better directed toward mathematics than poetry. And he was indeed wonderfully inventive, pioneering both vector analysis and, according to no less an authority than quantum pioneer Erwin Schrödinger, the math at the heart of quantum mechanics.[12]

Hamilton had been so excited after his electric insight that right there and then he'd carved into the stone bridge a single, simple line,

$$i^2 = j^2 = k^2 = ijk = -1.$$

It contained all that he needed to "calculate with geometry," as he put it—including calculating rotations in three-dimensional space. For this line of simple equations was a compact way of writing down the necessary algebraic relationships between his three imaginary numbers:

$$ij = k = -ji,\ jk = i = -kj,\ ki = j = -ik.$$

(You can see how this follows from the graffiti if you play with it algebraically: for instance, $k^2 = ijk \Leftrightarrow k = ij$.)[13] Using these definitions, if you had the patience to carry out 4-D multiplications such as

$$(a + ib + jc + kd)(w + ix + jy + kz),$$

you'd see that quaternions satisfy not only the law of moduli, but all the laws of ordinary real and complex multiplication—*except* the commutative law. In other words,

$$(a + ib + jc + kd)(w + ix + jy + kz) \neq (w + ix + jy + kz)(a + ib + jc + kd).$$

It's this last conclusion that was revelatory, and you can guess it without doing all the tedious multiplications, because the imaginary products themselves—ij and so on—are not commutative. This was a reluctant step for Hamilton, who had been working for years trying to fit first triples and then quaternions into the usual laws of arithmetic, so he deserves kudos for being brave enough to take it—and to recognize that he'd hit upon a brand-new kind of algebra.

THE REMARKABLE ALGEBRA OF WHOLE VECTORS AND QUATERNIONS

Hamilton identified two parts to a quaternion such as $P = w + ix + jy + kz$. He called the real number standing by itself a "scalar," represented here by w, and he named the imaginary part, $ix + jy + kz$, a "vector." The word "vector" comes from the Latin for "carrier," and the term "radius vector" had been used in astronomy for the movable line of sight that "carries" one point to another—say, from the eye to a star or planet. But as Hamilton would note later (in his *Lectures on Quaternions*), a "radius vector" is a scalar, a number representing only magnitude, whereas his term "vector" denotes magnitude *and* direction.

Hamilton's imaginary vector $ix + jy + kz$ looks very like the vector you might recognize as $x\mathbf{i} + y\mathbf{j} + z\mathbf{k}$, where the $\mathbf{i}, \mathbf{j}, \mathbf{k}$ are in boldface type because they are no longer fundamental imaginary numbers—rather, they're real "unit vectors," vectors whose magnitude, or length, is "unity" (which just means 1). So already you can see the genesis of today's vectors, although there's still a ways to go in the story before we see how this happened.

The quaternion symbol P encodes four pieces of information—the scalar w and the vector's three components x, y, z—so you can think of P as representing "one complex thought," just as Hamilton had foreseen in his anxious letter to De Morgan back in 1841.

If we crib the bold type from the future to distinguish Hamilton's vector from his scalar, we can write the quaternion P, with its imaginary-number "basis" i, j, k, as

$$P = w + \mathbf{p}, \text{ where } \mathbf{p} = ix + jy + kz.$$

(Hamilton sometimes put the imaginary numbers after the real number components as we do with vectors, so the order of the imaginary and real numbers in ix, jy, kz doesn't matter—you can think of it as xi, yj, zk if you prefer.) A "basis" is essentially just a set of independent unit quantities, one along each axis. More technically, these basis quantities are said to *span* the *vector space* in which the calculations take place, but these notions weren't yet properly formulated; the point here is that you can write the information in the vector $\mathbf{p} = ix + jy + kz$ just as well if you change from the unit imaginary basis (i, j, k) to the unit vector basis $(\mathbf{i}, \mathbf{j}, \mathbf{k})$, and write $\mathbf{p} = x\mathbf{i} + y\mathbf{j} + z\mathbf{k}$. That's because the key thing is that in each case the *components* have the *same* numerical values, (x, y, z)—because either way you're using a Cartesian coordinate system. And it is the components you need when you calculate with vectors.

In other words, you can think of a vector as a device for storing data—in the values of the components—while the basis is the device's hardware. Different devices have different designs—in this case a real basis, $\mathbf{i}, \mathbf{j}, \mathbf{k}$ or an imaginary one, i, j, k—but they both store the same information.

This is hindsight, though: the surprisingly acrimonious transition from Hamilton's vectors to modern ones is a story for chapter 8, but I mention it here for readers familiar with modern vectors.

If you *are* familiar with vectors, Hamilton's quaternion might seem strange: another case of apples and oranges, because adding a vector to a scalar is not something we do in college vector analysis. But at this stage of the story, we need to think of a vector simply as the imaginary part of a 4-D complex number, as Hamilton did—an algebraic expression that follows the rules of the imaginary numbers he carved on Broome Bridge.

With this in mind, now take a second quaternion,

$$Q = a + ib + jc + kd \text{ (or } a + bi + cj + dk) = a + \boldsymbol{q}.$$

Hamilton discovered that when you multiply out the product of P and Q, component by component, and when you group all the resulting terms as scalars or vectors, you get *two new* kinds of multiplication, indicated here by the modern dot and cross:

$$PQ = wa - \boldsymbol{p} \cdot \boldsymbol{q} + w\boldsymbol{q} + a\boldsymbol{p} + \boldsymbol{p} \times \boldsymbol{q}.$$

This is an amazingly compact expression, given that it actually contains twenty-two multiplications and eleven additions or subtractions. This extraordinary economy is one example of the power of quaternions and whole vectors (the ones represented by special notation such as the boldface type here, rather than by listing their components). Aside from the minus sign, computationally Hamilton's "scalar" and "vector" products—the $\boldsymbol{p} \cdot \boldsymbol{q}$ and $\boldsymbol{p} \times \boldsymbol{q}$, respectively—are just the same as the modern ones; so to see the compactness of quaternion (and whole-vector) multiplication, I've given the modern component forms of these products in the endnote, although you're likely familiar with them from math classes.[14] But you may not know that the bold type often used today for vectors is due to the idiosyncratic Englishman Oliver Heaviside. We'll meet him properly later, along with the urbane American Josiah Willard Gibbs, who gave us the dot for scalar products (also now called "dot products") and the cross for the vector (or cross) product.

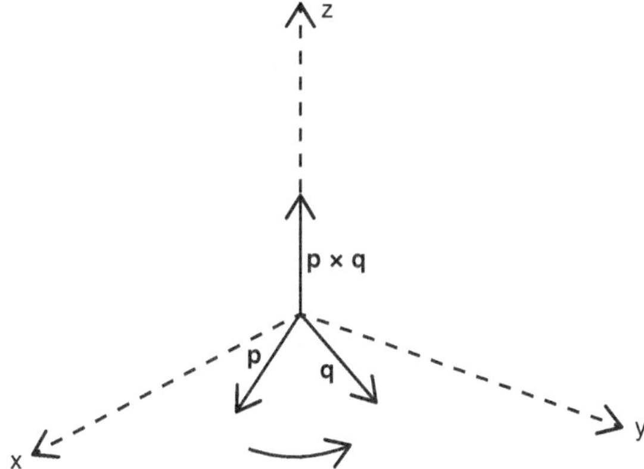

FIGURE 4.1. Right-hand rule for cross products. If you curl the fingers of your right hand in the direction shown by the arrow from $\boldsymbol{p}$ to $\boldsymbol{q}$, which in this case is anticlockwise, you get the direction of $\boldsymbol{p} \times \boldsymbol{q}$ from the direction of your thumb: it will point in the upward or positive z direction. To find the direction of $\boldsymbol{q} \times \boldsymbol{p}$, you have to turn your right hand upside down to curl your fingers clockwise, and your thumb will point downward, in the negative z direction. In this case $\boldsymbol{p}$ and $\boldsymbol{q}$ are in the x-y (horizontal) plane, but no matter what plane they lie in, their vector product will be perpendicular to that plane.

Hamilton himself hadn't designated these two new products as products of *vectors*; rather, he called them the "scalar part" and the "vector part" of the *quaternion* product. Nevertheless, for his two quaternions P and Q, Hamilton set the scalars w and a to zero and multiplied out the vector components just as we do today—except that instead of using boldface letters and a dot or a cross he wrote $S.\,PQ$ for the "scalar part" (our scalar product), and $V.\,PQ$ for the "vector part" (our vector product). The scalar product is commutative, but the vector product is not, and this is encoded in the mnemonic device called the "right-hand rule" (fig. 4.1). What this rule means is that when you curl your fingers in the direction from vector $\boldsymbol{p}$ to vector $\boldsymbol{q}$, your thumb points in the direction of the product vector $\boldsymbol{p} \times \boldsymbol{q}$. But if you reverse the order of the multiplication, your thumb points in the opposite direction. Which means that $\boldsymbol{p} \times \boldsymbol{q} = -\boldsymbol{q} \times \boldsymbol{p}$. And because vector products are noncommutative, so is the whole quaternion product PQ.

As you can see in the diagram (and in the previous endnote), the key thing about vector or cross products is that they multiply two vectors to get another vector; scalar or dot products multiply two vectors to get a scalar. (Which is why there's a math joke that asks, "What do you get when you cross a mountain-climber with a mosquito?" The answer is, "Nothing: you can't cross a scaler with a vector"—because the cross product can be applied only to two vectors, not to a scalar and a vector. This also highlights the biological definition of "vector," meaning an organism that "carries" pathogens from one animal or plant to another.)

This noncommutative vector multiplication was a Very Big Deal in the world of abstract algebra, a "curious, almost wild" revelation, as Hamilton put it. For it was the first time in four thousand years of recorded mathematical history that someone discovered a self-consistent algebraic system where *XY didn't* equal *YX*. And this meant there was a whole new algebraic world to explore, with who knew how many potential applications. That's why Hamilton is often described as "the liberator of algebra."

BREAKING THE RULES AND OPENING UP NEW WORLDS

Hamilton himself was initially puzzled and concerned by the fact that his quaternions led him to break such a time-honored rule. But before we see how he used his new mathematics to represent rotations in space, let me show you some of the other extraordinary new algebras and applications that his discovery of noncommutative multiplication opened up.

Today, for example, noncommutative algebras underpin quantum mechanics and general relativity. In quantum mechanics, Heisenberg's famous uncertainty principle says that you cannot *precisely* measure a quantum particle's position *X* and momentum *P* at the same time. But you *can* precisely measure first one of these attributes, say the position, and then the other. The trouble is that by the time you've made this second measurement, the position measurement you made first will no longer be quite so accurate, because subatomic particles are always moving or vibrating—so you'll get a different answer if you measure it again. This is noncommutativity in action, and it has a special equation (from which Heisenberg's principle follows):

$$XP - PX = i\frac{h}{2\pi}.$$

The h is Planck's constant; it represents the fundamental limit to measurement, and it is tiny: $6.6260693 \times 10^{-34}$ in units of kg m^2/second (or joule-seconds). Tiny or not, h is not zero, and that's the point here—for if it didn't matter in which order you measured the observable position and momentum, you'd have $XP - PX = 0$. This, of course, is the same as $XP = PX$, which is none other than the commutative law.

In general relativity, noncommutativity gives a measure of the curvature of space-time, so we'll revisit this when we come to tensors. Meantime, you might already be thinking that along with vectors, ordinary matrices offer a more familiar example of noncommutative multiplication.

MATRIX REDUX (AND A TIMELY NOTE ON THE TECHNOLOGY OF THE TIME)

Matrices, or arrays, are great ways of representing a lot of data concisely—so they, too, have something to do with vectors and tensors. Mathematicians had, in fact, been representing information in arrays and tables for thousands of years—but they hadn't been adding and multiplying these arrays as algebraic entities *in themselves* until Arthur Cayley spelled out how to do it, fifteen years after Hamilton announced his quaternion algebra.

Cayley, who was fifteen years younger than Hamilton, had spent the first eight years of his life in Russia (when his merchant father was based in St. Petersburg). He graduated top of his class at Trinity College, Cambridge, and taught there for several years on a temporary fellowship. A full fellowship required taking Anglican vows, so although he was a committed Anglican, taking Holy Orders was a step too far, and in 1846 he began training as a lawyer. He was almost as prolific as Euler, writing nearly three hundred mathematical papers in his spare time during the fourteen years he worked as a property lawyer as well as hundreds more afterward. One of his earliest papers, written before he left Cambridge for the law, was on quaternions—Cayley was the very first to publicly take up Hamilton's idea. He published this paper in 1845, barely a year after Hamilton had first announced his discovery in a paper he read to the Royal Irish Academy on November 13, 1843. And when Hamilton gave a series of four lectures on quaternions at Dublin University in 1848, Cayley was in the audience.[15]

Like everything, Cayley's work on matrices didn't arise in a vacuum, and there had been antecedents in the work of mathematicians such as Gauss and the lesser-known Ferdinand Eisenstein, who had been inspired to take up math after he met Hamilton, but who died tragically early of tuberculosis when he was just twenty-nine. There were also antecedents in Cayley's own work on determinants—which were developed before matrices, although today we learn about them as properties of matrices. Even in the ancient world, Chinese mathematicians had glimpsed the idea of matrices in the way they set out and solved simultaneous linear equations, using arrays and a form of what is now known as Gaussian elimination. ("Linear" equations have only x, y and so on, such as you get in the equations for straight lines—they don't include products or powers of the unknowns, such as the x^2 in the "quadratic" equation for a parabola, say, or a circle. Besides their ability to describe straight lines, linear equations have many useful applications, such as optimization—optimizing the accuracy of a machine learning model, for example, or optimizing business costs. They are also crucial in describing how to change from one coordinate system to another in particular ways—such as when you want to physically rotate or translate your Cartesian system, or the vectors in it, as we'll see later in figure 4.2. Such "linear transformation equations" will have an increasingly important role to play as this story progresses.)

The ancient Chinese "Gaussian elimination" method survives in the two-thousand-year-old *Jiuzhang Suanshu*, or *Nine Chapters on the Mathematical Art*, and it was way ahead of its time. Medieval Indian mathematicians did something similar—and in the seventh century Brahmagupta even used arrays to solve quadratic equations. All these early approaches were purely algorithmic—ideal for a computer! So, they were specifically computational—with an implied "do this then that" plan for each problem—rather than algebraic and general like symbolic matrix algebra (or, indeed, like Newton's and Leibniz's symbolic calculus algorithms). For Gaussian elimination doesn't use matrix multiplication or addition.[16] Cayley was using the same kind of convenient representation of simultaneous equations that the unknown author of *Nine Chapters* had hit upon—but because he had the benefit of symbolic algebra, this eventually led him to the theory of matrices. To begin with, though, he was simply trying to solve equations, just as his ancient predecessors had done—except that he was studying what his friend James Sylvester would soon dub "invariants."[17] More on invariants later, for they are fundamental to the idea of a tensor, and they're especially important in relativity theory. They also use the idea of linear transformation equations that I just mentioned.

In the early 1840s, Cayley began corresponding on the subject of transformations and invariants with George Boole. Today we think of Boole for his development of the "Boolean" algebraic logic that underpins computers, but that came a little later, after Hamilton's quaternions had opened up the possibility of completely new symbolic algebras. It's another example of the way the discovery of quaternions was important not only for itself, but also for helping to inspire a whole range of new algebraic discoveries, from matrices to symbolic logic. Anyway, Cayley was still a young professor of mathematics at Cambridge, and the self-taught Boole was running a school in Lincoln, although he would soon gain a professorship in Cork. The two men exchanged letters on their research, and Cayley began to wish they could meet in person—but, he lamented, there was not yet a rail link between the two towns.[18]

Cayley's lament is an interesting comment on the technology of the times—a reminder that the first public railways using steam locomotives to carry passengers as well as goods had been built barely a decade earlier, and they were still being expanded. Yet already they were creating new markets for manufactured products such as textiles, which in turn would drive new scientific and technological innovations. The Industrial Revolution was well underway—the invention of the steam engine had been part of it—and the railway project itself was sparking new applications of new scientific discoveries, such as the use of the brand-new invention of electromagnetic telegraphy to coordinate train traffic. The very existence of electromagnetism had only been discovered twenty years earlier, so this application highlights the ancient, two-way connection between economics and entrepreneurship on the one hand and science, mathematics, and technology on the other.

Unfortunately, it's a nexus that too often ignores the environmental impact of technology. Back then, at least there was an excuse, if ignorance counts as an excuse: it would be another decade before the pioneering American climate scientist Eunice Newton Foote, a distant relative of Sir Isaac, published her landmark paper on the heating effect of atmospheric carbon dioxide—which much of this exciting new nineteenth-century technology was spouting profusely. Not that Foote and other early climate scientists were thinking much about future climate change; they were focused on what Earth's climate was like in the past, and how to deduce its present temperature.[19] Some, however, were uneasy about this kind of "progress"—perhaps most famously the American writer Henry David Thoreau, who in 1845 withdrew to the woods by Walden Pond in Concord, Massachusetts, in order to live closer to nature. Like his poet friend Ralph Waldo Emerson, he was inspired by transcendentalism, New England's philosophical and literary back-to-nature movement—a kind of loose parallel to the English Romantic poets with whom Hamilton associated.

Cayley wasn't thinking about the industrial nexus with science, though—he was engaged in mathematics for the love of it. So it's not surprising that he soon realized matrices were more than just convenient methods of *representing* equations—they were algebraic structures in their own right, with their own algebraic rules.

For example, using algebraic symbolism you can write the simultaneous linear equations

$$2x + y = 7$$
$$x - 3y = 1$$

in matrix form like this:

$$\begin{pmatrix} 2 & 1 \\ 1 & -3 \end{pmatrix} \begin{pmatrix} x \\ y \end{pmatrix} = \begin{pmatrix} 7 \\ 1 \end{pmatrix}.$$

(By the way, $\begin{pmatrix} x \\ y \end{pmatrix}$ and $\begin{pmatrix} 7 \\ 1 \end{pmatrix}$ are examples of "column vectors"—where the information is written as a column rather than the rows I've been using so far, such as (x, y). It's the same information either way.) Writing these two equations like this takes the idea of "row-by-column" matrix multiplication for granted, but Cayley initially got the idea for multiplying any two compatible matrices—not just a matrix and a vector—when he was trying to find a neat, laborsaving way of representing multiple linear transformations. For example, if you want to rotate a coordinate frame and then translate the axes, you'll need transformation equations to rotate from x-y coordinates to x'-y' coordinates, say (as we'll see shortly, in fig. 4.2). Then you'll need to transform from x'-y' to x''-y'' in order to translate the axes. It turns out that on combining these transformations to get x-y in terms of x''-y'', the entries in the final matrix give the row-by-column definition. And this definition turns out to be another case where multiplication is not always commutative.

In other words, the information in a set of conventional linear equations can be represented in matrix form, as the ancient Chinese mathematicians had seen, but in Cayley's hands this led to the idea that matrices themselves could be multiplied, *regardless of what they represented*. So, he figured that they could be added, too: to add two matrices (of the same size), you just add the two numbers at the corresponding places, like this:

$$\begin{pmatrix} 3 & -1 \\ 2 & 0 \end{pmatrix} + \begin{pmatrix} 1 & 5 \\ -2 & 6 \end{pmatrix} = \begin{pmatrix} 4 & 4 \\ 0 & 6 \end{pmatrix}.$$

Cayley's close friend and colleague Sylvester, yet another mathematically gifted lawyer, had coined the term "matrix" in 1850—the same year, as it happens, that the first international telegraphic cable was laid, between Dover and Calais. You can imagine the excited talk about uniting the world via telegraphy—the same mix of entrepreneurial hype and genuine excitement we've witnessed more recently, over the connective power of the internet. (Unfortunately, the Dover cable failed—but others soon followed.) Sylvester, like Cayley, had studied at Cambridge, but as a Jew he wasn't entitled to graduate, thanks to the sectarian regulations that had led De Morgan to refuse to teach there. So, Sylvester and Cayley met in their capacity as lawyers—but they both worked on invariant theory and matrices in their spare time, and they both ended up as math professors, following their true calling.

While Cayley is generally credited as the founder of matrix algebra, other mathematicians soon developed more sophisticated matrix theory. In America, for example, father and son mathematicians Benjamin and Charles Peirce were pioneers of matrix and many other new, post-quaternion algebras, for Hamilton had opened the algebraic floodgates. Still, it is telling that Cayley had been motivated to find the rules of matrix algebra simply to carry out computations more efficiently: just as in ancient times, a lot of new math is created to solve practical problems or to facilitate computation. But once you can multiply matrices in one context, they, like vectors, can be applied to an amazing range of other practical computations.

IMAGE COMPRESSION, SEARCH ENGINES, MACHINE LEARNING, ROBOTICS: APPLICATIONS OF MATRIX AND VECTOR MULTIPLICATION

Once you know how to multiply matrices, sometimes you can "factorize" (or "decompose") them. One such factorization is called the "singular value decomposition" (SVD), and one of its modern uses is in digital image compression. The information—such as the location and color of each pixel—in the image you want to transmit is represented as a matrix with hundreds or thousands of rows and columns. The image matrix is then factorized into three matrices, with one of these factors (the "singular" one) containing the less interesting parts of the image—such as the sky or other background. Speaking very loosely, this less interesting detail can then be "factored out"—with the help of vectors, too—so there's much less information to transmit.

• • •

Speaking of huge amounts of information, in programming search engines, large arrays of data can be handled compactly via matrix and vector algebra. One of the earliest search engine methods used Boolean algebra to give exact matches to a user's query, via Boole's AND, OR, and NOT operations. But as Cornell computer science professor Gerry Salton realized in the 1960s, matrices and vectors allow users' queries to be matched partially and ranked according to relevance. (Fuzzy Boolean logic was later introduced in Boolean search engines to do this, too, and each approach has its advantages and disadvantages.) In the vector approach, information is stored as a matrix, with rows representing, say, key words, and columns representing different documents containing these key words. To simplify matters I'll assume the database has just three key words, which I'll call A, B, and C. Document 1 might contain no references to keywords A and B, say, but three references to C, so it would be represented as the vector $\begin{pmatrix}0\\0\\3\end{pmatrix}$, the first column of the matrix. Document 2 might have one reference to A, three to B, and two to C, so it would be represented as $\begin{pmatrix}1\\3\\2\end{pmatrix}$, and so on. Now suppose the user wants to search for information relating to B, so the "query vector" can be written as $\begin{pmatrix}0\\1\\0\end{pmatrix}$. (You don't want A or C, so you represent them by 0, and you do want B, denoted by 1.) Since each document is represented by a vector, its relevance to the query vector can be computed, via the scalar product (defined geometrically, as in the next endnote), from its vector's angle with respect to the query vector; the smaller the angle, the closer the match.[20]

Alternatively, in Google's PageRank algorithm—developed in the late 1990s by Stanford grad students Larry Page and Sergey Brin—a website is classified according to the number and rank of the other websites that link to it. The vector representing the proportionate number of links from website A to each of the other websites is the first column of a matrix (which I'll call M), and so on for the other websites. To find the rank of each website itself, it makes sense to start off with an equal ranking for all of them. This initial rank vector is then updated by multiplying it by the matrix of linking preferences. The process then keeps repeating, each time multiplying by M, so that by the nth step, the initial rank vector has been multiplied by M^n. The process keeps on iterating until there's no significant change in the rank vector. At this "equilibrium" point the website's PageRank vector is defined. With more website links coming on board, the process keeps on updating, as you can see when you search for something and note which websites come up first. But none of this would be possible without a definition of matrix multiplication.[21]

• • •

In machine learning, too, information is stored in strings and arrays—vectors and matrices, and even multidimensional arrays (which are, in fact, examples of tensors). In the real world it is generally only possible to collect a sample of all the possible data about a topic. Data science includes the art of extrapolating from this sample to make *predictions*—such as the likelihood of a person having a heart attack in the next year, or the likely load and price of energy at a given time—and *classifications,* such as whether a tumor is malignant or benign, whether an email is valid or spam, or, more recently and controversially, whether a face is yours or not. Programmers do this by fitting the data to a "model"—an algorithm or an equation that makes sense of the data. But sometimes it is too difficult or costly to devise an appropriately accurate model to fit the data at hand, and this is where machine learning comes into it. The idea is to program the computer with an algorithm that "trains" it to make predictions or classifications via a simple initial model; preliminary "training" data are fed in, along with the initial model and an error-testing routine. Making use of vector and matrix algebra, the programmer tells the computer how to estimate the accuracy of the initial model and to keep readjusting it until it reaches an optimum accuracy—analogously to the matrix multiplication iterations in the PageRank algorithm. The model has then been "trained," and the programmer can now use it to make predictions from new data.

Machine learning has expanded so rapidly that it is now behind a dizzying range of everyday things, from chatbots and recommendation algorithms to speech and image recognition, fraud detection, self-driving cars, medical diagnostics, and much more—for better and sometimes worse. (The "worse" includes not only misuse of AI by repressive governments or cyber criminals, and the disruptive changes to the way we live and work, but also the ethical issues surrounding the data used to train AI. These issues include using data without permission from their creators, and the racial, sexual, and gender biases that programmers—largely white men—have unconsciously programmed into training data, search engines, and other algorithms.)[22]

Matrices are used in countless other ways—including cryptography, where the noncommutativity of matrix multiplication (and of other algebraic structures) is currently being explored as an additional safety tool. One of the pioneers in this research was a sixteen-year-old Irish schoolgirl, Sarah Flannery. She admired Cayley so much that she named her algorithm after him.[23]

• • •

By now, though, if you've taken a linear algebra course, you might be thinking that a simpler, older, and more familiar application of matrix multiplication is rotations—and you are right! So, we are beginning to home in on Hamilton's preoccupation with rotations.

In robotics, for example, programmers need to know how to write instructions so that the computer can tell the robot how to move from one place to another. To raise and lower a robot arm, for instance, it must be rotated about a hinge (just like a shoulder or elbow joint), as in figure 4.2.

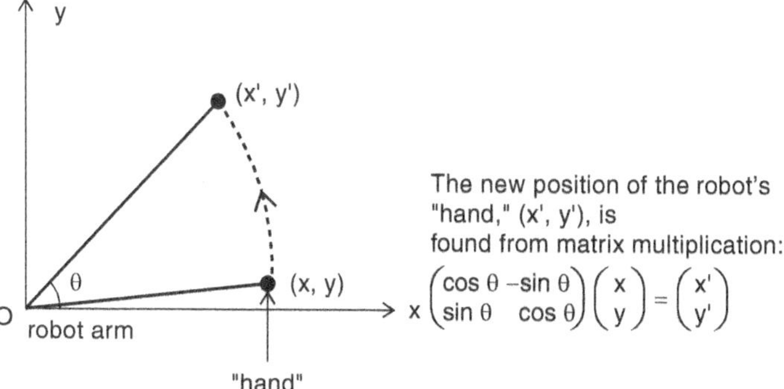

FIGURE 4.2. Rotating a robot arm about the origin of the $x - y$ plane, anticlockwise through an angle θ, using a rotation matrix. When you multiply out the left-hand side of this matrix equation, the resulting equations, showing how to write x' and y' in terms of x and y, represent a linear transformation. The details are not so important here, though: key is the *idea* of transformation equations.

In computer graphics, too, the position vectors of key features can be moved around in space via rotation matrices.

3-D ROTATIONS AT LAST!

For rotations in three-dimensional space, the matrix multiplications are a little more complicated than in figure 4.2, of course. But as with quaternions, too, matrices' noncommutative multiplications are right for the job, because two successive rotations in 3-D space generally don't commute, either. Try it with a book: Place it on a table with the front cover facing up and the spine toward you. Then, keeping its center fixed, rotate the book clockwise through 90°, so that the spine is on your left (this is a rotation in the horizontal plane of the table, say the x-y plane). Now turn the book over clockwise, rotating it through 180° until it's face down on the table. This time, you're rotating in the vertical x-z plane, so you've used all three dimensions of space—and when you're done, you'll notice that the book's spine is on your right. Now start again, but turn it over first, and then rotate it toward you: the spine will be on your left!

By the way, if you were to do the experiment with a featureless square box, or a ball with no markings, you wouldn't be able to discern this noncommutativity. That's because with rotations like these, spheres and cubes keep their shape and orientation on account of their symmetry about their centers. In other words, they are "invariant" under this sequence of rotations. This is a geometrical analog of the kind of invariants that Cayley was studying when he came up with matrix algebra.

Almost a century before Cayley, Euler had made a brilliant *geometrical* study of two- and three-dimensional rotation mathematics, using the spherical trigonometry pioneered by the ancient Greeks. Matrix algebra, on the other hand, is indebted to quaternions and their noncommutative multiplications—and quaternions had enabled Hamilton to give, at last, the very first purely *algebraic* computations for rotations in space.

First up, to represent the three-dimensional vector he wanted to rotate, he took just the vector part of a quaternion—for example, by setting $w = 0$ in $P = w + ix + jy + kz$, so that he had just the vector $\boldsymbol{p} = ix + jy + kz$.

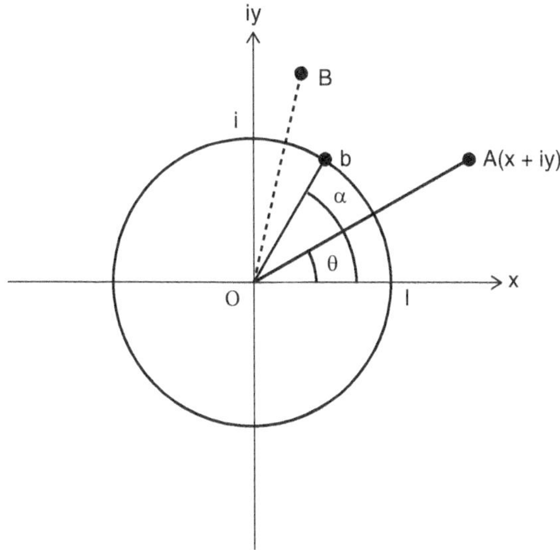

FIGURE 4.3. The "machinery" for rotating a complex number (cf. fig. 3.8). Represent the point A by the complex number

$$x + iy = r\cos\theta + ir\sin\theta = re^{i\theta}$$

To rotate OA about the origin through an angle of α, multiply it by the unit complex number $b = e^{i\alpha}$ to give the new point

$$B = (e^{i\alpha})(x + iy) = (e^{i\alpha})re^{i\theta} = re^{i(\theta+\alpha)}.$$

(The "machinery" has to be a unit vector if you just want to rotate the original vector, without changing its length as well, as in fig. 3.8.)

Then, for the algebraic "machinery" that will do the rotating, he chose a unit quaternion U encoding the required axis and angle of rotation. It's analogous to the way that in figure 4.3, the unit complex number

$$b = \cos\alpha + i\sin\alpha$$

is the "machinery" that rotates the line OA about the origin in the 2-D complex plane. Similarly, to get a rotation in three-dimensional space, the 3-D vector $\boldsymbol{p}$ has to be multiplied by the 4-D quaternion U—so this is where Hamilton's patient search for the right rules for multiplying his i, j, and k finally paid off. With the rest of the machinery in place—which I've

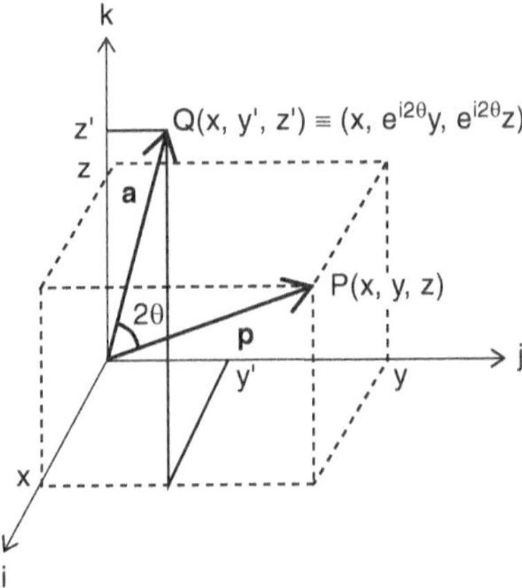

FIGURE 4.4. 3-D rotations using quaternions. When $\boldsymbol{p} = ix + jy + kz$ is rotated through an angle of 2θ about the *i*-axis, its new position vector is $\boldsymbol{a} = xi + e^{i2\theta}(yj + zk)$. I've chosen 2θ for the angle rather than θ because it will have an interesting connection with quantum mechanics, as we'll see shortly.

described in the next endnote—some nifty but simple calculations give the rotated vector. The endnote shows how to do this when rotating $\boldsymbol{p}$ about the *i*-axis, and you can see the result in figure 4.4.[24]

In math classes, you would likely have found a rotated vector using matrices, which is a good way to do it, as I've mentioned. But it turns out that quaternions are more efficient than matrices in more complicated situations where you need to combine several rotations around different axes, in order to orient an object in a particular way or to drive it via a smooth sequence of rotations—a satellite, for instance, or an airplane, a robot, your mobile phone screen, or a computer animation. Quaternions are more compact than matrices, and they are faster, because they require fewer calculations and use less computer power.

For example, in modeling, simulating, tracking, or guiding an aircraft or spacecraft, its orientation (or "attitude") is measured from three perpendicular axes—typically, along the length of the craft, across the width or wingspan, and the vertical up-and-down direction, with the origin at the center of gravity. Rotations around these axes are called, respectively, "roll," "pitch," and "yaw," and they are implemented mechanically by the yoke, throttle, and rudder (roughly speaking). To program simulations and tracking and electronical guidance systems, rotations about the three axes can be represented as three 3 × 3 rotation matrices—3-D versions of the example given in figure 4.2—and these are combined to give the required orientation of the craft. For instance, to adjust the yaw and pitch angles, you'd multiply the two relevant matrices—and if you count out all the row-by-column steps in multiplying two 3 × 3 matrices, you'll find there twenty-seven multiplications and eighteen additions. But if these rotations are represented by quaternions instead of matrices and roll, pitch, yaw angles, it turns out that there are only sixteen multiplications and twelve additions.

This is not the end of it, though. Unlike matrices, quaternions don't have the problem of "gimbal lock," which famously plagued the Apollo 11 mission to the moon. Gimbals are a set of rings that traditionally kept gyroscopes and other instruments steady during motion; there are three rings, each rotating around a different axis—but when two of them align, there are only two axes available for orienting the craft. This is reflected mathematically when the relevant rotation matrices are multiplied and you end up with an orientation matrix with too many zeroes, so that it can no longer include data for rotations about the missing axis.[25] So, the computer loses track of the orientation in the third dimension. This doesn't happen with quaternion representations.

Yet it is only relatively recently that quaternions have found their place in today's world. For instance, NASA first used them routinely to program rotations in guidance, navigation, and control systems in 1981. Their software includes vector, matrix, and quaternion algebra—for it is all marvelously useful, and different problems suit different approaches. Quaternions are especially useful both in orienting and guiding a spacecraft—for instance, they played an important role in simulating orbits for missions to the moon, and in tracking NASA's Mars Exploration Rovers—and in smoothly processing computer images from the raw data it collects.

• • •

In addition to all these technological applications, there's something else that makes quaternion rotations special—something mysterious and utterly surprising. They have an unexpected property that they share with the math describing the very building blocks of our world—electrons, protons, and neutrons. Even if you didn't feel like reading the endnote on the calculations leading to the rotated vector in figure 4.4, notice in the caption the $e^{i2\theta}$ term in the expression for the rotated vector $\boldsymbol{a}$. This is surprising, because it was found by multiplying the original vector $\boldsymbol{p}$ by the unit quaternion $U = e^{i\theta}$. It's a quirk of quaternion rotation algebra that turns θ into 2θ. And it's a quirk of quantum mechanics that although you can rotate an ordinary material thing, such as a ball or a planet, through 360° and get it back to its starting point, quantum math says that a particle such as an electron needs to "spin" through 720° to get back to its original state.

Not that the electron is *actually* spinning about itself, like Earth or a cricket ball—according to most physicists, electrons are best imagined as points, so they have no axis to spin around. Yet they have an angular momentum *as if* they were spinning. This bizarre and unexpected phenomenon was discovered through some ingenious putting of two and two together over several years, the kind of brilliant guesses and patient detective work that makes science so thrilling. So, I'm going to take a page or two to follow the trail of this extraordinary discovery, for it will also take us deeper into the fascinating connection between quantum mechanics and quaternions.

DISCOVERING "SPIN" AND THE QUATERNION CONNECTION

The first step toward "spin" was the self-evident fact that an electron could have orbital angular momentum, such as when it rotates not around itself but around the nucleus of an atom. But moving electrons generate tiny magnetic fields, as the Danish professor Hans Øersted had discovered long ago. (In 1820 he'd noticed the astonishing fact that a burst of electric current deflected the needle of a nearby magnetic compass, as if the current were magnetic as well as electric. This kind of set-up quickly became the basis of the first telegraphy systems, where the "compass" needle pointed to letters on a dial, the different letters corresponding to different amounts of current.) So researchers knew that electrons could be deflected by other magnetic fields, just as a magnet deflects iron filings. You can tell a lot about the properties of an electron's magnetic field by deflecting it in this way—so this is what Otto Stern and Walther Gerlach set out to do in Hamburg in 1922. Actually, they were experimenting with a beam of vaporized silver atoms—electrons would deflect too much for this set-up—but silver atoms have an unpaired electron that interacts with the field.

Stern's credentials included having been Einstein's first postdoctoral student a decade earlier, but now he and Gerlach were having trouble attracting sufficient funding for their experiment—in the early 1920s and the wake of World War I, things were still volatile in Germany. Then their equipment kept breaking down, until finally, after a year of struggle, they reckoned they'd taken a sufficiently good image of the deflected beam of silver atoms, which they'd captured on a photographic screen. But they didn't get the results you'd expect from ordinary physics, which suggested that the spatial orientations of the vaporized atoms' angular momenta would be randomly distributed, because of random fluctuations of their orientations during the heating process. In fact, they didn't get much of a result at all, at first. But as they stared at the plate in disappointment, two black lines began to appear, as if by some kind of "invisible ink" magic. Apparently, Stern's cheap brand of cigars had a high sulfur content, and as he was examining the plate, the sulfur in his breath mixed with the invisible silver atoms on the plate to produce black silver sulfide! These two black lines showed that there were just two possible values for the components of the silver atoms' angular momenta. In other words, the spatial orientation of angular momentum was quantized.

They weren't entirely surprised to find something like this—a discrete rather than continuous range of possible values—because Niels Bohr's early model of the atom was quantized. As quantum mechanics developed, however, limitations in Bohr's model came to light, so it wasn't clear just what Stern and Gerlach had found.

Then two young Dutch physicists, George Uhlenbeck and Samuel Goudsmit, suggested, in effect, that perhaps what Stern and Gerlach had measured were not *orbital* angular momenta but *spin* angular momenta—as if the electrons generating the magnetic fields *were* spinning around themselves. They'd come up with this idea by following up some anomalies in the atomic spectrum of hydrogen, whose atoms each have just one electron.[26] Using magnetic fields to manipulate the energy levels of the electrons, Uhlenbeck and Goudsmit figured out that the resulting spectral lines would fit this spin hypothesis if the electron's angular momentum depended on just two possible independent values—which is just what Stern and Gerlach had found.[27]

When Uhlenbeck talked about this strange idea with the legendary Dutch physicist Hendrik Lorentz—who will come into this story later in connection with relativity—Lorentz confirmed his fears, saying that if the electron producing this momentum really *were* spinning, it would have to be doing so at an impossibly high speed. So, the electron *isn't* spinning like a cricket ball—it just acts as if it is. Which seemed so bizarre that Uhlenbeck made a desperate visit to his and Goudsmit's mentor, Paul Ehrenfest, for they'd already sent him a paper about their hypothesis. But Ehrenfest had promptly sent it off for publication, telling the panic-stricken Uhlenbeck, "Well, it's a nice idea, though it may be wrong. But you don't have a reputation, so you have nothing to lose." Which wasn't much comfort for young Uhlenbeck.[28]

It all came together when the eccentric English physicist Paul Dirac began developing a *relativistic* quantum mechanical description of electron behavior. He found that to make an electron's wave equation relativistic, he *had* to include quantities that gave just the values of spin momentum that Uhlenbeck and Goudsmit had found. The German physicist Wolfgang Pauli had already made a start on the nonrelativistic mathematics of spin and had introduced three quantities now known as Pauli matrices, which Dirac adapted in his formulation. As Pauli spelled out explicitly, these three matrices—which help describe spin angular momentum components about the three spatial axes—relate to each other *in exactly the same way as Hamilton's unit quaternions i, j, k*: multiply any two of them and you get plus or minus the third—the same rules Hamilton had carved on Broome Bridge eighty-five years earlier![29]

What's more, the math says that if you want to rotate the spin axis of a quantum particle, the rotation equations are analogous to those for quaternion rotations. That's why electrons—and the other building blocks of matter—have that strange property where you need to rotate them twice to get back to where you started.[30]

As I've emphasized, though, subatomic particles don't really rotate like cricket balls; rather their spin aligns with a magnetic field, so if you could slowly rotate the field, you'd be rotating the spin axis (if not the particle itself). This is just what experimental physicists succeeded in doing, half a century after this astounding mathematical result. In 1975, two Australian physicists, Tony Klein and Geoff Opat, carried out an experiment that showed that this strange rotational behavior was physical—it wasn't just a mathematical artifact. Their experiment has some kinship with Thomas Young's famous two-slit experiment, which showed that the interference pattern between two beams of light was that of waves, not particles. Klein and Opat diffracted a beam of neutrons (traveling as a matter wave) into two parts; in this case, however, one part interacted with an external magnetic field that rotated the spin axis, and the other remained unchanged. The interference pattern made by these two beams showed that the spin had to be rotated through *two* cycles to get back to its original state. Which is just what the math predicted![31]

It was a stunning result, but before Klein and Opat could get their work into print, two other teams also confirmed this strange mathematical prediction.[32] Spin, whatever it was, was definitely real, and it behaved according to the same quirk of rotational math that Hamilton found when he finally managed to rotate vectors in three-dimensional space.

Today "spin" is a vital part of everyday life—in the magnetic resonance imaging (MRI) used for medical diagnoses, for example. MRI takes 3-D images of a patient's internal organs—so there's no need for invasive diagnostic surgery or radiation—and it does this by using magnetic fields to align the spins of hydrogen atoms, which are ubiquitous given the amount of water and fat in our bodies. The spins are then deflected via radio waves, different sequences of radio pulses highlighting different parts of the body; images are captured when the radio source is turned off and the spins return to their equilibrium state, giving off electromagnetic energy in the process.

• • •

Hamilton would be beside himself with joy if he knew that quaternions are related to something as fundamental as this strange electron spin property. And if he could see the glorious pictures of our universe that quaternions have helped bring to Earth, his poetic soul would thrill to the wonder and grandeur and mystery of it all—the kind of mystery he'd tried to capture in one of his favorite poems, *Ode to the Moon under Total Eclipse*. He'd written it when he was not quite eighteen, and he kept it till the day he died. In the poem, he questions the moon goddess on the "portentous gloom" as "crimson red" spreads over her "lovely brow" during the eclipse, asking if it be caused by some passing cloud from another world, or a wizard's magic spell? Or, perhaps, is it our Earth's "shadowy cone" that dims the moon to "our wondering eye?" This mix of poetic imagination and science is typical of Hamilton. He would have loved to know that his quaternions (and also what is now known as Hamiltonian dynamics, on which quantum mechanics was modeled) have helped bring to light some of the universe's most magical secrets.[33]

When he first discovered quaternions, though, Hamilton only glimpsed their wider possibilities. Yet right from the beginning he felt in his bones that quaternions had a vital role to play in physical applications—and if in his lifetime they never quite lived up to his hopes, never achieving the glamorous profile that they have today by association with NASA and quantum mechanics, still he had his share of glory. In 1848, for example, he was awarded medals for his discovery of quaternions, from both the Royal Irish Academy and the Royal Society of Edinburgh.

His long struggle to make sense of complex numbers, to expand them to three dimensions and to invent his 4-D quaternions, reminds us of the kind of deep thinking needed to make sense of even the simplest things we take for granted today. The same goes for De Morgan and Peacock's work on symbolic algebra, teasing out the laws of algebra that people had long been using without ever realizing they might apply to things other than just numbers. Today, it is mainly philosophers of mathematics who discuss such things as the relationship between numbers and symbolism, language, and meaning. Educational policymakers and funding bodies are focused mainly on how mathematics can be useful. So it is worth saying again that Hamilton had no idea just how useful his quaternions would prove in such high-tech applications as driving robots and spacecraft. Nor did he foresee what a ubiquitous role vectors would play in physics, engineering, and IT, for now they crop up just about everywhere you need to specify the spatial positions of objects or to store and handle data. Which goes to show that although modern math has its ancient roots in attempts to solve the kinds of practical problems the Mesopotamians were faced with—and although the need to solve practical problems is still a key driver of new mathematical insights—sometimes it is math for math's sake that leads to solutions of future technological and physical problems.

In fact, a couple of months after Hamilton told his friend John Graves about his quaternions, Graves invented "octonions"—not 4-D but 8-D complex numbers made from pairs of quaternions. They were just a curiosity, proof that there was even more to the algebraic world than numbers and quaternions—yet today they are being investigated for possible connections with the standard model of particle physics. This is the current model in which the building blocks of physical reality are classified in categories that share similar properties. For instance, matter is made of different kinds of quarks and leptons, the latter being particles with charge, such as electrons. Then there are the force-carriers or bosons, such as the photons that carry electromagnetic force, plus the Higgs boson, which carries mass.

The jury is still out on whether or not octonions will prove useful in this context—but as with the applications that flowed from Hamilton's noncommutative multiplication, once an idea is out there, you never know what it may inspire.[34] George Boole is another example; his system of algebraic logic evolved from the foundations of symbolic algebra laid by Peacock and De Morgan, and by Hamilton's quaternions. Little did Boole know, when he died suddenly and far too young after catching cold in a downpour of rain, that his system of expressing true and false statements in terms of 0s and 1s would eventually drive our computers and other electronic devices and gadgets.

Back in the 1840s, though, there was still a ways to go in developing the vector analysis that has broadened not just our technology but also our understanding of math, and of our universe. Over the next few chapters, we'll meet new generations of mathematicians and mathematical physicists, whose work developing Hamilton's vectors and quaternions will take us into yet more wondrous realms of reality.

First, though, we'll see that it always takes time for new ideas to prove their worth. And we'll meet an extraordinary outsider, who invented his own powerful vectorial system. His motivation was different from Hamilton's—he wasn't interested in rotations and complex numbers—but ultimately, we'll see that both approaches would help mathematicians to develop the sophisticated vector and tensor analyses that are so important today.

(5)

A SURPRISING NEW PLAYER AND A VERY SLOW RECEPTION

These days we know how useful Hamilton's quaternions are, but at the time they faced quite a battle for acceptance. In 1847, Hamilton gave his friend Robert Graves a humorous account of the discussion that followed his presentation of a paper on the subject, at a meeting of the British Association for the Advancement of Science. Incidentally, it was at a British Association meeting in 1833 that the term "scientist" was first introduced, in order to put scientific and mathematical researchers on the same professional footing as artists. The idea was William Whewell's, and he also coined the name "physicist" in 1840. Anyway, in 1847 Hamilton told Graves that while Dr. Peacock thought he'd given "a capital exposition" of his quaternion system, another member of the audience wondered about "the possibility of making mistakes in the use of my new calculus. In reply to which I disclaimed the power of setting any limit to the faculty of making blunders."[1]

There were a few more "gentle skirmishes" of this sort, continued Hamilton, and then the polymath Sir John Herschel gave quaternions a rapturous endorsement. Which was too much for the Astronomer Royal, George Airy. Airy had called Hamilton's mathematical prediction of conical refraction (which we saw in chap. 2) "perhaps the most remarkable prediction that has ever been made"—but he took a much dimmer view of quaternion algebra. Hamilton told Graves that after Herschel spoke, Airy rose to his feet "to speak of his own acquaintance with [quaternions], which he avowed to be none at all; but [he] gave us to understand that what he did not know could not be worth knowing." In other words, Hamilton added wryly, Airy seemed to assume that anything that to *him* seemed obscure or paradoxical *must* be erroneous.[2]

If someone of Hamilton's scientific standing was having trouble getting his system accepted, imagine how much harder it would be for a self-taught loner living in a remote country town. There must have been something vectorial in the air, for such a person actually existed: the German schoolteacher Hermann Grassmann. In one of those remarkable instances of independent parallel discovery that happen surprisingly often in science, Grassmann discovered a system analogous to Hamilton's at the very same time. Several others, too—notably the German August Möbius, of "Möbius strip" fame, and the Italian Giusto Bellavitis—had also been inching toward the idea. But it was Hamilton and Grassmann who ultimately succeeded—each unbeknownst to the other.

Four years younger than Hamilton, Grassmann was the son of a Pomeranian theologian, Justus Grassmann, who taught math and science at the local high school in Stettin—now spelled Szcezcin, because the area is in today's northwest Poland. Hermann and his eleven siblings were born in Stettin, but at school he'd shown no mathematical talent at all, unlike the prodigy Hamilton—Justus thought his son might find a job as a gardener or craftsman. Still, Hermann managed to get into Berlin University, where he studied philology and theology—he wasn't interested in becoming a gardener, no matter what his father thought. In fact, he'd set his eyes on becoming a teacher like Justus, so when he returned home from university, he began to study mathematics and science on his own. His father had written some math textbooks, and they evidently worked well, for soon Hermann passed the state teaching exams.[3]

After a year teaching in Berlin, he taught high school math, science, languages, and theology in his hometown for the rest of his life. Unlike George Boole, who also taught himself mathematics and then spent years as a schoolteacher, Hermann Grassmann never gained the recognition from his peers that might have led to a university position. It certainly wasn't for lack of merit: Grassmann may not have shown much talent at school, but he was a brilliantly original mathematician. And it wasn't for want of trying, either—he was quite ambitious, and put in extraordinary efforts to try to gain professional promotions.

He began trying to prove himself with an intensive study of the tides, and in the process he realized, like Newton, that physical phenomena acting in space are vectorial (not that either of them used that term). But while Newton used just the parallelogram rule and the idea that forces and velocities have both magnitude and direction, Grassmann became as entranced as Hamilton by the challenge of developing the underlying mathematics of vector algebra. And when he applied this algebra to the theory of tides, he found himself "astounded by the simplicity of the calculations resulting from this method."[4]

By 1840, he was ready to submit his 200-page dissertation on tidal theory to the Berlin examination committee that decided on teacher promotions. Unfortunately, but perhaps not surprisingly, the examiner failed to appreciate its revolutionary mathematical basis. Grassmann must have had extraordinary self-belief, working away in isolation, with no one to recognize his achievements. But he passionately believed in the power of his new method to simplify computations, and so, seemingly undaunted by the rejection of his paper on the tides—and by the consequent lack of promotion up the teaching ladder—in 1844 he published a monumental treatise on his new approach. He called it *Die Lineale Ausdehnungslehre*, or "The Linear Theory of Extensions."

GRASSMANN'S AMAZING *AUSDEHNUNGSLEHRE*

While Hamilton spoke of vectors, Grassmann used the German word *strecke*, which can be translated as "distance, route, line, or stretch"—and his basic geometric objects were indeed "lines," which could be "stretched" or "extended" to form planes, just as points could be extended into lines. This doesn't sound all that earth-shattering, but the radical and ingenious part of it was the way Grassmann built an *algebra* of geometrical lines and their extensions—just as Hamilton had been searching for an algebra of quaternions and 3-D rotations. And while Hamilton had been shocked to discover that multiplication of quaternions and vectors was not always commutative, so Grassmann reported that he was

> initially perplexed by the strange result that though the other laws of ordinary multiplication (including the relation of multiplication to addition) held, yet one could only exchange factors if one simultaneously changed the sign (i.e. changed + to – and – to +).[5]

This was a long-winded way of saying that multiplication of his geometric "lines" was not commutative—which goes to show the conceptual and linguistic importance of Peacock and De Morgan's work on identifying and naming the fundamental laws of arithmetic. In fact, Grassmann's talk of changing + to – was a way of saying that vector products are *anti*commutative. This is just what the right-hand rule for cross products shows (as we saw in fig. 4.1): if you curl your fingers from $\boldsymbol{p}$ to $\boldsymbol{q}$ and your thumb points upward, then if you curl them the other way, from $\boldsymbol{q}$ to $\boldsymbol{p}$, your thumb will point downward—which means that $\boldsymbol{p} \times \boldsymbol{q} = -\boldsymbol{q} \times \boldsymbol{p}$.

I mentioned earlier that the young Hamilton had studied Laplace's *Mécanique Céleste* and found an error in the logic of the parallelogram rule, and that he had discovered quaternions because he'd been intrigued by his friend De Morgan's work on symbolic algebra and by Warren's paper on the geometry of complex numbers. Grassmann also knew about the parallelogram rule, but otherwise he came to his theory of "extensions" by a very different route.

He learned about the power of algebra and calculus in physics from a textbook by the Italian-French mathematician Joseph-Louis Lagrange. It was called *Mécanique Analytique* (*Analytical Mechanics*), and it had been published half a century earlier—in 1788, the eve of the French Revolution; Laplace's *Celestial Mechanics* appeared a decade later. Lagrange and Laplace were not particularly interested in politics and managed to steer a neutral course through the Revolutionary years. In 1790 the French Academy of Science created a Committee on Weights and Measures—both Laplace and Lagrange were members—but when the Reign of Terror began in 1793, the fanatical new government closed the Academy. They did have the sense to keep the weights and measures committee going—minus a few members they deemed suspect, including foreigners[6]—and it is thanks to this committee that the metric system was created.

As for *Mécanique Analytique,* while Laplace had updated Newton's work on planetary motion—Grassmann built on Laplace for his theory of tides—Lagrange extended Newton's concept of motion in general. I mentioned that Johann Bernoulli, Émilie du Châtelet, and especially Leonhard Euler had begun translating Newton's work into the language of Leibnizian calculus, but Lagrange extended the theory of differential equations so that they finally became *the* language of mechanics—the language of motion and forces that we take for granted today. Lagrange was the only one of his parents' eleven children to survive beyond childhood, and he certainly did them proud.[7]

Grassmann acknowledged his debt to *Mécanique Analytique* in the foreword of *Ausdehnungslehre*—but he noted that in using his own new system, "the calculations often came out more than ten times shorter than in Lagrange's work." What he meant was a similar economy of expression as you get if you use quaternions for compound rotations. Grassmann's journey to vectors had begun not with Lagrange, however, but with his father's textbooks—and having such a father was one of the few pieces of luck he had in his long struggle for mathematical success.

As he noted in the foreword to *Ausdehnungslehre,* his father had defined a rectangle as the "geometrical product" of its length and width. This was novel in itself—the idea that you could multiply two geometric quantities *in themselves* to create a new geometric object—in this case, "multiplying" two *lines* to get a *rectangle.* By contrast, the traditional product in this context multiplies the *numbers* representing the length and width to get another number, which the ancients had long ago defined as the *area* of the rectangle. But Hermann took his father's idea into true vectorial territory by noting that you could do the same for all parallelograms, if you viewed the sides "not merely as lengths, but rather as directed magnitudes." And, of course, "directed magnitude" is another name for Hamilton's "vector."

You might remember from school that the geometric definition of the magnitude of the vector product $\boldsymbol{a} \times \boldsymbol{b}$ is $|\boldsymbol{a}||\boldsymbol{b}|\sin\theta$, where $|\boldsymbol{a}|$ means the magnitude of the vector $\boldsymbol{a}$, and similarly for $|\boldsymbol{b}|$. That's why there's yet another cross-product joke (from the television sitcom *Head of the Class*): "What do you get when you cross an elephant and a grape?" Answer: "Elephant grape sine-theta!" More seriously, in the case of a rectangle, $\theta = 90°$, and so $\sin\theta = 1$, which means you're back to the traditional rule where you find the area simply by multiplying the magnitudes of the sides. In other words, the role of $\sin\theta$ is hidden for rectangles, but not for other parallelograms.

Grassmann called this kind of geometrical multiplication an "outer product." In three dimensions, it differs slightly in its interpretation from Hamilton's (and the modern) vector product, as you can see in the endnote, but computationally it is the same.[8] Grassmann also defined what we, following Hamilton, call the "scalar product" of two vectors, but which he called an "inner product." He defined his inner product using just the same geometrical definition you may have learned for the scalar product, $\boldsymbol{a} \cdot \boldsymbol{b} = |\boldsymbol{a}||\boldsymbol{b}|\cos\theta$—although he didn't have this modern vectorial notation, of course.

While we use Hamilton's terms "vector" and "scalar" in vector analysis today, we'll see later that Grassmann's "inner" and "outer" products would also eventually find a modern home, as the names for products of n-dimensional vectors and tensors, respectively. Hamilton had told De Morgan he saw no reason you couldn't find an algebra of what he called "polyplets" (by analogy with "triplets") of the form $(a_1, a_2, \ldots, a_n)$, but he had stopped with quaternions, where $n = 4$ (or 3 when the focus is on the vector part). Grassmann's approach also focused on three dimensions, but he presented it in an abstract way that was readily adaptable to any number of dimensions—in fact, he applied it to pioneer the new theory of n-dimensional linear algebra.

Grassmann's algebra obeyed all the usual laws except the commutative law for outer products—like Hamilton, he knew it was important to establish these rules for his new algebraic system. But it went further than Hamilton's system in that it contained not just vectors but the germ of tensor algebra. At this point in the story, though, it must be said that Grassmann's self-taught mathematical style was rather impenetrable.

Hamilton's wasn't much clearer in his expanded 800-page *Lectures on Quaternions,* which he published in 1853. For a start, there was the fact that both he and Grassmann were trying to articulate brand-new ideas that their peers would find difficult to digest anyway—the fundamentals of vector analysis, linear algebra, and more: in Hamilton's case, three-dimensional complex numbers and rotations, and in Grassmann's, n-dimensional vector spaces and prototensor algebra. They both began to develop vector calculus, too (of which more in the next chapter). But the originality of their work meant that each of them had to invent new words to describe what they were doing—dozens more opaque and ultimately superfluous words than the few that finally caught on (notably the "vector," "scalar," "inner," and "outer" products that I just mentioned). Both men also had a penchant for an excess of philosophizing—Hamilton's philosophy of time, and Grassmann's philosophical approach to the fundamental concepts of mathematics. For instance, Grassmann was a devotee of dialectics, and he applied it in categorizing concepts as "equal-different," "real-formal," and so on.

It helped Hamilton's case that he already had a reputation for mathematical genius, whereas Grassmann was an unknown country schoolteacher—it's amazing that he managed to find a publisher at all for *Ausdehnungslehre.* Even Hamilton had trouble publishing his *Lectures,* as he told a friend shortly afterward:

> It required a certain *capital* of scientific reputation, amassed in former years, to make it other than dangerously imprudent to

> hazard the publication of a work which has, although at bottom quite conservative, a highly revolutionary air. It was a part of the ordeal through which I had to pass, an episode in the battle of life, to know that even candid and friendly people secretly, or, as it might happen, openly, censured or ridiculed me, for what appeared to them my monstrous innovations.[9]

He must have been disappointed at the book's reception. Even John Herschel, who had been so excited when he heard Hamilton lecture on his new discovery back in 1847, wrote to him saying the *Lectures* would "take any man a twelvemonth to read, and near a lifetime to digest." He grasped it a little more easily when he came back to it six years later, but after three chapters he gave up in despair over the distracting, idiosyncratic philosophy. He advised Hamilton that if he focused on explaining his method's mathematical algorithms and terminology more clearly, then readers might be "better prepared to go along with you in your metaphysical explanations."[10]

Readers were even less prepared for Grassmann's strange interweaving of philosophy and mathematics, because the fact that he was self-taught in math showed up in his lack of fluency in the mathematical language used by established mathematicians. In those days before the internet, every time a new paper was published Grassmann would have had to travel to Berlin to find an academic library—if indeed he ever heard there *was* a new paper to follow up—and even today, the fastest train route from Szcezcin to Berlin is almost a two-hour journey each way. It's another example of why mathematical progress at that time tended to arise in those countries and institutions that were able to foster well-connected communities of scholars.

Nevertheless, as soon as *Ausdehnungslehre* was published in 1844, Grassmann sent a copy to the venerable Gauss, the leading German mathematician and perhaps the best in the world at that time. Gauss replied rather dismissively, saying that he'd been working on the same ideas for half a century and had published some of them in 1831. I mentioned his pioneering geometrical representation of complex numbers, which neither Grassmann nor Hamilton had known of when they made their discoveries, but, as we'll see in chapter 10, Gauss also initiated what is now known as "differential geometry." Grassmann's work on "inner" and "outer" products would also contribute to this branch, but in 1844 it seemed to Gauss that there was nothing new in Grassmann's dense jungle of words. In his reply Gauss said that he was very busy, and that it would take time to familiarize himself with Grassmann's "peculiar" terminology—but apparently, he never got around to it.

Undeterred, Grassmann took a trip to Leipzig to meet Möbius, Gauss's former student, who had published some early vectorial ideas. It's another reminder of the state of technology in the mid-nineteenth century, for there were no telephones or email to make long-distance communications possible without having to leave home. The upside of having to travel, of course, is the personal contact, and Grassmann's visit was a convivial meeting of mathematical "kindred spirits," as Möbius recalled. Grassmann followed it up with a letter diffidently asking his new friend if he would review *Ausdehnungslehre*, since he would be well able to judge the work for both its weaknesses and "whatever merits the book may contain."[11]

It must have been a nail-biting wait for a reply. In fact, poor Grassmann had to wait four months, because Möbius simply couldn't get his head around such an innovative and eccentric book. In his reply he said he'd made many attempts to read it, but each time he couldn't get past the philosophy. Still, as he told his colleague Ernst Apelt, when he "forced" himself to skim through it, he felt that Grassmann's systematic presentation of the fundamental ideas of geometric addition and multiplication—what we know today as the algebra of vectors—had something about it that might be good for the development of mathematics. So, as he told Grassmann in his reply, he'd asked a mathematician familiar with philosophy to review it (someone he referred to as Drobisch). But just in case, he suggested that it might be best for Grassmann to review it himself.[12]

No review was forthcoming from Möbius's philosophical colleague, and Grassmann was not so bold as to send out a review himself. Instead, he sent a copy of his book to Johann Grunert, the editor of *Archiv der Mathematik und Physik*. Like everyone else, Grunert found it too difficult, but he kindly suggested that Grassmann write his own review-summary. And so he did, although it was hardly more enlightening than the tome itself. But at least he'd managed to introduce his work to his mainstream colleagues—his was the only review *Ausdehnungslehre* received. It was simply too "strange," too abstract and unintuitive, as Apelt told Möbius—and Apelt had only read Grassmann's review! He certainly wasn't about to start on the book. Heinrich Baltzer did try to read it after Möbius recommended it to him, but whenever he tried to enter into Grassmann's thought-processes, he would "become dizzy and see sky-blue before my eyes." Möbius replied saying he knew the feeling![13]

And so it went. But Grassmann was a battler. Like Hamilton, he was convinced that his method would make many computations in physics and geometry much simpler, and in 1845 he published a paper on electrodynamics—the mechanics of moving electric particles and changing electromagnetic fields (as we now call them)—which utilized his new "outer product." Electromagnetism was a new and exciting topic, as mathematicians attempted to make sense of a growing number of experimental results—this was twenty-five years after Hans Øersted discovered the very existence of electromagnetism, but twenty years before James Clerk Maxwell's theory of the electromagnetic field. Maxwell's theory will prove a turning point in the story of vectors, but in the meantime, Grassmann realized that his new system could better describe a result published by the French pioneer of electromagnetism André-Marie Ampère, back in 1826. It concerned the strength and direction of the magnetic force exerted by a small electrical circuit—so you can see why vectors or Grassmann's "directed lines" might be the ideal language for it. But the formula that Grassmann deduced from published experimental results gave some subtly different results from those given by Ampère when it came to certain untestable aspects of this electromagnetic phenomenon. The details do not matter here—except to say, first, that there is still contention over which of the two formulae, Ampère's or Grassmann's, is the better experimental fit; and, second, that Grassmann's vectorial approach fitted better with Maxwell's, of which more soon, and so Grassmann's version won out over Ampère's when it finally achieved recognition in the 1870s, in the wake of Maxwell's triumph.[14]

Luckily, despite his unpolished mathematical idiosyncrasies and innovative results, the gifted young outsider did find early supporters in people of the caliber of Möbius, whose fledgling vectorial work was important in projective geometry and the study of invariants. Möbius also prefigured the field of topology with his remarkable "Möbius band": a twisted strip of paper joined at the ends that has only one side, as you can see if you trace a line along the surface. Another of Gauss's former students, Johann Listing, also discovered this phenomenon, around the same time. Listing coined the term "topology," which today refers to the study of *general, invariant* properties such as the number of sides of a surface, or the number of holes in an object: for instance, a donut and a teacup each have one hole, and if they were both made of plasticine or dough, you could turn one into the other without cutting or gluing. In other words, topology doesn't focus on the *specific* shape of the donut or teacup but on properties that don't change under this kind of continuous "plasticine" deformation—and it will prove useful in describing curved surfaces and space-times. As for Möbius, he evidently saw something special in Grassmann's work even if he himself couldn't read much of it, for he wrote to Grassmann in early 1845 to tell him about an academic essay competition. It was just the sort of collegial kindness the isolated schoolteacher needed—and the essay topic was right up his alley.

In chapter 3, I mentioned Newton's pioneering role in expressing physical quantities in terms of both magnitude and direction, but Newton's nemesis Leibniz also has a part to play in the story. It has to do with the power of symbols. "It is worth noting," Leibniz said, "that notation facilitates discovery. This, in a most wonderful way, reduces the mind's labors." We've seen this in calculus, where the Leibnizian dy/dx notation for differential calculus is better adapted to computations and conceptual understanding than Newton's $\dot{y}$—and John Wallis had seen it in Thomas Harriot's full use of symbols in algebra itself. But Leibniz wanted a similar laborsaving language for geometry, too.[15]

He expressed the idea in a 1679 letter to the Dutch physicist Christian Huygens, pioneer of the wave theory of light and many other things—including the pendulum clock. (Though it seems a quaint relic today, the pendulum clock was a great advance in accurate timekeeping. In their experiments on falling motion half a century earlier, Galileo had had to measure time by a dripping "water clock" and Harriot had used his pulse.) Leibniz suggested to Huygens that a new form of algebra was needed, one that is "distinctly geometrical or linear and which will express situation directly as [ordinary] algebra expresses magnitude directly." (By "situation" he meant relative position in space.) Descartes had spearheaded the use of coordinates to describe such positions, but Leibniz was looking for something more clearly geometrical—something that we, with hindsight, might imagine as a language of vectorial arrows or Grassmannian directed lines, independent of coordinates. Leibniz himself didn't find such a system, and his letter to Huygens was published only in 1833—by which time Möbius and Bellavitis had already taken tentative steps in the vectorial direction. But the 1845 essay competition was designed to give mathematicians the challenge of answering Leibniz's suggestion more fully, and the prize was to be awarded in 1846, in honor of Leibniz's two hundredth birthday. As it turned out, the challenge was so great that Grassmann was the only one who entered. It was his first piece of luck after publishing *Ausdehnungslehre*, for his prize-winning paper was published in 1847, with an appendix by Möbius, whose fame and connections made it more likely that other mathematicians would take notice of his protégé.

Not that the essay was much more readable than *Ausdehnungslehre*, although it did help to bring Grassmann's book into the wider mathematical community. But when Hamilton finally heard of it, sometime around 1850, such was its reputation for difficulty that he thought he might have to learn to smoke in order to get through it! When he began to read it in the fall of 1852, however, he told De Morgan it was "a *very* original work," which "the Germans, if they think me worth noticing, will perhaps set up in rivalship with mine, but which I did not see till long after my own views were formed and published." Three months later, he told De Morgan he was still reading Grassmann "with great admiration and interest," although some days he found it hard to read in German. He was amazed, naturally enough, that Grassmann had, "with the most obvious and perfect independence," found such a similar system to his quaternions.[16]

Two days later, his enthusiasm had dimmed somewhat, but he wrote to De Morgan again, saying that he considered Grassmann "a great and most German genius." But he was having trouble grasping all of Grassmann's different kinds of multiplication (he had many more than the two that have come down to us in vector-tensor algebra). "His *outer* products I think I *do* understand, and that is saying something for a person who has not learned to smoke. And even his *inner* products . . . I can swallow pretty well [for they] have much analogy to my '*scalar parts*' of a quaternion, and his 'outer products' to my '*vector parts*.'" (As I mentioned, Hamilton's "scalar and vector parts" of a product of quaternions are our "scalar and vector products" of vectors.)[17]

De Morgan was fascinated by Hamilton's progress with *Ausdehnungslehre*, and a week later he received another update. Hamilton said that he'd been reading with his daughter looking over his shoulder, "amused at the folly of philosophers," when he noticed that while he himself had got the idea of adding "lines" from John Warren, Grassmann seemed to have worked it out himself. (He didn't know about Justus Grassmann—and even today no one knows how Justus came to his idea of geometrical algebra.) However, Hamilton continued, Grassmann took 139 "ostrich-stomach-needing" pages to get to the point—the actual algebra of directed lines (or vectors). Hamilton had conceived this early, he said, and had taken it as the starting point of his 1848 public lectures on his new system. Presumably he thought he was thereby sparing his listeners the need for a cast-iron "ostrich" stomach to digest this new mathematics. Still, the published version of those lectures was tough going, because Hamilton took so much care to explain each fundamental step in his new algebra that he employed an excess of strange terminology—including all the new words that he, like

Grassmann, had coined. But he did have a conversational style, and he also illustrated his new vector algebra with practical examples—particularly from astronomy and his work as Ireland's royal astronomer.[18]

Hamilton had begun expanding these lectures into a book before he'd heard of Grassmann, and the resulting *Lectures on Quaternions* was published not long after his series of letters to De Morgan. Hamilton used Greek letters to distinguish vectors from numbers, which are usually signified by Latin letters—and on page 62 of his preface, when distinguishing "ordinary algebra," where $ab = ba$, from his new noncommutative vector algebra, where $\alpha\beta = -\beta\alpha$, Hamilton acknowledged that Grassmann had come up with his own "species of non-commutative multiplication for inclined [or directed] lines in a very original and remarkable work." But he made it clear that his own work was independent of the "profound philosopher" Grassmann's, and he noted that Grassmann had not been able to extend the two-dimensional complex plane to all of space.

This is the nub of the difference between the two discoveries. Hamilton's quaternions were designed largely to handle algebraically three-dimensional geometrical operations such as rotations, and they have found some extraordinary applications as I've mentioned. Grassmann's system was more abstract, which made it less useful than vectors and quaternions in immediate applications in 3-D physics, but more useful in the later generalizations that would ultimately lead to modern tensor analysis.

Not that Hamilton—or Grassmann—foresaw all this at the time, of course. Neither did anyone else. Despite Hamilton's earlier fame, in the 1850s and early 1860s not even his initial idea of representing complex numbers as ordered pairs of real numbers had taken off, let alone his quaternions. Grassmann's ideas had made even less headway. Yet he and Hamilton never gave up. I mentioned earlier Hamilton's joyous 1859 letter to a new colleague (Peter Guthrie Tait), saying of their progress in applying quaternions to physics, "*Could* anything be simpler or more satisfactory? Do you not *feel*, as well as think, that we are on a *right track*, and shall be *thanked* hereafter? Never mind *when*. . . ." Grassmann felt similarly: in the 1862 version of *Ausdehnungslehre*—a reworking in a more mathematical, less philosophical style, plus some new material—he ended the foreword by saying, "I remain completely confident that the labor which I have expended on the science presented here and which has demanded a significant part of my life as well as the most strenuous application of my powers, will not be lost." He hoped not to seem arrogant, but he also hoped his ideas would eventually bear fruit, no matter how long they might lie dormant.

And lie dormant they did, for the second version of *Ausdehnungslehre* fared little better than the first.

CREATIVE FERMENT

Mathematical creativity is not unlike any other form, and in art, music, and literature, new forms and styles often take a while to catch on, too. But things that work gradually become accepted, especially when a range of new artists and art forms begins to meld or cross-pollinate. Indeed, while Hamilton and Grassmann had been working out a new way of handling geometry using algebraic operations and symbolism, another mathematical revolution had been taking place in geometry.

Euclidean geometry was beautifully rigorous and had reigned supreme for more than two thousand years. Trouble was, it was a science of everyday "flat" space: the pages of a book, for example, or a slate, a tabletop, a patch of sand—any surface on which Euclid could draw such things as parallel lines that never meet and triangles whose angles add up to 180°. But it was clear that on a *curved* surface such as a ball or the globe of Earth, two parallel lines *do* meet. Check out two nearby lines of longitude as they cross the equator: they start off parallel, but as you move north, they gradually move closer together until they meet at the North Pole. So, while Hamilton and Grassmann were discovering that you could ditch the commutative law for multiplication and still have a consistent algebra, other mathematicians were developing new "non-Euclidean" geometries by abandoning Euclid's so-called parallel postulate.

Chief among them were Gauss in Germany, Janos Bolyai in Hungary, and Nicolai Lobachevsky in Russia, and their discoveries will have ramifications for the story of tensors—a story that includes Einstein's now-famous curved space-time. Meantime, as non-Euclidean geometry filtered into the mainstream in the 1860s, so people became more interested in noncommutative algebras—and also in Hamilton's algebraic treatment of complex numbers as ordered pairs of real numbers, moving on from their "Euclidean" geometric representation on the Argand plane.

It would take yet another sixty years for mathematics, and mathematicians, to become sophisticated enough not only to invent tensor analysis, but also to revisit *Ausdehnungslehre* and to recognize in its very abstractness new tensorial tools. But that was far too long for Grassmann to wait, and in the 1860s and 1870s he turned mostly away from mathematics; the few papers he wrote were not up to his earlier standard, as if his heart had gone out of it. Instead, he put his prodigious mental energy into writing textbooks and papers on subjects related to his teaching—German, Latin, math, music, religion, and botany. Like Hamilton, he'd also learned Sanskrit, and in the 1870s he made a 1,123-page translation of the *Rig Veda* and published his own even longer work on the Hindu classic—exhausting labor that deservedly earned him membership in the American Oriental Society and an honorary doctorate from the University of Tübingen, in 1876. He did all this, the textbooks and the philology, as well as much of his math and physics, while teaching full time and raising eleven children (and burying four of them)—he'd married in 1849, five years after he first published *Ausdehnungslehre*.

For some political context, I should add that as well as courting and marrying in 1848–49, Grassmann had joined his brother in publishing a weekly political newspaper in those years—Grassmann wanted to see Germany unified under a constitutional monarchy, but at the time, the likes of Otto von Bismarck opposed constitutional reform. It would take several territorial wars and two decades before unification of a new German empire was achieved—with Bismarck as chancellor. While Grassmann was producing his newspaper, the most pressing political issue in Ireland was the growing movement for Roman Catholic independence—but Hamilton was a staunch Anglican (the national religion of Ireland at that stage, for the country was still part of the United Kingdom). He put his energy not into politics but into the promotion of science, through his role as president of the Royal Irish Academy, and his active contributions to the British Association for the Advancement of Science.

• • •

Hamilton never gave up his pursuit of quaternion applications. Aside from his *Lectures on Quaternions,* he published more than a hundred papers on the subject. Despite his fame and mainstream mathematical prowess, there were many who thought he was deluding himself by devoting the last twenty years of his life almost solely to quaternions. There was a widespread feeling that quaternions were brilliant, but they weren't all that useful.

Hamilton's fortunes began to change, however, with the rise of a new generation, and the entrance of Maxwell and Peter Guthrie Tait in the next chapter of our story. Tait became a fierce supporter of quaternions, and it was largely through his and Maxwell's work that others would develop the modern vector analysis taught in college today. Poor Grassmann's ideas would continue to prove too general—and too difficult to disentangle from his "strange" philosophy and his "peculiar" terminology—to be a direct influence. We'll check in with him again later, but even so, it is vector analysis and non-Euclidean geometry that will initially lead us to tensors.[19]

(6)

TAIT AND MAXWELL

Hatching the Electromagnetic Vector Field

Although Hermann Grassmann's work was difficult to penetrate and excessively abstract, it is also partly an accident of history that William Rowan Hamilton, not Grassmann, is the direct link to the vector analysis taught today. For surely it was fate that brought Peter Guthrie Tait and James Clerk Maxwell together as schoolboys.

They were both born in Scotland in 1831—twelve years before Hamilton announced his quaternion breakthrough—and from the age of ten they both attended the prestigious Edinburgh Academy. Not that it was a happy experience for the motherless young Maxwell at first. He was an eccentric child, being raised by his doting and unconventional father, a barrister and the Laird of Glenlair, the family estate some ninety miles southwest of Edinburgh. So when he arrived at the big city school, Maxwell was an odd figure, with a quirky sense of humor, a hesitant way of speaking, and homemade clothes that were comfortable rather than fashionable—and for all these transgressions of the class code he was mercilessly bullied. It didn't help that his academic performance was indifferent: as a self-directed learner he initially found school boring; this, together with his slow speech and shyness, led to the nickname "Dafty," so you can imagine how that singled him out as a boy "waiting" to be punched.[1]

Fortunately for him, and for the future of science, his gentle classmate Lewis Campbell befriended Maxwell, so that eventually he was able to settle down and become one of the top students, along with Campbell and Tait. Tellingly, it was a productive and friendly rivalry, and Tait and Maxwell soon became firm friends, too. Like Maxwell, Tait had lost a parent—his father, who'd been secretary to a duke—and had moved to Edinburgh from the country, although his hometown of Dalkeith was only eight miles away.

The Edinburgh Academy was a relatively progressive school, and Tait, Maxwell, and Campbell excelled there. Then, at sixteen, they all went on to Edinburgh University, although after a year Campbell transferred to Oxford—he became a classicist and an Anglican minister, and later a wonderfully empathetic biographer of his childhood friend Maxwell—while Tait transferred to Cambridge, where he outshone all his peers, graduating as the Senior Wrangler of 1852. The Senior Wrangler was the student with the best marks in the Tripos, a grueling series of around sixteen increasingly difficult mathematics examinations, held over eight days. Mathematics was seen as important mental preparation for all students no matter their ultimate profession—and if you managed to be one of the highest scoring wranglers you were assured of a good career pathway. Tait was so excited that first he telegraphed the news to his family, and then he wrote to his erstwhile Edinburgh teachers, "I'm all in a flutter, I scarcely can utter, etc, as the song has it: I AM SENIOR WRANGLER!" He also took out the prestigious Smith's Prize, for the top student in an even more complex series of math exams—and all up it was such a fine achievement that his and Maxwell's old school held a special celebration in his honor.[2]

After graduating, Tait won a fellowship at Cambridge—one of the benefits of being a top wrangler, for it gave him a job and teaching experience that would presumably lead to a permanent lectureship later, at Cambridge or some other leading institution. For the moment, though, he reveled in his newfound security, free of the grind of exam preparation. He even treated himself to a copy of Hamilton's *Lectures on Quaternions*, hot off the press—he had no idea what it was about, but he was intrigued by the strange name "quaternion." He read the first six chapters while away on a summer holiday shooting trip, but once back at Cambridge, his teaching and research took up all his time, and he laid *Quaternions* aside. (Maxwell, by contrast, refused to hunt or fish, such was his love of animals. He was especially brilliant with horses, and he always had a much-loved dog by his side.)

While Tait was at Cambridge, Maxwell had stayed on at Edinburgh—apparently to please his father—but the excellent, broad-ranging teaching there proved the perfect training-ground for the practical and philosophical skill with which he would soon transform physics. Then, after finishing his Edinburgh degree in 1850, he enrolled at Cambridge to further his scientific studies. It took a while to settle in, but soon the shy, awkward young Dafty made "a troop" of new friends, for the Cambridge crowd found him "genial and amusing," able to converse in a witty and erudite way on any subject at all. "I never met a man like him," one new acquaintance recorded after such a conversation, while another later reminisced, "Everyone who knew him at Trinity [Maxwell's college at Cambridge] can recall some kindness or some act of his which has left an ineffaceable impression of his goodness."[3]

While Maxwell delighted in his new social life, he was not so impressed with the Cambridge curriculum of the time, with its emphasis on technical proficiency rather than deep thinking. At the end of the year as he was preparing for exams, he expressed his frustration in a long poem, "A Vision of a Wrangler." Here's a taste of his humor, and his view of the exam system:

> In the grate the flickering embers
> Served to show how dull November's
> Fogs had stamped my torpid members
> Like a plucked and skinny goose.
> And as I prepared for bed, I
> Asked myself with voice unsteady,
> If of all the stuff I read, I
> Ever made the slightest use.[4]

This is but the second of twenty-four increasingly erudite verses centered on a fantastical incarnation of his evil new alma mater, as he expresses his outrage at a system that demands slavish rote study and then rewards its best students with cushy academic jobs—these wranglers then becoming part of the elite who keep the system going. The poem ends with a serious spiritual declaration, that it is better to silently contemplate the "glories of Creation" than to mindlessly study the detached mathematical symbols that physicists use to represent nature. He was twenty-one when he wrote this extraordinary poem; in twelve years' time, however, he would show just how wondrous the symbolic representation of nature can be. Later, he would also become an examiner and moderator himself, and he would play a leading role in making the Tripos questions—and therefore the Cambridge syllabus—much more relevant to the exciting new science being discovered.[5]

In November 1853, however, as his final exams drew near, he wrote another long poem—"Lines Written under the Conviction That It Is Not Wise to Read Mathematics in November after One's Fire Is Out!"—in which he speaks of the soul-crushing exam pressure on young people, and the folly of mistaking academic prizes and degrees for wisdom. So mostly he preferred to study what interested him instead of cramming—Tait was amazed at how unprepared Maxwell was for the Tripos. Still—thanks to Maxwell's "sheer strength of intellect," as Tait put it—he managed to graduate as Second Wrangler, tying with the Senior Wrangler (E. J. Routh) for the Smith's Prize. His father expressed his pride in a touchingly supportive letter: "[Your cousin George] came into my room at 2 am . . . having seen the Saturday *Times*, received by express train, and I got your letter before breakfast. . . . As you are equal to the Senior Wrangler in the champion [Smith's] trial, you are but a very little behind him." As for Maxwell's former school math teacher, he was "beside himself" with excitement, just as he'd been with Tait's success.[6]

You can get a feel for the difficulty of the Smith's Prize exams from question 8 on the February 1854 paper, which Maxwell sat: it asked for a proof of what is now known as "Stokes's theorem"—the Irish-born mathematician George Stokes had been setting the Smith's Prize papers for the past few years. The 1854 paper was the first time this theorem appeared in print—today it's in every undergrad calculus textbook, and it relates a "line integral" to a "surface integral." Many students today find it difficult enough to apply, let alone prove. No one knows if Maxwell, or any of his fellow examinees, proved it, either, but I suspect not, for Maxwell later credited the full proof to William Thomson, the future Lord Kelvin.[7]

Line and surface integrals will play a role in our story, so in case you haven't met them before, let me give you an idea of what they are. The ordinary integral calculus we learn at school integrates a function with respect to the independent (horizontal) variable, often x or t—so when you integrate, you're adding up the areas of all those skinny rectangles under the curve, as we saw in figure 2.1. By contrast, a simple example of a "line integral" is figure 2.3b, where I integrated with respect to a small segment ds of a line—a circle in that case—to find the circumference. In other words, in a "line integral" you're adding up an infinitesimal number of little lengths rather than areas—so it's an exact version of the circumference approximations Ahmes and Archimedes made thousands of years ago. A "surface integral" does a similar thing over a whole surface—but it needs a "double integral" so that you can integrate in two dimensions. For example, in figure 2.3a I could have found the area of the circle more easily using a surface integral, as you can see in the endnote.[8] But the crucial point here is the *idea* that you can integrate not just with respect to the horizontal axis, but also along curved lines and across surfaces—and through volumes, too, if you use a triple integral.

As for Maxwell, in ten years' time he would make good use of the theorem he first met in the Smith's Prize exam, but in the meantime, and despite his outstanding exam results, he wasn't awarded a Cambridge fellowship. Apparently, the powers-that-be thought he was too careless—a mathematical failing he openly acknowledged, and one I'm sure many of us can identify with. So, the brilliant if sometimes slapdash Maxwell graduated without a job, while a few months later Tait was promoted to professor of mathematics at Queens College, Belfast; his copy of *Quaternions* still lay unfinished, waiting for him to find the time to study it. Which suggests there was an upside to Maxwell's predicament, given that his father was comfortably off: between stints as a private tutor at Cambridge and volunteering at the newly founded Working Men's College, he had time to contemplate the rapidly developing new science of electromagnetism, at home in the peace of Glenlair—a beautiful estate that still looks much as it did in Maxwell's time (thanks largely to the efforts of its present owner, Duncan Ferguson).[9]

One of the first things Maxwell did in his newfound freedom was write to William Thomson, whose work on the math of electricity had already made an impression on him. Maxwell said that he wanted to get himself up to speed on the state of electromagnetic research, and he asked if Thomson would advise him on where he should start.

This bold approach was not entirely out of the blue, for he and Thomson had met several years earlier. It was at an Edinburgh meeting of the British Association in 1850, where nineteen-year-old Maxwell—still shy and awkward, with his regional accent and hesitant way of speaking—had bravely stood up to make a point at question time, about a paper on optics and vision that had just been presented to the meeting. (Maxwell was also a pioneer of the study of vision and color, and he would work out how to create the first-ever color photograph of an everyday object—a tartan ribbon, proud Scot that he was.) According to one attendee, the audience was "half-puzzled, half-anxious and perhaps somewhat incredulous . . . as they gazed on the raw-looking young man who, in broken accents, was addressing them." So, you can imagine Maxwell's surprise and relief when, after the meeting had finished, he was approached by a rising star in the world of mathematical physics: twenty-six-year-old Thomson, who wanted to know more about his ideas.[10]

Now, in 1854, Thomson replied generously to Maxwell's letter, and with his advice, Maxwell set himself the task of reading up carefully on all the electromagnetic results known so far. He would have studied some of the relevant math and physics at university, but in the next few pages I'll take you on a brief historical journey through the key things he needed to know before he could transform the subject—and thereby inaugurate, with

Maxwell and his color top. Cambridge, Trinity College, Add.P.270a; by kind permission of the Master and Fellows of Trinity College, Cambridge.

Tait's help, a brand-new approach to physics: the *vector field*. Or rather, Maxwell would take an approach that others had intuitively glimpsed and breathe such new life into it that it was, indeed, brand-new.

MAXWELL'S JOURNEY IN 1854—AND A SURFEIT OF INTEGRALS

Hans Ørsted's 1820 discovery of the magnetic effect of an electric current could not have been observed before 1800, when Alessandro Volta invented the first electric battery. Before then, only *static* electricity could be stored, in an early form of a capacitor called a Leyden jar. Static electricity is often generated by friction, as when you brush your hair in a darkened room and see sparks of electricity. The brushing knocks electrons out of the atoms in your hair—although no one knew about electrons back then.

But often it is enough for two materials just to touch each other for electrons to be exchanged—as in what Maxwell called "the electricity of kissing." (In a related vein, capacitors—sources of charge—are embedded in most mobile phone screens, so that when you touch the screen electrons flow between it and your finger: the resulting change in capacitance is interpreted as a touch command.)

So, the first mathematical studies of electricity were about the effects of static charges—charges that tend to stay put rather than freely moving in a current. In particular, in 1785 Charles Augustin Coulomb did his famous experiment with a torsion balance and showed that the electric force from a charged particle drops off in inverse-square proportion to the distance from the source—just like Newton's law of gravity for the force exerted by a stationary object's mass. It was an astonishing coincidence that the mathematical laws of two such different phenomena should have the same form.[11]

At around this time, Joseph-Louis Lagrange brought his powerful mathematical mind to bear on a new formulation of the theory of gravity—and just like the inverse square law, it would soon prove useful in describing electricity, too. Lagrange's formulation involved the concept of a mathematical "potential"—an idea related to "potential energy" and the "work" done by a force. For instance, it takes effort—work—to lift a weight against gravity, but once it's up there it has the potential to expend energy doing more work if it falls. You can see this kind of application in the huge demolition balls hoisted on cranes or in the falling water that drives a turbine. To lift up the demolition ball, though, you need to apply a force to it, à la Newton's second law of motion; the "work" done by this force is defined as "force times the distance the object is moved." In the endnote, I've given the math of "work" and how it leads to the potential—but the main point is that the definition of work involves an integral.[12]

Today, "voltage" (named in honor of Volta) might be a more familiar term for electric potential, and its role is similar to that of temperature in heat and pressure in fluids. If there's a pressure difference between the ends of a pipe filled with water, then the water flows from the high-pressure end to the low-pressure end. Similarly heat flows from a hotter region to a colder one. When things change like this, derivatives come into play, so you won't be surprised to find that if V is the potential associated with a force, then the force's components are:

$$\frac{\partial V}{\partial x}, \frac{\partial V}{\partial y}, \frac{\partial V}{\partial z}.$$

The curly d's indicate "partial" derivatives, which show that V is changing all through space, not just in one direction as with the familiar $\frac{dy}{dx}$.[13]

The application of Lagrange's idea to electricity—and to magnetism, too—was the result of the work of several mathematicians over several decades, beginning with Pierre-Simon Laplace, who developed potential theory more fully than Lagrange. Then came the ubiquitous and versatile Carl Friedrich Gauss, the remarkable self-taught mathematician George Green—whose day job at the time was being a miller, although a decade later, at the age of forty-three, he graduated from Cambridge as Fourth Wrangler—and Laplace and Lagrange's student Siméon-Dénis Poisson. When Mary Somerville and her husband visited Paris in 1817, Laplace invited them to his country house—she was one of the few in Britain who had studied and understood his *Mécanique Celeste*. She found Laplace very kind and attentive, and Poisson an entertaining and vivacious dinner guest.

Sophie Germain, however, had a rather different experience with Poisson, for she had the temerity to compete in the same playing field as he. She anonymously entered a pioneering paper on the theory of vibrating surfaces—a brand-new topic opened up to mathematics through a physical discovery (the patterns made when sand is scattered on a plate set vibrating by a sound)—in an essay competition sponsored by the French Academy of Sciences. It was such a difficult problem that, just as with Grassmann's entry in the Leibniz centenary competition, Germain's was the only essay submitted. It is likely her identity was an open secret—but it's true that her self-education showed, and her derivations weren't completely correct, and so her first effort was rejected. She reworked her paper and submitted it again—prizes were only awarded if the standard was high enough;

otherwise, the competition remained open. Perhaps because she was an outsider, Poisson, who was one of the judges, behaved rather badly, by severely criticizing her second essay (the other judges gave it an honorable mention) and then lifting its conclusion and working on it himself. Undeterred, she kept on improving her paper and finally won the prize with her third attempt, in 1816. I mention it here not just as a shout out to Germain and her persistence despite the obstacles for women in science, but also because the math of vibrating surfaces includes some similar equations and expressions to those being developed for gravity, electricity, and magnetism around the same time.

For instance, Laplace had shown that the potential V obeyed a really neat equation:

$$\frac{\partial^2 V}{\partial x^2}+\frac{\partial^2 V}{\partial y^2}+\frac{\partial^2 V}{\partial z^2}=0.$$

Poisson later showed that in physical applications the right-hand side is not always zero—it depends on what and where you're measuring. Either way, Maxwell described the physical meaning of the expression on the left-hand side as the "concentration" of the potential, and mathematically it is known today as the "Laplacian" of V, where the Laplacian is the "differential operator"

$$\frac{\partial^2}{\partial x^2}+\frac{\partial^2}{\partial y^2}+\frac{\partial^2}{\partial z^2}$$

(or "Laplace's operator," as Maxwell called it).

As I mentioned in chapter 2 in connection with $\frac{d}{dx}$, the term "operator" just means that the expression directs you to "operate on"—in this case, take the second (partial) derivatives of—some function that is yet to be inserted. In physics, the functions you want to operate on have a physical meaning, such as the potential, but mathematically speaking, any suitably differentiable function will do. The idea of something that "operates" when a function is "inserted" will become especially important when we come to tensors.

Meantime, in the wonderful way that nature and mathematics seem intertwined, the Laplacian appears not only in potential theory but in the theory of heat, vibrations, and waves: instead of using the potential V, you can take the Laplacian of the temperature T, say, as in the "heat equation," or the stretching distance u in a wave, as in the "wave equation." The significance of this for our story, though, is that Lagrange, Laplace, and the other pioneers simply used Cartesian coordinates and components. They had the Newtonian idea of a vector, but whole-vector calculus had yet to be invented.

Once Tait settled down to reading Hamilton's *Quaternions*, however, he would begin to apply the potential and Laplacian in full vector form—and Maxwell would then use these vector quantities to unprecedented advantage. But that was still fifteen years in the future. In 1854, Maxwell was still reading up on what came before he entered the ring, and Tait was still busy with his new academic duties in Belfast.

• • •

As if all this new calculus of static electricity wasn't challenging enough, when Øersted discovered the existence of electromagnetism—which involved *moving* charged particles in a flowing electric current—things became a whole lot more complicated.

In 1821, André-Marie Ampère used Øersted's strange discovery to pioneer telegraphy—and he also began trying to quantify electromagnetic effects experimentally and then express them mathematically. He made amazing progress, single-handedly initiating the mathematical study of electromagnetism (or as he called it, "electrodynamics"). Maxwell dubbed him the "Newton of electricity," because he was such a great experimenter *and* mathematical theorist. In particular, he used line and surface integrals to quantify the relationship between electric current and the magnetic force it produces in various experimental set-ups—but when he died in 1836, he still hadn't managed to find a completely general description of electromagnetism.[14]

At the same time as Ampère began his electromagnetic research, Englishman Michael Faraday used Øersted's discovery to develop a rudimentary prototype of an electric motor. Then, in 1831, the year Maxwell and Tait were born, Faraday discovered the inverse effect to Øersted's: he showed that you could *generate* an electric current by moving a magnet through a coil of wire. (It is the relative motion that induces the current, so you can just as well turn the coil instead of the magnet, and we'll see this set-up shortly, in fig. 6.2.) In yet another example of independent codiscovery in science, the American Joseph Henry discovered the same thing, some months later.

Faraday demonstrated this remarkable phenomenon with a prototype generator, which others would later develop into commercial propositions. The rest is history, for better and worse, for by the late 1800s the world began to go electric. In the 1880s, for example, Thomson built a house that was one of the first in the world to include electric light, while in Germany and Italy, Einstein's father and uncle were running an innovative technological company making and installing generators, bringing electricity to businesses, public buildings, streets, and even households for the very first time. Today, of course, we rue the fact that most large generators have been coal-fired—in the absence of a natural force such as falling water or wind, you needed heat to produce steam, in order to turn the huge coils that generate the current. Still, few of us would want to do without electricity, so we owe thanks not just to the modern scientists and engineers designing sustainable, climate-safe ways of producing it, but also to the nineteenth-century experimentalists and mathematicians who were trying to figure out just how it all works.

When Maxwell first set out on his electromagnetic journey, sorting out the math of it had become rather tricky. It had to do with the advantages and pitfalls of using analogies in science—in this case, between gravity, hydrodynamics, and electricity. I've mentioned already the mathematical parallels between gravity and static electricity, but hydrodynamics—the study of how fluids behave as a result of forces—was important because in those early days people tended to visualize electricity as a fluid. That's why they used the word "current" to describe the flow of "charged fluid," whatever it was: the electron wouldn't be discovered until the very end of the nineteenth century. But Maxwell realized, as very few had done, that analogies could lead you astray—indeed, electricity turned out not to be a fluid at all. Still, he knew that initially, at least, this analogy had offered a timely and powerful way into the mathematics of the new phenomenon of electromagnetism. The challenge was to use the math without confusing it with the physical reality.

For instance, another term that has evident links with fluids is "flux." The *Oxford English Dictionary* defines it as "a process of flowing or flowing out." But in the study of fluids in the eighteenth and early nineteenth centuries, "flux" acquired a more specific meaning, as the amount of fluid flowing through a given surface—such as the cross-sectional area of a pipe—in a given time. You can think of it as counting the number of molecules of fluid leaving the pipe each second—and you can measure this number using mass or volume. In figure 6.1, I've shown the flux using volume.

Once you make a quantitative definition like this, you can define it mathematically. For fluid flowing through a cylindrical pipe, the flux turns out to be the scalar product of the cross-sectional area of the pipe and the velocity of the fluid (as you can see in the caption to fig. 6.1). In the case of flux through a surface of any shape, not just a circle with its neat area formula, the definition involves a *surface integral*. That's because you're effectively adding up (integrating) the amount of liquid flowing across each little element of area on the surface.

All this suggests that flux is a property of any vector, not just the velocity of a fluid. And sure enough, just as Coulomb's law of static electricity and Newton's law of gravity have the same mathematical form, and just as Lagrange's potential and the Laplace-Poisson equations can be applied to both electricity and gravity, so it turned out that flux integrals could be applied to other things than fluids—including gravity, electricity, and magnetism. Gauss was one of the pioneers in developing this math, so the laws relating the gravitational and electrical flux through a closed surface to the enclosed amount of mass and charge, respectively, are called "Gauss's laws." In fact, you can deduce Newton's and Coulomb's laws from Gauss's laws—and there's a Gauss's law for magnetic flux, too. The main point here, though, is that all these early applications of flux involved *integrals*. But physically, the really interesting things happen when these electric and

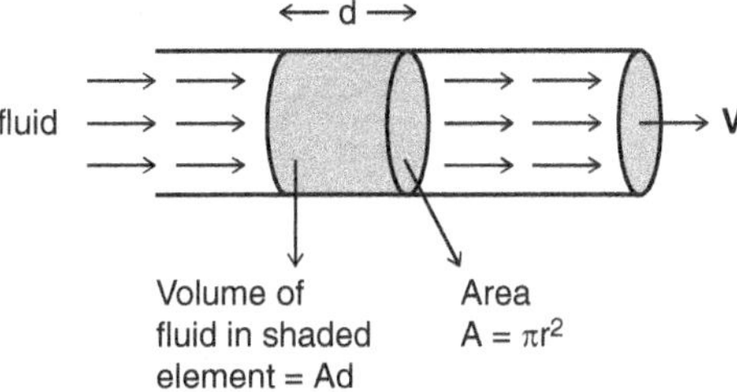

FIGURE 6.1. Flux of fluid flowing through a pipe. Here the velocity of the fluid (shown by the arrows) is constant. Its direction is perpendicular (or "normal") to the surface shown at the end of the pipe, so in this case all the water flows through that surface. The small cylindrical section of fluid that I've highlighted in the middle has volume $V = Ad$, i.e., the circular base with area A $(= \pi r^2)$ times the length d. So, the fluid flows through the shaded surface at the end of the pipe at a rate (volume per unit time) of $A\frac{d}{t}$. But $\frac{d}{t}$ is distance/time, which is the speed, so the velocity vector is this speed in the direction shown. Which means the amount of fluid flowing through the end of the pipe in a given time—that is, the flux—is Av. This formula makes sense, because the faster the flow and the larger the area, the more volume (or number of molecules) of fluid passes through the end of the pipe in a given time.

It's more complicated when the pipe outlet is partially blocked, or at an angle to the flow. For instance, if the flow direction is at an angle θ to (the normal to) the surface it's flowing through, you need the component of the vector $\boldsymbol{v}$ in that normal direction. That's because flux is the amount of fluid flowing through—that is, *across*—the surface. So, the formula for flux becomes the scalar product, $\boldsymbol{A} \cdot \boldsymbol{v}$ $(= Av\cos\theta)$. Making $\boldsymbol{A}$ a vector means specifying that the direction of a surface is defined to be that of its normal. When your surface is curved or irregularly shaped, however, you need to use vector calculus to find the area. Then the general flux formula is $\iint_S \boldsymbol{v} \cdot d\boldsymbol{A}$. But I'm mentioning this just for completeness—we won't need it again.

magnetic fluxes *change*, for that's when electromagnetic effects come into play. So, these changing fluxes hold the key to practical applications of electromagnetism, as in the schematic setup for electric motors and generators in figure 6.2.

Exactly how this happened, and how it could be expressed mathematically, is what a host of researchers were trying to find out, including Ampère and Augustin Cauchy in France, Green and Thomson in Britain, and the German physicist Wilhelm Weber. Maxwell read up on them all and admired their various discoveries and hypotheses. Yet none of these luminaries had managed to find a complete theory of electromagnetism.

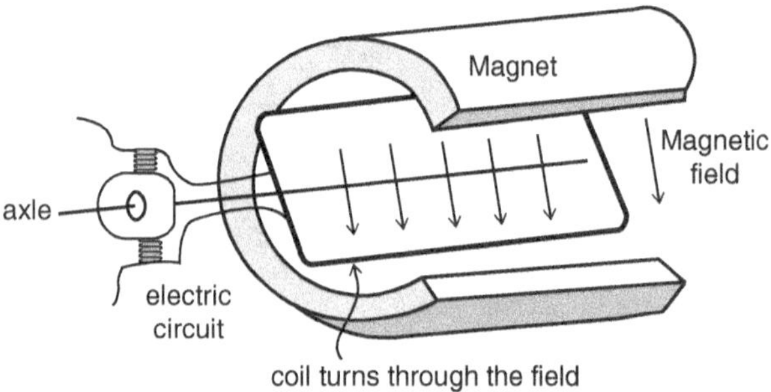

FIGURE 6.2. Electric motors and generators: As Øersted discovered, an electric current can produce a magnetic force, and it had long been known that two magnets attract or repel each other (opposite poles attract). So, when a current-carrying coil is placed in an external magnetic "field"—shown here by the arrows between the magnetic poles—its own magnetic field reacts to the external one. This deflects the coil and turns it—and this turning effect (torque) is harnessed in electric motors. As the coil turns, the angle it makes with the external magnetic field changes—which means the magnetic flux through the coil is changing (cf. fig. 6.1). So, if you mechanically turn the loop first, then the changing magnetic flux through the loop induces the electromagnetic field needed to produce an electric current—and this is the basis of electricity generators.

Now we begin to see where Maxwell's training in philosophy comes to the fore—not to mention his genius. While analogies between different physical phenomena can certainly be very useful, they come with hidden assumptions—and Maxwell realized that his forerunners' choice of mathematical tools, with all those integrals, embodied the biggest assumption of all. For most researchers assumed that gravity, electricity, magnetism, and electromagnetism all acted at a distance—instantaneously leaping directly from one material body to another, one charged particle, or one magnet, to another, as well as from a moving magnet to a wire or from a current-carrying wire to a magnet. Maxwell was the only one who recognized clearly that the flux and other integrals I've been talking about here were assumed to embody action-at-a-distance, because they focused only on the points, lines, and surfaces that formed the limits of integration or *boundaries* of the integral. For instance, the work done by a gravitational or electric force $\boldsymbol{F}$ in moving an object from point r_1 to point r_2 is given by $\int_{r_1}^{r_2} \boldsymbol{F} \cdot \boldsymbol{dr}$; when you carry out such an integral, you need only the limits of integration—you don't need to know anything about the path between r_1 and r_2.[15] So you can see that this type of integral was the right tool for those who believed all the action happened at the two points, interacting instantaneously from one point to the other. To take another example, you only need the limits of integration in (the modern version of) Newton's proof that the gravitational attraction between two spherical objects, such as the sun and a planet, acts as if all their mass is concentrated at their centers. You can see why this suggested action at a distance: it's as if nothing gravitational happened in between these two central points. Maxwell realized that, similarly, all those electromagnetic integrals his predecessors were using took no account of any processes that might be going on not just at the boundaries, but also *within* the intervening or surrounding space. Perhaps that was why no one had managed to crack the electromagnetic code.

But all this was about to change. After his year of careful study, Maxwell was ready to set aside the works of the mainstream mathematicians, and to take up an idea suggested by a rank outsider.

A BRILLIANT IDEA

Faraday had made his experimental discovery of electromagnetic induction back in 1831, the year Maxwell was born. As far as mathematical theory went, however, the self-taught Faraday hadn't been able to follow all that fancy mathematical talk of line and surface integrals, Gaussian flux laws, and so on. He'd left school at thirteen, taking up an apprenticeship with the kindly bookbinder George Riebau, who had fled France during the Revolution. Faraday could barely read and write when he first arrived at Riebau's workshop—his family belonged to a fundamentalist Christian sect that had little interest in learning because the Bible was its authority, and he'd often played truant from school and its harsh schoolmaster. But Riebau encouraged young Faraday to practice after work by trying to read some of the books they were binding—and in this extraordinary, painstaking way, Faraday had become enchanted with electricity, thanks to an entry in the newly bound *Encyclopaedia Britannica*. He set up a rudimentary

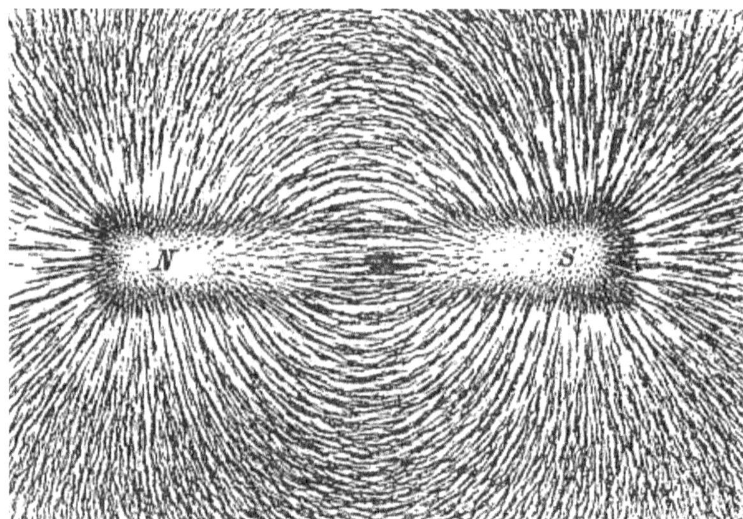

FIGURE 6.3A. Iron filings aligning around a magnet. Newton Henry Black and Harvey N. Davis, *Practical Physics* (New York: Macmillan, 1913). Wikimedia Commons, public domain.

laboratory at the back of Riebau's shop, attended public lectures, and eventually learned enough to impress the Royal Institution's leading light, Humphry Davy—and from there he worked his way from lab sweeper to Davy's assistant until, ultimately, he became Britain's leading expert on the practical, experimental side of electromagnetic research.

So, while Faraday didn't know much math, he certainly knew more than most about the physical nature of electromagnetism—and this led him to develop a unique approach to the question of what was going on in the space between electric charges and magnets. He first got the idea when he noticed something that may well have fascinated you, too, if you took a school science class: the neat way that iron filings line up in loops around the poles of a bar magnet. That's because each little filing orients its own tiny poles to those of the bar magnet, as in figure 6.3a—and as Maxwell later explained, *lines* of filings are formed because each time a new filing is added, it must also align itself end to end with its neighbor, for the attracting magnetic force is concentrated at the poles near the ends of each little magnet.[16]

The traditional, "action-at-a-distance" way of looking at this was that if you had just one iron filing present—a single little "test" filing, as in figure 6.3b—then the force between it and the magnet would interact instantaneously. In other words, the magnetic force would act *only* between the magnet and the filing, and it would do so instantaneously. But Faraday reasoned that even when you'd taken all the other iron filings away, the force from the magnet was still acting in all the same places as it was before—even if you could no longer see its effects. In particular, there must be force acting all the way *through* the space between the magnet and the "test" filing—just as there'd been when all the filings were present. To describe this idea, Faraday came up with the concept of "lines of force," in direct analogy with the lines of iron filings. It was as if the filings made manifest the invisible force emanating from the magnet, just as light makes a watermark visible.

The way iron filings line up around a magnet brings to mind something like a field of wheat. When a breeze blows across the field, each little stalk bends in response to the force of the wind. Similarly, Faraday proposed that both electric and magnetic forces were conveyed through "fields" emanating from charged particles and magnets—they did *not*, he declared, act at a distance. It was a simple, utterly brilliant idea. Trouble was, he didn't

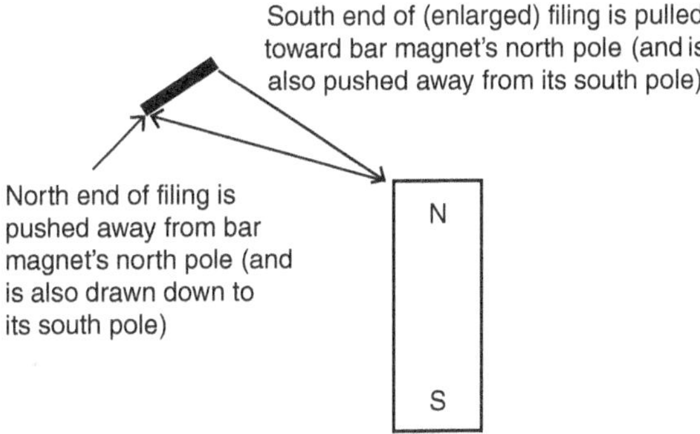
South end of (enlarged) filing is pulled toward bar magnet's north pole (and is also pushed away from its south pole)
North end of filing is pushed away from bar magnet's north pole (and is also drawn down to its south pole)
N
S

FIGURE 6.3B. The forces that align each individual filing (each of which is like a tiny bar magnet). Faraday also did the same thing for electricity, placing a little electric "test" charge at various spots in the space around a central charge and noticing how the force acted in each place. In this way he built up a picture of the electric and magnetic fields.

know how to turn it into a predictive, mathematical theory that could be tested against further experiments. So, while he'd gained immediate mainstream acclaim for his undeniably skillful experimental research, his concept of electric and magnetic fields was all but ignored.

WHAT MAXWELL DID WITH FARADAY'S FIELD

In fact, before Maxwell came on the scene, William Thomson was virtually the only established theorist to try to put some math into Faraday's idea. When he was just seventeen, Thomson had made a mathematical comparison between electrostatic effects and heat flow, adapting Joseph Fourier's mathematical study of heat. So, Maxwell began his own electromagnetic career by extending Thomson's approach to fluids, making the analogy between the streamlines in a river and Faraday's lines of force around a magnet, and around an electric charge.

In early studies of heat and fluid flow, the likes of Fourier, Newton, and Euler had, in fact, intuited the idea of a field, albeit it an implicit way. But they did have the idea that *each* particle of fluid has a velocity (and a temperature), which is similar to Faraday's idea that there are electric or magnetic forces acting on hypothetical unit test charges and filings placed at *each* point in the space around an electrically charged body or a magnet. Actually, Faraday had a clearer idea about lines of force than he did about the field, and it was Thomson who first defined the concept in physics: whenever there is a definite magnetic force at every point in a space, he'd said in 1851, you have a "field of magnetic force," or more simply, "a magnetic field." And similarly, of course, for an "electric field."[17]

Maxwell called his paper—his first on electricity—"On Faraday's Lines of Force," and he presented it to the Cambridge Philosophical Society in two parts over the winter of 1855–56. (In 1855 he'd finally secured a fellowship at Cambridge!) When it was later published, he sent a copy to Faraday. At the age of sixty-four, Faraday had grown weary of the mainstream's rejection of his field idea, so you can imagine his joy on finding that twenty-four-year-old Maxwell had brought it to life at last. Faraday replied warmly, and the two men became friends.

Maxwell had also been nursing his adored father, who died in April 1856. He poured his grief into a poem and then carried on as best he could as the new Laird of Glenlair, with all its attendant duties. (And its pleasures: Maxwell had always got on well with the estate's tenants, and he and his father had set up a program for those who wanted to improve their reading.) A few weeks later, he finally landed a professorial appointment, at Marischal College in Aberdeen, beating Arthur Cayley to the job—Cayley was trying to get back into academia after his years as a lawyer—and Tait, too.[18] Two years later he married the college principal's daughter, Katherine Dewar. Like Hamilton's wife, Helen, Katherine was strictly religious; Maxwell, like Hamilton, had a broader view of spirituality, although they were both committed Christians.

Many other papers followed in the next decade as Maxwell honed his conception of Faraday's field until, at last, he was ready to publish his radical field theory of electromagnetism. It includes the ideas of potential and flux that I've mentioned—but it contains relatively few integrals. Instead, there are many derivatives and "partial differential equations"—equations that involve derivatives with respect to x, y, and z, such as the derivatives of the potential in the Laplacian I showed earlier.

He wasn't the only one to use differential equations, of course—people had been using them in physics ever since Newton. It's just that he was clearer than anyone else about when to use each kind of calculus, and why. He explained his preference for derivatives in electromagnetism carefully, noting that while line and surface integrals focus only on such things as the distance between the two interacting particles and the "electrifications or currents in these bodies"—which was fine if you assumed that electric and magnetic forces acted at a distance—partial differential equations show how things *change* through the *whole* space around the bodies. So, he suggested, such equations are the natural tools for expressing the way changing electric and magnetic forces and fluxes produce electromagnetic effects. The field, whatever it was made of, mediated these changes through space—just as the field of wheat mediates a breeze rippling through it like a wave.[19]

• • •

Maxwell was the first to put Faraday's discovery of magnetic induction—the inducing of electric current by a relatively moving magnet—into symbolic mathematical form. Moving the magnet through a coil of wire, or turning the coil through a magnetic field, changes the magnetic flux—or what Faraday thought of as the number of "lines of force" from the magnet through a loop of wire, as you can see in figure 6.2; Maxwell had to find the right equations for relating this changing magnetic flux to the amount of current induced. This is known today as "Faraday's law."

Then there was "Ampère's law," the mathematical version of Øersted's discovery, in terms of the magnetic force produced by a changing electric current. Maxwell extended Ampère's work on this by defining current more carefully, and then showing how this relates to changing electric flux.[20] I spoke before about flux as a surface integral, and Ampère had used surface (and line) integrals, too, for he had worked in the action-at-a-distance tradition. So, Maxwell needed not only to complete Ampère's results but also to rewrite these integrals as derivatives, in order to express the field concept.

There was much more to describe and define—including the Gauss-Coulomb laws for static electricity and magnetism, which Maxwell now had to rewrite in field language. For even when you have a charge or a magnet just sitting still, the inverse square law shows how the force emanating from each of them changes with distance ("static" means the force doesn't change in *time*)—so derivatives are a natural choice of language for these laws, too. In the magnetic field around a bar magnet, you can see these changes in the alignment of iron filings, the filings bunching up near the poles where the force is strongest but spreading out as it weakens with distance—as in figures 6.3a and 6.3c.

The nub of how Maxwell changed from traditional action-at-a-distance integrals to partial differential field equations lies in the "fundamental theorem of integral calculus," which you may remember from school, although I'll spell it out in the next endnote. You'll also get a hint there of the marvelous use Maxwell made of Stokes's theorem, which had featured in the Smith's Prize exam—and you'll see how he used what is now, post-vectors,

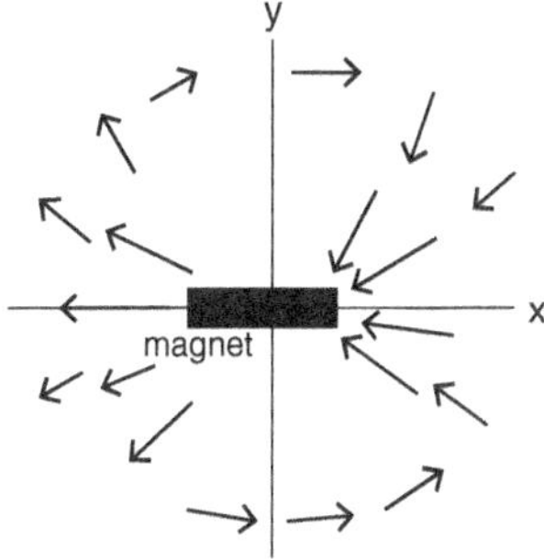

FIGURE 6.3C. In hindsight, you can see how the physical lines of filings shown in figure 6.3a can be mathematically represented as a vector field. The arrows point in the direction of the force at that point, and their length indicates the strength of the force—so they are longer close to the magnet's poles, where the force is strongest.

known as the "divergence theorem," pioneered by Gauss, Green, and the Russian mathematician Mikhail Ostragradsky.[21]

We'll see Maxwell's equations in the next chapter, but the point here is that first, in his landmark paper on the electromagnetic field—which he presented to the Royal Society at the end of 1864 and published in its *Philosophical Transactions* in January 1865—he formally united electricity and magnetism for the first time.

Second, like Gauss, Stokes, et al., Maxwell used only the component form of Stokes's theorem and the divergence theorem (he didn't state them in the compact whole-vector form you see in textbooks today). So, in his 1865 paper, he found relationships between the *components* of the various quantities he'd defined to describe the electromagnetic field. In other words, he had not yet formulated his theory in full vector form. He did talk about quantities having both direction and magnitude, though, so he *had* created a vectorial electromagnetic theory. With Tait's help, he would get to Hamiltonian whole-vector calculus in just a few years' time.

The year 1865 is a momentous one in American history as well as in physics and math history, for the bloody Civil War finally ended. It is fitting that one of the men who would help transform Maxwell's equations into modern (post-Hamilton) vector notation is Josiah Willard Gibbs, whose father had long agitated for the abolition of slavery. During the war, young Gibbs was completing his engineering doctorate at Yale—the very first American engineering doctorate. It would be another decade before he discovered Maxwell, and almost another still before he made his mark in the history of vectors. So, part of what I've been trying to show throughout this story is just how long it takes for ideas to develop and find their best form. I think that knowing something of this long journey can help students who are struggling to grasp an idea, by showing that earlier mathematicians had struggled, too. But it can also help show the power of specific ideas that we might now take for granted—and in taking them for granted, we might miss some of that power.

For instance, Maxwell chose to represent the behavior of electromagnetism using differential equations, but these are mathematically equivalent to the integral equations that the early electromagnetic pioneers might have developed had they succeeded in their quest—in fact, the integral form of Maxwell's equations is often used today. (This equivalence is where the "fundamental theorem" in the previous endnote comes into it.) The difference is in the interpretation, although today, the idea of a *field* is paramount in physics, no matter which form of calculus you use—for as Faraday, Thomson, and Maxwell showed, a "field" is just the distribution of values of a particular quantity through space. For example, a record of the temperature of the air around you right now is a "scalar field"—a set of numbers representing the magnitude of the temperature at each point in space—while the electric and magnetic forces acting on little unit test charges or filings placed throughout the field are "vector fields," because at each point these forces have both magnitude and direction. (In other words, the difference between a vector and a vector field is that a vector is defined at a given point, and a vector field assigns a vector to *each* point in the relevant space.) But it was Maxwell's choice of differential calculus that enabled him not only to define electromagnetic fields mathematically, but also to make a surprising and far-reaching discovery—one that has revolutionized the way we communicate with each other, and with the universe itself.

AN UNCANNY DEDUCTION

When Maxwell published his electromagnetic field equations in 1865, after ten years of hard work and deep thinking, he had formally united electricity and magnetism. Yet magnificent achievement that it was, this was not the end of it, for he also deduced from his field equations the reason that electromagnetic effects do *not* act at a distance. They are propagated as transverse waves, the same kind of wave that Thomas Young had identified for light via that famous two-slit experiment.

What's more, Maxwell found his theoretical electromagnetic waves had just about the same speed as light, too. (I say "just about" because of the difficulty in making accurate measurements at the time.) It was a coincidence too delicious to ignore—and it proved that Maxwell's hunch about derivatives was as profound as it was simple. The original transverse "wave equation," a differential equation, had been derived a century earlier to describe the way a plucked string vibrates—but this remarkable new mathematical parallel suggested, as Maxwell put it modestly, "that light itself (including radiant heat, and other radiations if any) is an electromagnetic disturbance in the form of waves propagated through the electromagnetic field according to electromagnetic laws." In one fell swoop he had united not just electricity and magnetism but light, too. He'd even pointed to the possible existence of "other radiations," a tantalizing prediction that set the likes of Heinrich Hertz on the path to discovering radio waves.

As Einstein later wrote, imagine Maxwell's feelings when he made the connection between electromagnetism and light! "To few men in the world has such an experience been vouchsafed," he added. Maxwell himself had privately expressed his excitement in his typically understated way, in a letter to his cousin in January 1865: almost as an aside he'd added, "I have also a paper afloat, with an electromagnetic theory of light, which, till I am convinced to the contrary, I hold to be great guns." He was not given to boasting, so you can sense how thrilling it must have been when he made his uncanny deduction.[22]

Maxwell's momentous discovery of the electromagnetic nature of light was only possible because of his clever choice of mathematical language.[23] Not that it met with rapturous acceptance at first. The problem for the mainstream was that Faraday's conception of the field had been expressed with meticulous data but no equations, and now Maxwell's theory was expressed *entirely* in terms of equations—with no physical model for conceiving what the field actually *was*. It was the very same response that had dogged Newton's theory of gravity. In various earlier papers, Maxwell had indeed suggested several mechanical models that might explain what the field was and how it transmitted electromagnetic effects, but he knew there was no way of proving the existence of any of them. The mathematics of the field's *consequences*, he believed, should be enough—just as Newton had known it was enough that his equations accurately described the effects of gravity, even though he hadn't explained gravity itself.

Still, Maxwell kept on thinking about how he could express his fields more clearly. At the beginning of his 1865 paper, he'd said that the first step in finding a theory of electromagnetism had been to "ascertain the strength and direction of the forces acting between the bodies"—just as Newton had done when setting out to define force in the first place. It was a vectorial approach, but not yet a full vector theory, for Maxwell hadn't yet studied quaternions. But that was about to change—with help from his old friend Tait.

QUATERNIONS AT LAST

While Maxwell had been developing Faraday's field idea, Tait had been studying the mathematics of heat. If T is the temperature, the "heat equation" relates the Laplacian of T to $\frac{\partial T}{\partial t}$, so it shows how the temperature changes through space as it changes in time—through your morning coffee, for example, as it cools while you carry it from the kitchen. Fourier had published the heat equation in 1822, and then used it in the founding document of climate science, his 1827 paper, "On the Temperatures of the Terrestrial Sphere and Interplanetary Space." Thomson had made the first electric field analogy by adapting Fourier's work on heat, as I mentioned, and today its formula showing how things diffuse has myriad uses, from engineering to policymaking (for ideas can diffuse, too). Anyway, in 1857, Tait suddenly remembered something he'd read in Hamilton's *Lectures on Quaternions* on that hunting trip four years earlier. Hamilton had cast the Laplacian into vector form, by defining a "vector operator" that he expressed like this (except he still used ordinary d's instead of the modern curly ones that were only beginning to gain ground):

$$\lhd = i\frac{\partial}{\partial x} + j\frac{\partial}{\partial y} + k\frac{\partial}{\partial z}.$$

If you've studied vector calculus, you'll recognize this as

$$\boldsymbol{\nabla} = \frac{\partial}{\partial x}\boldsymbol{i} + \frac{\partial}{\partial y}\boldsymbol{j} + \frac{\partial}{\partial z}\boldsymbol{k}$$

in today's notation. (As I mentioned in the prologue, there are other modern ways of denoting vectors, such as hats or bars on top of a letter or squiggles underneath, but boldface is both arresting and easy to type, so it is common in modern textbooks and papers.) Either way, though, you're dealing with a Cartesian coordinate system, so we'll see shortly that you essentially get *the same results,* no matter whether you use the unit imaginary numbers i, j, k or the unit vectors $\boldsymbol{i}, \boldsymbol{j}, \boldsymbol{k}$ as the basis of your vector operator.

Tait immediately set to work applying this new kind of operator in his work on heat and electricity, turning Hamilton's symbol $\lhd$ into ∇ in the process. Tait also linked this symbol with a name, "nabla," which his assistant William Robertson Smith suggested, because "nabla" is the Latin transliteration of the Greek word for an ancient Assyrian harp, which had an inverted triangle shape similar to ∇. (Later, Gibbs will replace Hamilton's "nabla" with "del"—apparently because ∇ also resembles an upside-down version of the Greek letter "delta." Both names are used today.) As for the boldface type in the modern representation of $\boldsymbol{\nabla}$, it will make its way into history when the iconoclastic Oliver Heaviside eventually enters our story. But I'm introducing it here because I think it makes it easier for modern readers to see, as Tait certainly did, that Hamilton's operator *itself* is a vector.

This means, for example, that it can turn an ordinary (scalar) function, such as the potential, into a vector. Earlier I mentioned that if a force has a potential V, the Cartesian components of the force are $\frac{\partial V}{\partial x}, \frac{\partial V}{\partial y}, \frac{\partial V}{\partial z}$. These are just the components you get if you insert V into $\nabla = i\frac{\partial}{\partial x} + j\frac{\partial}{\partial y} + k\frac{\partial}{\partial z}$

or into the modern vector version, $\boldsymbol{\nabla} = \frac{\partial}{\partial x}\boldsymbol{i} + \frac{\partial}{\partial y}\boldsymbol{j} + \frac{\partial}{\partial z}\boldsymbol{k}$. So instead of writing the force by listing its components,

$$\boldsymbol{F} = \left(\frac{\partial V}{\partial x}, \frac{\partial V}{\partial y}, \frac{\partial V}{\partial z}\right),$$

you can write it much more simply and transparently as

$$\boldsymbol{F} = \boldsymbol{\nabla} V,$$

or $F = \nabla V$ as Tait, and later Maxwell, wrote it.

It's simpler because it's a more compact way of representing $\boldsymbol{F}$, saving time and space when you are writing it out or programming it. And it's more transparent because it expresses the force as a whole vector, rather than listing its components separately. This makes it easier to see that the problem is about force and potential as physical concepts, rather than just a list of numerical (component) values. This is a subtle point, and it will take some of the players in our story a long time to understand it.

Physics aside, since $\boldsymbol{\nabla}$ is a vector, you can also form its scalar product with itself, writing the Laplacian as $\boldsymbol{\nabla} \cdot \boldsymbol{\nabla}$, or ∇^2 as Hamilton wrote it, and as we also do today. So, for example, Laplace's equation becomes

$$\nabla^2 V = 0,$$

which is neater than $\frac{\partial^2 V}{\partial x^2} + \frac{\partial^2 V}{\partial y^2} + \frac{\partial^2 V}{\partial z^2} = 0$.

You can also take the scalar product of nabla and another vector—and you can take the vector or cross product, too. We'll see more of this in the next chapter. It will enable Tait to use these nabla operations to write Stokes's theorem and the divergence theorem in terms of whole-vector calculus—and he'll be the first to do this. First, though, he threw himself into Hamilton's book. He was so excited about the role that quaternions and their vector calculus operators could play in physics that he asked if Hamilton would mind corresponding with him. Hamilton, of course, was delighted to find a mathematical kindred soul. At the beginning of their correspondence in 1858, he told Tait of that marvelous day when the secret of quaternions revealed itself in a flash at Broome Bridge, "which," he added, "my boys have since called the Quaternion Bridge."[24]

By 1859 Hamilton felt comfortable enough to share quite personal feelings with Tait, as he did with De Morgan—including recounting the despair he'd felt when his first love Catherine Disney had married someone else. (It was years later that she told Hamilton she had loved only him, and that her parents had pressured her into a more "suitable," ultimately unhappy marriage.) But as he told Tait, "Those days are over: happily? Yes, so far as getting a little more sense, and less sensibility, is concerned."[25] It seems he'd read Jane Austen's *Sense and Sensibility*! At any rate, at the age of fifty-four, Hamilton was content.

Twenty-eight-year-old Tait seemed content, too. He had married Margaret Porter two years earlier, and he was also reveling in his work. As well as his teaching in Belfast, he published his first quaternion paper (on waves) in 1859. Then, in 1860, he won a promotion, to professor of natural philosophy at his and Maxwell's old university, Edinburgh. Maxwell had applied for the post, too, but rightly or wrongly Tait was seen as the better lecturer. J. M. Barrie was one of Tait's students, before he became famous as the author of *Peter Pan*, and he claimed that Tait was the most superb demonstrator ever. He recalled that during classes Tait's "small twinkling eyes had a fascinating gleam in them. . . . I have seen a man fall back in alarm under Tait's eyes." Yet, Barrie continued, those eyes could also "be merry as a boy's"—especially when "he turned a tube of water on a crowd of students who would insist on crowding too near an experiment." But Maxwell had his student fans, too—for instance, the future astronomer David Gill described how Maxwell would stay behind with interested students for hours after the lecture. As for Tait and Maxwell themselves, such professional rivalry had no effect on their friendship, judging by their letters to each other. After all, they'd been friends and rivals since school.[26]

We'll see something of their quirky correspondence in the next chapter, for Tait is about to become famous for his work on quaternions, and Maxwell is about to pepper him with questions. He will use the answers in his *Treatise on Electricity and Magnetism*—and this extraordinary book will set the scene for the final step on the road to modern vector calculus.

(7)

THE SLOW JOURNEY FROM QUATERNIONS TO VECTORS

Tait was fascinated by the way quaternions—or rather, the vector part of a quaternion—enabled him to write physics equations so much more compactly, especially when he applied Hamilton's operator ∇ ("nabla"). Everyone else—from Newton to Maxwell—had been writing out separate equations for each component of a vectorial quantity such as force or velocity. Hamilton himself had focused on math rather than physics, so he was delighted by the new uses Tait was finding for his quaternions—and Tait had traveled from Belfast down to Dublin so that he and Hamilton could talk in person. They also exchanged dozens of letters, including Hamilton's joyful letter hoping that someday he and Tait would be thanked for their efforts. Hamilton did receive recognition but no real thanks for his part in the story—but when he died in 1865, just a few months after Maxwell published his trailblazing electromagnetic field theory, thirty-four-year-old Tait was ready to carry on in his stead.

First up, he wrote a 37-page obituary for his friend and mentor—it appeared in the *North British Review* a few months later, in 1866. He opened by urging his readers to pause amid the "din of controversy" and "the battle of intellectual giants" that dominated academic life, and "seek repose in the contemplation of something far more elevated and much more subtle: the character and works of a man of *genius*." Yes, he continued, the word "genius" has become overused, because it refers to something far more creative and original than mere cleverness—but Hamilton *was* a genius of the highest order, fit to rank alongside "those of the grandest of all ages and countries, such as Lagrange and Newton." For Hamilton contributed far more than quaternion analysis, marvelous though it was. His early papers on optics and dynamics, Tait said, showed "a mastery over symbols, and a flow of mathematical language (if the expression can be used) almost unequalled."

In 1867, Tait published his own *Elementary Treatise on Quaternions*, the first (relatively) accessible textbook on vectorial methods. In fact, aside from the notation—he follows Hamilton in using Greek letters for vectors, and S and V instead of the modern dot and cross for scalar and vector products—it reads not unlike a modern vector analysis text. Yet with Hamilton gone and a new collaborator on the scene, his passion for applying quaternions was . . . not dampened, because he published several more research papers on the subject over the next five years, but somewhat curtailed. For his new collaborator *hated* quaternions!

I'm speaking of William Thomson—Maxwell's early mentor, and the future Lord Kelvin. He and Tait had met in 1861, a year after Tait moved from Belfast back home to the University of Edinburgh. Thomson was Belfast-born but Glasgow-raised, and he'd become a professor at the University of Glasgow when he was only twenty-two. Yet despite some remarkable mathematical papers—including the one on Faraday's field idea, written when he was just seventeen—Thomson would soon become rich and famous for his technical rather than his mathematical achievements. In the late 1850s, for example, he'd been the chief technical advisor during the dramatic laying of the very first telegraph cable under the Atlantic—he'd been on board ship when the first cable broke, and again during a terrible storm that threatened to derail the whole project. Eventually the cable was laid, and Thomson, who was a director of the North Atlantic Telegraph Company, sent the first telegraphic signals from Europe to America. He also invented many profitable electrical instruments and helped pioneer the science of thermodynamics—the Kelvin temperature scale is named after him. With so many irons in the fire, it's amazing he still managed to find time to collaborate with Tait: they were writing a new physics textbook, *A Treatise on Natural Philosophy*. "Natural philosophy" was the name still being used for theoretical physics, and it was the subject in which they were both professors.

From Tait's point of view, however, the amount of time Thomson actually devoted to their project was rather small. "I am getting sick of the great Book," Tait had written to him in 1864. "If you send me only scraps . . . at rare intervals, what can I do? You have not given me even a hint as to what you want done in our present chapter about statics of liquids and gases!" Thomson was then away in Germany, and Tait went on to say that although he'd almost finished a chapter on the kinetics of particles, he certainly didn't feel like paying forty-five times more postage to send it overseas when in all probability Thomson wouldn't get around to reading it till he was back in Scotland.[1]

Tait had also been working on a chapter on kinematics for his *Elementary Treatise on Quaternions*, which he'd been writing at the same time. He told his readers that although quaternion calculus may be difficult to learn, it is worth mastering because it offers "the most extraordinary advantage to the advanced student, not alone as aiding him in the solution of complex questions, but as affording an invaluable mental discipline." But Thomson was having none of it. He used vectorial component equations, of course, but he saw no need whatsoever for the whole-vector calculus that Hamilton's nabla operator allowed. He made the good point that you carried out computations with components anyway—in the scalar product, for example, when you calculate $\boldsymbol{a} \cdot \boldsymbol{b} = a_1b_1 + a_2b_2 + a_3b_3$. Or when you show that $\nabla^2 V = 0$ by calculating $\frac{\partial^2 V}{\partial x^2} + \frac{\partial^2 V}{\partial y^2} + \frac{\partial^2 V}{\partial z^2}$, or in physics and engineering problems when you "resolve" a vector into its components (as in fig. 8.1)—to take just two more examples. So, for their joint project, Tait reluctantly deferred: quaternions were banished.

Despite their disagreements, Thomson noted that Tait was always ready to "brighten" their intense collaborative mathematical labor with "delightful quotations" from Shakespeare, Dickens, and Thackeray. Sometimes the two men worked together in Tait's small home study, by the light of a gas lamp that cast eerie shadows on a wall already dimmed by tobacco smoke. Those dingy walls also contained a list Tait had written in charcoal of the greatest scientists then living: Hamilton was first, followed by Faraday (who died in 1867). Tait and Thomson had discussed their friend Maxwell—"a rising star of the first magnitude," as Thomson put it—but thought him still too young to make the hallowed list.[2]

At last, Thomson and Tait's *Treatise on Natural Philosophy* was ready for the publisher. It saw the light of day in 1867 and would prove very influential because of its cutting-edge content, including the emerging science of thermodynamics. It even included a discussion on calculating machines for sophisticated mathematical computations. Charles Babbage and Ada Lovelace had brought the real possibility of such a machine to attention in the 1840s, but Babbage's "engine" never quite got off the ground, so Thomson described his own theoretical designs. The book was colloquially known as T and T'—pronounced "T and T-dash"—and the nicknames stuck for Thomson (T) and Tait (T') themselves, too.

Maxwell soon picked up a handle of his own: dp/dt. This came courtesy of a form of the second law of thermodynamics in Tait's 1868 book, *Sketch of Thermodynamics*: $\frac{dp}{dt} = JCM$, where JCM are James Clerk Maxwell's initials. (We don't need this formula, but the symbol p stands for "pressure," J for "Joule's equivalent" (the mechanical equivalent of heat), and C for "Carnot's function," while M is related to the amount of heat, usually written today as dQ/dV.) It's fitting, because Maxwell also helped pioneer the kinetic theory of gases—Tait likely crafted this form of the equation as a tribute. Maxwell's wife, Katherine, had helped with some of his experiments on heat and color, keeping the fire stoked and regulating the temperature, and making and recording experimental observations. She reminds us of the many unsung wives who have helped their famous husbands.[3]

While quaternions and whole-vector calculus are missing from *T* and *T′*, there are plenty of component-form equations. Tait must have had great delight in rewriting some of them, in a fraction of the space, in his *Elementary Treatise on Quaternions.*

Since some of these applications were to electricity and magnetism, he immediately recommended that Maxwell study the last twenty or thirty pages of the book, on the power of nabla. His letter suggests that Maxwell already had some knowledge of this "∇ business," as Tait put it—and in fact Maxwell had used ∇^2 as shorthand for the Laplacian operator in his 1865 paper. But now he was planning a monumental treatise on electricity and magnetism, surveying all the known experimental and mathematical results on which he'd built his final theory, which he then intended to compare with the action-at-a-distance electromagnetic theories that still dominated mainstream physics. And as he read Tait's book, he was excited by the possibilities inherent in its development of quaternion calculus.

In fact, he was so excited that when he reviewed Thomson and Tait's "great Book" for the journal *Nature*, he couldn't help but lament the lack of discussion on even the simplest rules of vectors—especially since one of the authors was "an ardent disciple of Hamilton." Thomson, of course, was *not* the ardent disciple! He thought quaternion and vector calculus was nothing but a shorthand for equations that had already been devised using components—as Maxwell had done when he built up his electromagnetic field equations in his 1865 paper. But now, in the early 1870s, Maxwell had discovered vector calculus, and he told Tait that Thomson didn't seem to realize it was "a flaming sword that burns every way"—all through space—whereas the component form was a brute-force Cartesian "ram, pushing [only] westward and northward and (downward?)."[4]

Maxwell also kept his and Tait's childhood friend Lewis Campbell up to date, telling him, "I am getting converted to Quaternions," and mentioning that he was using them in the *Treatise on Electricity and Magnetism* he was writing. He explained that since the operator ∇ was called "nabla," its application was a "Nablody"—Maxwell loved puns and nonsense! With

Tait he was equally playful, but he was also able to be much more technical, and he'd told him he was "dabbling in Hamilton" because "I want to leaven my book with Hamiltonian *ideas* without casting the operations into Hamiltonian *form* for which neither I nor[,] I think[,] the public are ripe. Now the value of Hamilton's *idea* of a vector is unspeakable"—that is, unspeakably good![5] He explained what he meant by this in a new paper, "Remarks on the Classification of Physical Quantities," which he presented to the London Mathematical Society on March 9, 1871. (Incidentally, at the very next meeting of the society, on April 13, the president opened the proceedings with a brief but glowing and well-deserved elegy for De Morgan—Hamilton's old friend, and a pioneer of modern symbolic algebra—who had died on March 18, aged sixty-four.)

First Maxwell spoke of the need to classify the burgeoning number of physical quantities that had arisen because of the progress of science—pointing out that if this classification is done through *mathematical* analogy, it can be of great help in solving new problems whose equations have the same form as that of a known problem. We saw this earlier with the wave equation analogy between vibrating strings and Maxwell's electromagnetic waves, and with Thomson's use of Fourier's analysis of heat flow as an analogy for describing Faraday's field lines, or "lines of force," and with Maxwell's similar streamline analogy. Maxwell mentioned Thomson, but he went on to say that Hamilton had provided an even more fundamental type of classification than analogy, by dividing physical quantities into scalars and vectors. This innovation was so fruitful, Maxwell said, that the importance of quaternions could only be compared with Descartes's invention of coordinates. He was spot on, given that quaternions and vectors (and their tensor extensions) are fundamental to the way modern physicists, engineers, and data scientists describe and analyze the way objects occupy physical and digital spaces. It's fascinating to go back in time and "witness" a brand-new idea, such as Hamilton's, as it comes into being—and to observe, with hindsight, just who did and who didn't recognize its significance.

Maxwell went on to define two new subcategories of vectors: forces and fluxes. Forces, he said, are vectors related to length or distance—as in line integrals—and fluxes (such as magnetic induction) are vectors related to areas, or surface integrals. For his 1865 paper he'd used the component form of the divergence theorem and Stokes's theorem to relate these integrals to derivatives, as I outlined in chapter 6. By 1871, however, he'd discovered the power of Hamilton's nabla operator, ∇, for defining such physical quantities in terms of whole vectors and vector fields. We'll see these nabla operations shortly, and they are so important, Maxwell said, that he gave them special names. You might already know them: "divergence," "curl," and "grad"—but it was Maxwell who gave them to us, so it's worth spending a little time seeing why he chose them.

THE POWER OF NAMES

The great power of mathematics lies in its visual symbolism, but to *think* about what you see you also need to be able to say it in your head. Maxwell had first tried out the idea in a letter to Tait back in the late autumn of 1870: "Here are some rough hewn names. Will you like a good Divinity shape their ends properly so as to make them stick?"[6]

First up, he suggested, "the result of ∇ applied to a scalar function might be called the slope." That's by analogy with the slope, $m = \frac{dy}{dx}$, of a straight line $y = mx + c$, because similarly you're taking derivatives when you apply ∇ to a scalar function, such as the electric potential (or voltage), or the potential of a force. We often think of scalars as just numbers, but functions can be scalars if they take numerical values that don't depend on the coordinates you're using. To take another example, pressure and temperature are scalars, so a function of pressure and/or temperature is a scalar function. Anyway, as I showed in chapter 6, $F = \nabla V$ (and its modern vector equivalent $\mathbf{F} = \nabla V$) means that the force has components $\frac{\partial V}{\partial x}, \frac{\partial V}{\partial y}, \frac{\partial V}{\partial z}$, analogous to the straight-line slope $\frac{dy}{dx}$. Today we read $\mathbf{F} = \nabla V$ aloud as "F equals grad V," where "grad" is short for "gradient," which is another name for "slope." So, Maxwell's term did sort of stick! It shows how a scalar function changes through space.

Next, Maxwell spoke of what happens when you applied ∇ to vectors.

I've generally been using bold type for vectors to make it easier for modern readers, but as I've noted it's essentially the same in quaternion form: the difference between the modern vector basis and Hamilton's imaginary number basis only becomes apparent in the scalar product. That's because $i^2 = -1$, by the definition of an imaginary number, whereas $\mathbf{i} \cdot \mathbf{i}$ is *defined* to equal 1, and similarly for $\mathbf{j}$ and $\mathbf{k}$. (At least they will be so defined shortly, when we meet Heaviside.) So, whenever you take the scalar product of two vectors, such as

$$v = v_1 i + v_2 j + v_3 k \text{ and } w = w_1 i + w_2 j + w_3 k$$

in Hamilton's notation, you're multiplying the basis quantities too. Which means that using Hamilton's notation for scalar products, which Tait and Maxwell used, too, you have

$$S.\, vw = v_1 w_1 i^2 + v_2 w_2 j^2 + v_3 w_3 k^2 = -(v_1 w_1 + v_2 w_2 + v_3 w_3).$$

So, the only difference between Hamilton's scalar product and the modern one is that minus sign out front: $S.\, vw = -\mathbf{v} \cdot \mathbf{w}$. Similarly, the scalar product of the nabla operator and v is $S.\, \nabla v = -\boldsymbol{\nabla} \cdot \mathbf{v}$. Maxwell suggested we read this as the "convergence" of the vector field v, but you might recognize $\boldsymbol{\nabla} \cdot \mathbf{v}$ as the "divergence" of $\mathbf{v}$—a name introduced later, because when you lose the minus sign, your vectors diverge instead of converging (as in fig. 7.1). So here again, Maxwell's name did essentially stick.

In his *Treatise*, Maxwell gave the diagram shown in figure 7.1b to illustrate convergence. He used the idea to show that (what is now called) the divergence of "the electric force on a test charge"—which he called the "electromotive intensity," $\mathfrak{E}$, and which is now called the "electric field," written as $\mathbf{E}$—is equal to 4π times the electric charge density, ρ. (As he told Campbell in his Nablody letter, he'd run out of Greek letters, Hamilton's choice for vectors, and was "heretically" using German capitals such as $\mathfrak{E}$ instead.) Maxwell wrote this equation in the component form given in the caption to figure 7.1. But he also gave a *whole-vector* version, which, ignoring the annoying minus sign in Hamilton's scalar product, he wrote like this:

$$\rho = S.\, \nabla \mathfrak{D}.$$

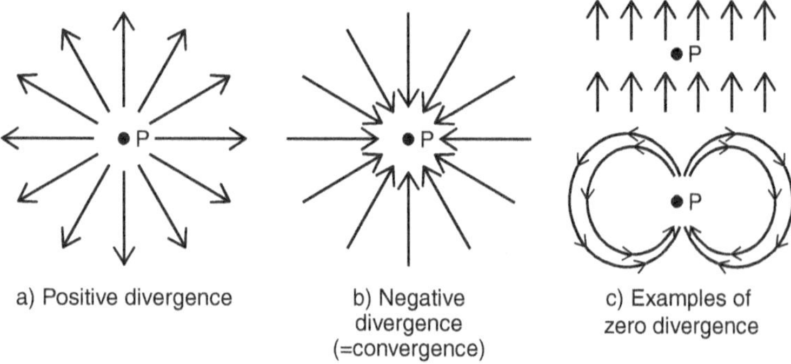

FIGURE 7.1. Divergence. For example, Coulomb's law tells us that if there's a positively charged particle at P, and you place a unit positive test charge at various points in the field around P, the electric force is in the direction shown by the arrows in (a). As you move away from P, the force drops off according to the inverse square law, so the arrows would get smaller. If the test charge is negative, it is attracted to the positive charge at P, so you have case (b). In volume 1 (art. 77) of his *Treatise on Electricity and Magnetism*, using Gauss's idea that Coulomb's law is related to flux, Maxwell wrote this law as

$$\frac{\partial X}{\partial x}+\frac{\partial Y}{\partial y}+\frac{\partial Z}{\partial z}=4\pi\rho,$$

where the left-hand side of the equation is the divergence of the electric force (whose components he wrote as X, Y, Z), and ρ is the electric charge density of the field. I showed how Maxwell used the idea of flux to derive this equation in an endnote to chapter 6. The lower diagram in (c) looks like the lines of force around a bar magnet, shown in figure 6.3a—and as we'll see, another of Maxwell's equations is that the divergence of the magnetic field is zero.

Since he'd already defined $\mathfrak{D} = \mathfrak{E}/4\pi$, this is just

$$4\pi\rho = \nabla \cdot \boldsymbol{E}$$

in modern notation. ($\mathfrak{D}$ is Maxwell's so-called "electric displacement," but it's the equations we need here, not the physical details.)[7]

We'll see shortly just how and why this modern transformation in notation happened, but modern readers will likely have some knowledge already of vector calculus, and some will know Maxwell's equations, too—so I'm trying to relate the historical process to what you may already know. The key thing, though, is that in each of these notational forms, the striking thing is the physical quantities: the charge density, and the diverging electric field. By contrast, when you write the same equation in the component form shown in figure 7.1, the emphasis is on the math rather than the physics. That's why Maxwell wrote to Tait saying that contrary to the view of "unbelievers" such as Thomson, "the virtue of the 4nions [his shorthand for quaternions] lies not so much as yet in *solving* hard questions, as in enabling us to see the *meaning* of the question and its solution."[8]

Maxwell used the symbol $\mathfrak{B}$ for what he called the magnetic induction; today we often call this simply the magnetic field vector, ***B***. For a stationary magnet, all the field lines begin at the magnet's north pole and end at its south, as in figure 7.1c, so there's no net magnetic flux through a closed surface around the magnet—which means there's no divergence of the magnetic field vector ***B***. (To see this, imagine the bar magnet enclosed in a large ball. The field lines loop from one end of the magnet to the other—they don't branch out separately and pass one-way through the ball's surface, the way they do through the rotating loop in the electric motor/generator in fig. 6.2.) This is another way of saying that, unlike a charged particle such as an electron or proton, which is either negative or positive, natural magnets always have two poles. No one has detected a naturally occurring magnetic monopole, although they've been created in the lab at the virtual quantum level; they also arise theoretically in string theory, following from the work of Paul Dirac in 1931. (My colleague Tony Lun and I are among those who have long been exploring the existence of gravitomagnetic monopoles. It's another example of the use of analogies between different topics because of similarities in the equations.[9])

Maxwell gave his magnetic divergence equation in component form—rather than specifically saying that the divergence (or convergence), the *scalar product* of nabla and $\mathfrak{B}$, equals zero: $S.\,\nabla\mathfrak{B} = 0$, or $\nabla \cdot \boldsymbol{B} = 0$ in modern notation. But he did give an equivalent equation in whole-vector form, which we'll see in a minute.[10] First, though, he had to find a name for the *vector product* with nabla.

As we saw in the right-hand rule for vector products in figure 4.1, if you curl your fingers in the direction from one vector to another, your thumb points in the direction of their vector (or cross) product. It turns

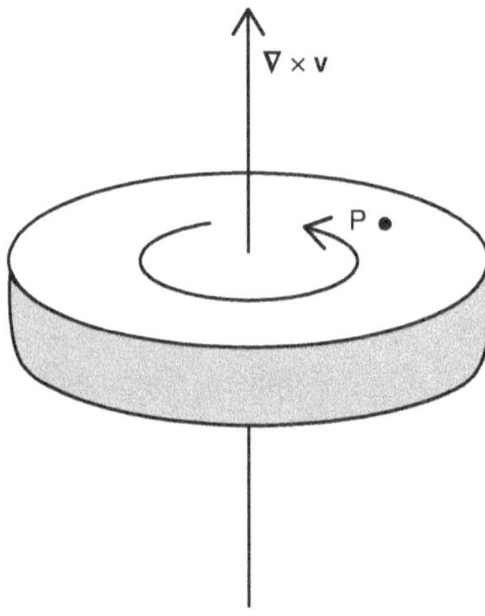

FIGURE 7.2. The curl of the velocity of the rotating disk at any point *P* lies in the direction shown.

out that this curling motion is also physically significant, in the sense that if your fingers indicate the motion of a rigid body rotating around a fixed axis through your thumb, the vector product of nabla and the velocity of any point on the body lies along this same axis (as in fig. 7.2). So, the vector product with nabla has something to do with rotations. In his letter to Tait, Maxwell suggested several possible names for this operation, including "twirl"—which, he thought, was "sufficiently racy," although perhaps "too dynamical for pure mathematicians." So, "for Cayley's sake," he said, what about the name "curl." Cayley had developed the mathematics of "scrolls" or "skew surfaces"—like the twisted ruled surfaces you might have seen as string sculptures—and, of course, you "curl" a scroll when you roll it up. In articles 25 and 26 of his *Treatise,* Maxwell settled on the name "curl" for the operation $\nabla\times$, albeit "with great diffidence"—and this name, too, stuck.

In his electromagnetic field equations Maxwell expressed one of his curl operations like this:

$$\mathfrak{B} = V.\,\nabla\mathfrak{U}$$

($\boldsymbol{B} = \nabla \times \boldsymbol{A}$ in modern notation), where his $\mathfrak{A}$ and the modern $\boldsymbol{A}$ represent the *vector* potential of the magnetic field $\boldsymbol{B}$. (The *scalar* potential described in chap. 6 has a physical interpretation, related to work and potential energy—and an example is the voltage. The *vector potential of the electromagnetic field*, however, is essentially a mathematical quantity that is related to the magnetic field via Stokes's theorem.)[11] As you may know from math class—and as Maxwell knew from Tait's work—the divergence of a curl is always zero. So, this equation was, in fact, equivalent to $S.\,\nabla\mathfrak{B} = 0$ (or $\nabla \cdot \boldsymbol{B} = 0$).[12]

Maxwell also had other curl equations in the *Treatise*, and together they led to the famous curl equations in the modern form of Maxwell's equations that we'll see in the next chapter. Meantime let's pause for a moment, for it might be hard to appreciate in hindsight just what a breakthrough it was when Maxwell showed how to write his theory of electromagnetism in whole-vector form. Its ramifications for our story will be huge, especially for physics—and, therefore, for the way we understand our universe.

THE POWER OF WHOLE VECTORS

On October 19, 1872, while he was checking the proofs of his *Treatise*, Maxwell explained to his old friend Campbell that his publisher, Clarendon Press, managed to average around nine sheets in thirteen weeks. It's another reminder of the technology of the times, when each page in a book had to be typeset manually, one painstaking letter at a time. As for proofreading, it is always laborious, but Maxwell told Campbell that their mutual friend Tait "gives me great help in detecting absurdities."[13]

Maxwell's *Treatise on Electricity and Magnetism* finally appeared in 1873. In the same year, he gave a special lecture at London's Royal Institution, in honor of James Watt—the eighteenth-century Scottish engineer who made the original Newcomen steam engine so much more efficient. Thomas Newcomen had designed his engine to pump water out of mines, but Watt's improved engines had much wider application, helping power the Industrial Revolution. Maxwell spoke of the increasing complexity of our social and employment interactions because of all this new industry, and urged his listeners to keep in mind, "amid the rattle of machinery and the press of business," the simpler, more "noble" aspects of life.[14]

Statue of James Clerk Maxwell, erected in Edinburgh in 2009. His faithful dog Toby lies at his feet, and he's holding the color wheel that helped him understand the nature of color vision and color photography.

As for the virtues of whole vectors, in an 1873 *Nature* review of Tait's new book about quaternions, Maxwell pointed out that while calculations are vital to the job of mathematics, a mathematician is much more than a mere calculator (or "dry drudge," as Newton had put it); rather, creative mathematicians invent new, laborsaving methods. But while Thomson believed quaternions and vectors were just a shorthand for equations already worked out, Maxwell saw that "Quaternions, or the doctrine of Vectors, *is* a mathematical method, but it is a *method of thinking*"—it is not just a creative laborsaving device for calculations.[15]

He'd made the same point, with fey humor, when Tait was awarded the Royal Society of Edinburgh's Keith Medal in 1870, for two of his quaternion papers—including one on the rotation of a body around a fixed *point* rather than around an axis as in figure 7.2. (It was a difficult topic, and the pioneering female mathematician Sonia Kovalevsky would become famous in the late 1880s for her own award-winning analysis of such motions. Unlike Tait, she used components, not vector calculus methods, for their usefulness was still being contested.) In his speech in honor of Tait, Maxwell said that a mathematician often gets so tired with all his calculating that he has no energy for thinking—whereas "Tait is the man to enable him to do it by thinking, a nobler though more expensive occupation, and in a way by which he will not make so many mistakes as if he had pages of equations to work out." Tait was delighted by Maxwell's wry take![16]

At the beginning of his *Treatise*, Maxwell put this more succinctly, saying that in "physical reasoning" rather than calculating, you want to see the object you're studying for *itself*, as a whole, rather than focusing on its coordinates and components. He made this point over and over, in his publications and his letters—that with whole vectors you can see the physical meaning of what you're doing, rather than getting lost in the calculations.[17] Aside from Tait, though, few seemed to be listening.

Thomson was perhaps the most famous dissenter. He didn't even accept Maxwell's theory of electromagnetism in component form, likening it to mysticism—all that math, with no mechanical model explaining in concrete terms how electromagnetic waves propagate! Brilliant as he was—and much as he recognized Maxwell's wide-ranging scientific genius—Thomson never realized the power of the mathematical "vector field" that now underlies so much of physics, and which has also found applications in biology and other sciences. Even in the 1890s, after Heinrich Hertz had produced the radio waves that confirmed Maxwell's theory, Thomson advised a colleague against using the term "vector." It contributes nothing, he said, to the simplicity of the geometry of a problem, adding bizarrely, "Quaternions came from Hamilton after his really good work had been done; and, though beautifully ingenious, have been an unmixed evil to those who have touched them in any way, including Clerk Maxwell."[18]

Stokes, too, refused to teach Maxwell's theory in the 1890s, and young Einstein's professors also refused—partly because the theory had only recently begun to make its mark in Germany, thanks to Hertz and other German-speakers; so Einstein played truant in order to study it for himself. In a few years' time, he would incorporate Maxwell's equations into his famous 1905 theory of special relativity—and he'd speak of vectors, although he'd use only component forms. A decade later, however, he would write Maxwell's equation in both modern whole-vector form *and* tensor form, in his paper introducing the general theory of relativity.

We'll see more of Einstein's work later. The point here is that it took a long time for Maxwell's theory of electromagnetism, and his praise for vector calculus, to find widespread acceptance. He published the theory in 1865, and in vector-calculus form in the *Treatise* in 1873, but it must have seemed to him that no one really cared.

THE TIDE BEGINS TO TURN

In 1873, Maxwell (and Tait) turned forty-two—and the up-and-coming William Kingdon Clifford turned twenty-eight. Clifford had been a child prodigy like Hamilton. He graduated Second Wrangler from Cambridge: like Maxwell, he'd preferred to study what interested him rather than cramming for the Tripos—he was especially interested in the cutting-edge algebraic work of Arthur Cayley and James Sylvester. Also like Maxwell, Clifford was quite an athlete: Maxwell was a superb horseman and a strong if inelegant swimmer (Tait, on the other hand, preferred golf), but Clifford was a man of prodigious strength—he could pull himself up on a gymnast's bar with just one arm. He knew of Maxwell and Tait's work, and the three men became friendly, having met at various meetings of the British Association for the Advancement of Science and the Royal Societies of London and Edinburgh.[19]

In 1877 Clifford gave a series of lectures on quaternions at University College London—the secular university where De Morgan had taught, in protest at the Anglican vows needed for a position at Oxford and Cambridge. In the same year, Clifford took a controversial stand in the fierce debates around science and religion that were then taking place, writing atheistic articles suggesting that we need firm evidence before we believe something—faith alone was not enough.[20]

In 1878, he gave his own mathematical treatment of vector algebra in his book *Elements of Dynamic*: although he was a pure mathematician, he emphasized the importance of vectors in dynamics, the study of motion due to forces. (In the process, he replaced Maxwell's negative "convergence" with our modern term "divergence.") In his review of Clifford's book, Tait praised Maxwell's "great" *Treatise* as perhaps the first, outside the specialist quaternion literature, to promote the use of vector methods in physics—which is why I've spent quite a bit of time on Maxwell, for his *Treatise* is a turning point in the story of vectors. Tait went on to say that Clifford "carries the good work a great deal farther, and (if for this reason alone), we hope his book will be widely welcomed." Tait had as mischievous a sense of humor as Maxwell, and he used it to take a swipe at Clifford's use of "Dynamic" instead of "Dynamics"; more seriously, he lamented Clifford's introduction of too many other idiosyncratic and unnecessary new terms. But he especially regretted that Clifford hadn't made sufficient use of the nabla operator: it was as if he'd gone quite a way along the quaternionic road and then suddenly stopped.[21]

Still, in his various writings through the 1870s Clifford had done something special, something that made him a worthy successor not only to Hamilton, Tait, and Maxwell, but to Grassmann, too. He was one of the very few who had closely studied both Hamilton *and* Grassmann, and he set out to create a new algebra that united the two approaches. He called it a "geometric algebra," and he based it around a "geometric product" that united Hamilton's scalar product (Grassmann's inner product) and Grassmann's outer product (which was a more general version of Hamilton's three-dimensional vector product). Clifford's geometric product looks like this in modern notation (but like Maxwell and Tait, Clifford used Hamilton's S and V for scalar and vector products, and none of them used modern boldface for vectors):

$$\boldsymbol{ab} = \boldsymbol{a} \cdot \boldsymbol{b} + \boldsymbol{a} \wedge \boldsymbol{b},$$

where $\wedge$ is called a "wedge"—in three dimensions $a \wedge b$ is equivalent to the cross product $a \times b$. Hamilton's full quaternion product is also a combination of these two types of vector multiplication, as I showed in chapter 4 (w and a are the scalars in the quaternions P and Q, respectively, and $\boldsymbol{p}$ and $\boldsymbol{q}$ are the vector parts):

$$PQ = wa - \boldsymbol{p} \cdot \boldsymbol{q} + w\boldsymbol{q} + a\boldsymbol{p} + \boldsymbol{p} \times \boldsymbol{q}.$$

But it applies only in three dimensions.

For physics in the everyday 3-D world, Clifford, like Maxwell, found that the separate scalar and vector products were far more useful than the full quaternion product—and this was another step toward the break away from quaternions to modern vector analysis. But Clifford was a pure mathematician at heart, and pure mathematicians love to push the math as far as they can, for the fun of it—as well as for inklings of future applications. For such a mathematician, the interesting thing about quaternions is that aside from the fact that their product isn't commutative, they obey all the usual laws of algebra—associative, distributive, and so on—and, as in ordinary algebra, you can divide with them, too. In mathematical terms this means a quaternion q has an "inverse" q^{-1}, defined so that $qq^{-1} = 1$, where the 1 here is a quaternion, $1 + 0i + 0j + 0k$.[22] In the quaternion framework for representing rotations, for example, the inverse reverses the direction of the rotation. Inverses are also important if you want to solve equations. You might be familiar with this idea from matrix equations such as $AX = B$, where the solution is $X = A^{-1}B$, if the matrix A does have an inverse.

But here's the thing: there's no such inverse using just vector products—you need the full quaternion product. Similarly, Clifford's full geometric product has an inverse. And just as quaternions belatedly proved so efficient for calculating complex rotations in three-dimensional space, so modern mathematicians have realized that the geometric product enables many of the things you can do with vectors to be done more efficiently.[23]

In a sad irony, the undervalued Grassmann had also been trying to unify his system and Hamilton's—in 1877, the very year he slipped out of the world altogether, at the age of sixty-eight. He was gone but not forgotten, as

Clifford was the first to show, with his own unification of the two systems. Not that many people back then were paying attention to Clifford, either.

A decade later, the Italian mathematician Giuseppe Peano would also discover Grassmann's *Ausdehnungslehre,* and he would publish his own summary and extension of it—including the first modern, axiomatic definition of a vector space. This takes the idea of vectors beyond intuitive models such as arrows or lists of components, defining them abstractly instead—as elements of a *vector space* whose structure is defined by analogy with that of real numbers. As we saw with the earlier work of Hamilton and his friend Augustus De Morgan on the laws of algebra, this structure means that when adding vectors, and when multiplying them by a scalar, the closure, commutative, distributive, and associative laws hold, and identity and additive inverse elements exist. We don't need the details for our story, though—we just need the *idea* that vectors are more abstract than those school-level arrows. Besides, few were listening to Peano, either. But his and Clifford's work exemplified the change in Grassmann's posthumous fortune that was beginning to take place as mathematicians sought ever more rigorous definitions.

A CROSS-CULTURAL INTERLUDE

Aside from quaternions, there are some interesting loose connections between Clifford, Tait, and Maxwell, which help put their work and times into a broader scientific and cultural context.

Clifford was a friend of the venerable novelist and critic George Eliot and her partner George Lewes—Clifford and his literary wife, Lucy, often attended their famous Sunday afternoon gatherings of writers, scientists, philosophers, and interested friends. Like Lewes, Eliot took an interest in science and math. She had studied geometry in the 1850s at an adult class "for the ladies"—in 1882 Tait would grumble to Cayley that he was "virtually forced" to give a course of lectures for "ladies," but presumably the problem was with his workload, not the women. Now, in the 1870s, Eliot was reading up on the new, non-Euclidean geometry that we'll see more of in chapter 10, and in which Clifford also made his mark.[24]

As for Lewes, he was a philosopher, critic, biographer, and a marvelous raconteur. In 1873 he was finishing a new book, *Problems of Life and Mind*. It was a forensic, empiricist exploration of the possible biological basis of consciousness, part of "the new psychology" of the time, which was bringing a more modern scientific approach to long-standing debates on the "mind-body problem"—the contested relationship between consciousness (mind) and the brain (body). The nineteenth century had been an extraordinarily creative one for mathematics and physics, as we've seen, but as Maxwell had told his students back in 1856, physics is the simplest science, and it tells us nothing about emotions, spirituality, or other "higher" aspects of our lives. Lewes wanted to explore these "higher aspects" by applying to the biological study of psychology the kind of empiricism and scientific rigor developed in physics.

Yet by contrast with physics' dramatic progress, advances in biology had been problematic. It is one thing to break a chemical compound or an electric circuit into its constituent parts, but in biology such a reductionist approach had dire sexist and racist consequences. For instance, the discovery that women had smaller brains than men was immediately assumed to prove the cultural expectation that they were unfit for strenuous intellectual work. Somerville and Germain suffered greatly under this assumption, and although they eventually succeeded against the odds, most women did not even try. The same pseudoscientific approach helped entrench racism. Even that most brilliant star in biology's nineteenth-century firmament—the theory of evolution by natural selection—was being co-opted into "proving" cultural tropes about the inferiority of the poor, of women, and of non-Europeans. Charles Darwin was a friend of Eliot and Lewes; so was Herbert Spencer, whose name became linked with the worst kind of "social Darwinism" but who was perhaps more a "liberal" utilitarian. He was also a philosopher and a pioneering sociologist, although Maxwell thought he'd pushed the social evolution analogy too far, and he and Tait thought Spencer's metaphysics showed an inadequate grasp of physics.[25]

Lewes, however, was on top of the latest physics and referred pertinently to Tait's and Maxwell's scientific writings in his *Problems of Life and Mind*. Like the good scholar he was—and today his work is still seen as relevant to the mind-body debate—he asked Tait to check the proofs. When Tait finally managed to find the time to read them, he sent his comments to Lewes's (and Eliot's) publisher, Edinburgh-based John Blackwood, adding, "I found it pretty stiff reading, far worse than double its amount of analytical [mathematical] formulae: but it is very interesting indeed, and will thoroughly rile the so-called Metaphysicians." Unfortunately, Blackwood decided it was *too* stiff a read, and so Lewes had to look around for another publisher.[26]

It wasn't just as Lewes's referee that Tait knew Blackwood, for the two men sometimes played golf together. After one such round, Tait showed Blackwood a poem Maxwell had written him, summing up the President's Address at a recent British Association meeting in Belfast—Maxwell and Clifford had attended but not Tait. The big topics of the meeting were evolution, and the scientific status of atoms—this was in 1874, twenty-five years before the electron was discovered and almost four decades before the nucleus of the atom. Maxwell supported the idea of atoms, but he was not a materialist, and with his wicked wit he made great fun of endless "British Ass" discussions among "our witless nobs." You can see the tenor of it in the opening lines:

> In the very beginning of science, the parsons, who managed
> things then,
> Being handy with hammer and chisel, made gods in the likeness
> of men. . . .[27]

Eventually we got commerce, the poem continues, and earthly power, and finally atoms "supplanted both demons and gods," and became "the unit of mass and of thought"—"structureless germs" that enabled us to "inherit the thoughts of beasts, fishes and worms." The poem has a go at both science and religion—or rather, at the current debate between the more extreme proponents of either side—and Blackwood loved it. So much so that he offered to publish it in his famous *Blackwood's Magazine*, which had long published Lewes, and had also given George Eliot her start as a fiction writer. In fact, *Blackwood's* had recently published her masterpiece,

Middlemarch—many novels at the time first appeared in serial form in magazines—so Maxwell's anonymous poem was in illustrious company. He'd been taken aback by the invitation—he told Tait that the private "paroxysms" of British Asses should not be "promiscuously stereotyped"; but in the end the idea was such a hoot that his "better ½" agreed to it.[28]

Although his poem lampooned both parsons and materialists, Maxwell had already made important contributions to the discussion on evolution, suggesting, for example, that the finite size of molecules would limit the amount of genetic information that could be passed down. (Darwin had originally believed that heredity was absolute, that everything was passed down as a whole—hence Maxwell's gibe about the thoughts of fishes and worms.) Still, although he'd rejected much of formal religion, Maxwell believed that God created all matter—and Clifford, who was an atheist, thought Maxwell had been too quick to limit the evolutionary role of atoms.[29]

In 1878, the year he published his major work on vectors, Clifford also published *Lectures and Essays,* with articles about matter, the mind-body problem, and a scientific approach to ethics. Maxwell was asked to review it, but he found it a delicate task—made more difficult than usual by the vague ill health that had been troubling him for a while now. He told Campbell there were many things in Clifford's *Lectures* that needed "trouncing," but that he wanted to express his criticisms gently, for "Clifford was such a nice fellow."[30]

All this debate—the back and forth, the witty barbs and careful criticisms—highlights the great scientific ferment of the times, and the broad participation by all the leading lights, no matter their primary field. On the subject of vectors, however, this ferment is about to erupt into debates as heated as any of those on more obviously controversial topics such as evolution. Tragically, Maxwell and Clifford will not get to play their parts.

FAREWELL MAXWELL AND CLIFFORD

When Maxwell came to write his careful review of Clifford, his mind simply would not work—and at last he knew his illness was serious. The doctor declared it abdominal cancer and gave his patient only a month to live.

Maxwell died at the age of forty-eight, in late 1879. Everyone who knew him was devastated at the loss, so early, of so bright and gentle a spirit. His "grand simplicity," as Campbell put it, touched everyone, and the tributes poured in. Tait's obituary for him is primarily a sober biographical summary, although you can see glimpses of the affection and admiration he had for his childhood friend. But as Tait's own biographer and former student Cargill Knott would later put it, "Tait had an unstinted admiration for the genius of Maxwell, a deep love for the man, and a keen appreciation of his oddities and humour."[31]

A moving funeral service was held at Maxwell's college, Trinity. Then he was taken home to Glenlair, and he still lies in the lonely churchyard nearby.

• • •

Clifford was gone, too, and far too soon, for he was only thirty-three when he died of tuberculosis earlier in 1879. Years later, the playwright George Bernard Shaw told Lucy Clifford that her husband had been smarter than anyone except Albert Einstein—adding drily, "cleverer even than ME!"[32]

In Campbell's tender biography of his childhood friend Maxwell, the last page includes a tribute to both him and Clifford, which was published in the *Pall Mall Gazette*. It is a fitting way to end this chapter, before we meet in the next some of the young men who carried on their work. "Maxwell's name stands," the article said, "by the unanimous consent of all who have any voice in such matters, in the very foremost rank." He had a rare yet influential genius, a quality he shared "with Faraday in an earlier generation . . . and with Clifford, a fellow mathematician . . . whose intellect was in more ways than one akin to Maxwell's own." Had they lived longer, who knows what else they might have accomplished.

(8)

VECTOR ANALYSIS AT LAST—AND A "WAR" OVER QUATERNIONS

Maxwell wasn't fated to see the achievements of the young men who were inspired by his *Treatise on Electricity and Magnetism*. This small band of devotees has been dubbed "the Maxwellians," and it includes George FitzGerald, who deduced from Maxwell's equations a theoretical way of detecting electromagnetic waves; his friend and collaborator Oliver Lodge; and Heinrich Hertz. Lodge and Hertz independently succeeded in generating these waves in the lab, for the very first time—spectacularly confirming Maxwell's theory in the process. Hertz presented his results in 1888, and poor Lodge—who had been preparing to present his own findings at the upcoming British Association meeting—discovered he'd been scooped.[1]

By 1901, the technology had improved so much that Guglielmo Marconi and his team were able to send the first long-distance telegraphic signals, soaring across the Atlantic Ocean rather than under it as William Thomson's had done—and the wireless era was born. In fact, Marconi's signal was sent from a transmitter designed by Maxwell's former student Ambrose Fleming.

For our story, though, the key Maxwellian is an eccentric telegrapher, Oliver Heaviside—for he is the one who extended Maxwell's whole-vector approach and turned it into modern vector analysis.

By the 1880s, many thousands of miles of cables had been laid, and telegraphy had become a glamorous new technology, akin to information technology in the early 2000s. And much like the recent popularity of IT courses, Thomson reported on the "epidemic" of young students, in his lab classes at the University of Glasgow, who wanted to train as telegraphic engineers. Young Heaviside, however, never went near a university. The downside of the nineteenth-century explosion of new technology was the erosion of work for craftspeople, including Heaviside's wood-engraver father, whose skills now had to vie with new photographic techniques for reproducing images. His mother tried to make ends meet by running a small school, but Heaviside grew up on the edge of poverty—and with a loathing for the mean streets of London with their cheating tradespeople and their boozy pubs.[2]

Following two of his older brothers, he started work at sixteen as a telegrapher for his uncle, Charles Wheatstone, who had codesigned one of the earliest electrical telegraph systems back in the 1830s, and who had married Heaviside's maternal aunt. A year or so later, in 1868, Heaviside landed a job working on the new Anglo-Danish telegraph cable, testing the signal speed along the line—and he became so fascinated by the technicalities of electricity and its measurement that he studied all he could find about it. He found Thomson's early theoretical work on telegraphic signal transmission particularly helpful, so already he was no ordinary telegrapher.

His deep thinking paid off when he published a paper that proved to be his ticket to academic respectability. It was on the best way to use a Wheatstone bridge—a device for measuring electrical resistance, named for his uncle, who had improved and popularized the original design by Samuel Christie. You needed to be able to measure resistance accurately so you could regulate your signals and make sure your cables wouldn't overheat. Thomson was so impressed with Heaviside's paper that he personally congratulated him.

Maxwell was impressed, too. Young Heaviside had had a rough start in life—made worse by scarlet fever that had left him partially deaf—but he had oodles of confidence and had sent Maxwell a copy of his paper. It had been published in February 1873, when Maxwell was in the final stages of seeing his *Treatise* into print, but he added in at the last minute a reference to the paper along with a brief outline of its key result. When Heaviside browsed through the *Treatise* later that year, he was blown away—and not just because his own work scored a mention. Rather, he was awed by the marvelous depth and breadth of Maxwell's book: "I saw that it was great, greater, and greatest" he enthused, and so he set out to master it.[3]

At the same time, he was doing well in his trade. Over the years he'd gained a great deal of electrical knowledge—helped also by collaborations with his brother Arthur—and his career seemed to be on the up and up. But suddenly, in 1874 when he was twenty-four, he quit his job and returned to live with his parents. No one knows quite why. It's been suggested that his prickly personality may have made life as an employee difficult. But it also seems that he'd fallen in love with electrical theory—and with the "heaven sent" Maxwell—and wanted the time to study and write papers.[4]

Whatever the reason, Heaviside, a lifelong bachelor, lived an increasingly isolated life devoted to writing increasingly sophisticated articles on electrical and telegraphic theory, especially for the weekly trade magazine *The Electrician*. His first paper on Maxwell's theory, "On the Energy of Electric Currents," appeared in 1883.

He also wrote on mathematics—including vector analysis, which he first discovered in Maxwell's *Treatise*. It is Heaviside who introduced the clear, bold typeface for vectors, and who got on the wrong side of Tait by dismissing quaternions as useless for practical purposes. He wrote facetiously, in his book *Electromagnetic Theory*,

> "Quaternion" was, I think, defined by an American schoolgirl to be "an ancient religious ceremony." This was, however, a complete mistake. The ancients—unlike Professor Tait—knew not, and did not worship, Quaternions.[5]

Heaviside is often hilarious, even in the middle of a technical passage! He's also often wise—as when he pauses midway in an account of Maxwell's electromagnetic innovations, taking aim at scientists who criticized them in "a carping and unreceptive spirit." He was referring to the complaint that Maxwell presented his final theory purely mathematically, with no physical model for how electromagnetism was supposed to get from one place to another.

Like Maxwell, Heaviside focused on vectors and their scalar and vector products, rather than on quaternions—but he went further by discarding all reference to vectors as the imaginary part of a quaternion. "I never understood it," he said. "The reader should thoroughly divest his mind of any idea that we are concerned with the imaginary $\sqrt{-1}$ in vector analysis."[6] After all, as I pointed out earlier, the components of a vector are the same whether you use an imaginary or real basis. So, it is thanks to Heaviside that Hamilton's imaginary i, j, k became today's $\mathbf{i}, \mathbf{j}, \mathbf{k}$.

By making $\mathbf{i}, \mathbf{j}$, and $\mathbf{k}$ real unit vectors whose square is 1, rather than the -1 that comes from squaring imaginary numbers, you also get rid of the minus sign in Hamilton's scalar product that I showed in the previous chapter. This unnecessary sign had annoyed Maxwell no end, too, but it was Heaviside who tossed it aside. "How could the square of a vector be negative?" he asked—adding drolly, "And Hamilton was so positive about it." His wonderful insouciance is perhaps a reflection of his outsider status: speaking loosely, the physicists didn't take him seriously because he hadn't been to university, and the practitioners didn't take him seriously because they thought theorizing was for nobs.[7]

The Maxwellians were notable exceptions to this rule—and so was Thomson. At the 1888 meeting of the British Association in Bath—where the disappointed Lodge presented his discovery of electromagnetic waves in the shadow of Hertz's prior announcement in Germany—there was also discussion about how to interpret Maxwell's theory. Lodge, FitzGerald, and Thomson all knew and admired Heaviside's work on this, and they promoted it so well that the official write-up of the meeting noted that "everyone expressed regret at the absence of Mr. Heaviside." Heaviside himself, loner that he was, had no intention of ever attending a scientific gathering. But at last, the physicists were taking notice of him.[8]

HEAVISIDE FAMOUSLY CHANGES THE FORM OF MAXWELL'S EQUATIONS

The discussion in Bath concerned the physical interpretation of Maxwell's mathematical quantities and equations. We've seen that Thomson was one of those who'd dismissed Maxwell's theory as having impressive math but no physical model for how electromagnetic waves propagated—but Heaviside had been asking a different question. As a telegrapher, he was interested in the problem of sending signals—which he saw as electrical energy—without distortion. Maxwell *had* written on the energy of the electric and magnetic fields, which Heaviside turned from $\mathfrak{E}$ and $\mathfrak{B}$ to the modern $\boldsymbol{E}$ and $\boldsymbol{B}$; but Maxwell apparently preferred the mathematical elegance of the "potentials" to a more direct emphasis on the physical fields.

In particular, and as I showed in chapter 7, he chose to represent the magnetic field by a vector potential:

$$\mathfrak{B} = V.\,\nabla\mathfrak{A}$$
$$(\boldsymbol{B} = \nabla \times \boldsymbol{A} \text{ in modern notation}),$$

where his $\mathfrak{A}$ and the modern $\boldsymbol{A}$ represent the vector potential of the electromagnetic field. It's a neat way to do it mathematically, because then you can use vector calculus identities, such as the divergence of a curl is always zero, to deduce further equations—in this case, that the divergence of $\boldsymbol{B}$ is zero: $\nabla \cdot \boldsymbol{B} = 0$, or as Heaviside wrote it, $\text{div}\,\boldsymbol{B} = 0$. (Like Clifford, he made the change from Maxwell's "convergence" to the modern "divergence." And he wrote $\nabla \times \boldsymbol{A}$ as $\text{curl}\,\boldsymbol{A}$.)

Similarly—and using Newton's dot notation as shorthand for the time derivative—Maxwell had represented the electric field as

$$\mathfrak{E} = V.\,\mathfrak{G}\mathfrak{B} - \dot{\mathfrak{A}} - \nabla\Psi,$$

or in modern symbols,

$$\boldsymbol{E} = \boldsymbol{v} \times \boldsymbol{B} - \frac{d\boldsymbol{A}}{dt} - \nabla\phi,$$

where Ψ (or ϕ) is the electric potential or voltage, and $\boldsymbol{v}$ is the velocity of any charged particle whose motion is contributing to a magnetic field (as per Øersted) that in turn contributes to the electric field (as per Faraday). If the electric field is due to a changing magnetic field alone, then $\boldsymbol{v} = 0$.

Heaviside allowed that sometimes the mathematical potential approach could be useful for calculations, but he wanted to get straight down to the physics of electromagnetic energy. When he did include a potential in an equation, he introduced the shorthand "pot"—just as he used "div" and "curl" for the divergence and curl operators—adding mischievously that in this case, "pot . . . has no more to do with kettle than the trigonometrical sin has to do with the unmentionable one. It seems unnecessary to say so, but one cannot be too particular." Quite! But in sorting out Maxwell's equations in relation to the transmission of energy, Heaviside "murdered" the potentials, as he told FitzGerald—and with good reason. In the *Treatise* (and in his original 1865 paper), Maxwell had derived his wave equation for light—or for any electromagnetic signal—in terms of the vector potential $\mathfrak{A}$ (or $\boldsymbol{A}$), and Heaviside had shown that this was unphysical when it came to analyzing the transmission of electromagnetic energy.[9]

So now Heaviside's main goal was to recast Maxwell's equations so that the physical fields, $\boldsymbol{E}$ and $\boldsymbol{B}$, would be the main event—just as they are today (generally speaking, at least).[10] What he didn't know was that in an 1868 paper, Maxwell *had* derived a wave equation directly in terms of the field. It's beautifully simple, and it's a pity he didn't take this approach further. Instead, it is Heaviside, that acerbic self-taught loner, who brought this approach into the mainstream.[11]

In fact, you may have read in popular books and blogs that in Maxwell's papers, there were twenty component-form "Maxwell's equations," and that it was Heaviside who reduced these to the four beautiful whole-vector equations that are famous today in textbooks (and on T-shirts):

$$\nabla \cdot \boldsymbol{E} = 4\pi\rho$$

$$\nabla \cdot \boldsymbol{B} = 0$$

$$\nabla \times \boldsymbol{E} = -\frac{\partial \boldsymbol{B}}{\partial t}$$

$$\nabla \times \boldsymbol{B} = \frac{\partial \boldsymbol{E}}{\partial t} + 4\pi \boldsymbol{J}$$

(where ρ is the charge density and J is the current density).[12] This claim on behalf of Heaviside is not quite the true picture, however. First, though, let's look again at what these famous equations mean. The first two are divergence equations, which relate to the flux flowing through a closed surface around the charge or magnet, as illustrated in figures 6.1 and 7.1. These two equations are vector forms of the Gauss-Coulomb laws. The third and fourth equations relate to Faraday's and Ampère's laws, and they involve the curl of the fields. As I showed in figure 7.2, the curl operation relates to rotations. So, the $\nabla \times \boldsymbol{E}$ equation encodes the experimentally observed fact that a changing magnetic field induces a current that flows *around* a loop, while the $\nabla \times \boldsymbol{B}$ equation shows that an induced magnetic field "rotates" or circles around a current-carrying wire. (If you scatter iron filings on a board through which the wire passes, you'll see the filings line up in circles.) The derivatives on the right-hand sides of these two equations show that the electric and magnetic fields must be changing in time if they are to induce each other.

It's an extraordinarily elegant and powerful set of equations.

The story of how we went from Maxwell to these four beautiful equations is a little more complicated—as these things usually are—than the popular legend that has grown up around the previously long-neglected Heaviside. Aside from the fact that Heaviside wrote "div" and "curl" instead of using the dot and cross, as far as I'm aware he never wrote these four equations all together: just as Maxwell's main field equations were interspersed with equations for various applications or for defining various terms, so were Heaviside's. That's how it goes with original research, for you are presenting your definitions, justifications, and applications as you go. He did write his version of the two curl equations together—he called them "duplex equations"—and he sometimes wrote the two divergence equations together, but even then, the two pairs of equations were interspersed among dozens of other equations. What's more, Heaviside didn't write these four equations in quite the modern form: he tweaked them to make them more symmetrical, as you can see in the endnote.[13]

Much more important than these details, however, is the fact that in his *Treatise* Maxwell had, in fact, reduced his twenty component equations to just five whole-vector ones. Notation aside, these are the two equations for $\boldsymbol{E}$ and $\boldsymbol{B}$ in terms of potentials, which I showed earlier: the first is equivalent to $\nabla \cdot \boldsymbol{B} = 0$, as we saw, and the second is equivalent to $\nabla \times \boldsymbol{E} = -\frac{\partial \boldsymbol{B}}{\partial t}$ (as I'll show in the next endnote). Then there's the $\nabla \cdot \boldsymbol{E}$ equation I gave in chapter 7, plus an equation for the curl of $\boldsymbol{B}$, which he wrote like this:

$$4\pi\mathfrak{C} = V.\nabla\mathfrak{H}.$$

He defined $\mathfrak{H}$ to be proportional to $\mathfrak{B}$ for a purely magnetic field, and he defined $\mathfrak{C}$ to be the conduction current plus the derivative of the electric displacement $\mathfrak{D}$ (which is proportional to $\mathfrak{E}$ as we saw briefly in the previous chapter)—so this means, in modern notation with appropriate units, that his curl equation is, indeed,

$$4\pi\boldsymbol{J} + \frac{\partial \boldsymbol{E}}{\partial t} = \nabla \times \boldsymbol{B}.$$

Maxwell also gave an additional equation, for the mechanical force created by the electric and magnetic fields—and this, as Heaviside said, is what engineers need when designing dynamos and motors. It is essentially what is now called the Lorentz force law—named after the same Lorentz we met in chapter 4 in connection with electron spin.

So, the novel thing that Heaviside did—Maxwell's 1868 paper aside—was to rewrite Maxwell's equations in terms of $\boldsymbol{E}$ and $\boldsymbol{B}$ rather than the potentials (and to change those hard-to-read Gothic letters to clear boldface ones). This brings out, in those four famous equations, the beautiful symmetry between the electric and magnetic fields, and it was a masterstroke by Heaviside. In yet another case of independent codiscovery, Hertz had done something similar. (Tragically, Hertz never got the chance to take his work much further. He died of blood poisoning in 1892, when he was just thirty-six years old—so he never lived to see the extraordinary uses his radio waves were about to find, just as Maxwell never got to see the spectacular way in which Hertz had confirmed his prediction of those waves.) Still, this change from potentials to field vectors follows easily from Maxwell's version of the equations. I've already shown that Maxwell had whole-vector equivalents of the $\nabla \cdot \boldsymbol{B}$, $\nabla \cdot \boldsymbol{E}$, and $\nabla \times \boldsymbol{B}$ equations; getting the $\nabla \times \boldsymbol{E}$ equation from Maxwell's equation for $\boldsymbol{E}$ is also relatively straightforward, as you can see in the endnote.[14] That's why Heaviside never claimed that these were his equations: they're all there, *plus* the force equation, in those five whole-vector equations in Maxwell's *Treatise*.[15]

Heaviside did bring out the symmetry, and he put the focus firmly on the $\boldsymbol{E}$ and $\boldsymbol{B}$ to make the equations more practical—but his "murdered" potentials are very much alive and well in mathematical physics today. The vector potential $\boldsymbol{A}$ may not be physical—it is $\boldsymbol{E}$ and $\boldsymbol{B}$ that are directly measurable—but as Heaviside himself said, they often enable calculations to be done more simply. This computational use of potentials is often known today as "gauge theory": the idea is to choose a "gauge"—an equation in terms of $\boldsymbol{A}$ and ϕ—that simplifies the calculations without changing the values of $\boldsymbol{E}$ and $\boldsymbol{B}$. In other words, the electric and magnetic fields remain *invariant* under "gauge transformations"—analogous to the way a rotating ball looks the same no matter which way you turn it. Gauges are now used to simplify calculations not just in electromagnetism, but in quantum theory and relativity, too.

VECTOR ANALYSIS: THE MATHEMATICAL LEGACY OF HEAVISIDE AND GIBBS

I've spent time untangling the reality from the myth about who wrote Maxwell's equations, because aside from illustrating how difficult it was for people to appreciate Maxwell's quaternionic whole-vector equations, it shows something that is still often misunderstood about how science evolves. Newton's laws, to take another example, are never written in the form he wrote them in *Principia*, because new mathematical techniques, notations, and understandings enabled others to write them in their modern vector calculus form. But the content is still Newton's. And so it is with Maxwell's equations—and he came much closer to the modern form than Newton did. So, Heaviside's most important mathematical innovation is not so much his reformulation of Maxwell's equations per se, as his development of modern vector analysis.

In the first volume of his extraordinary book *Electromagnetic Theory*, Heaviside set out in detail all the rules of vector analysis—breaking vectors away completely from their original role as the imaginary part of a quaternion. After all, he said, "we live in a world of vectors"—something

Newton was the first to make clear, as I showed in chapter 3. So, Heaviside continued, "an algebra or language of vectors is a positive necessity."[16] And he went on to show, in his many papers on telegraphy and related practical electromagnetic theory, just how powerful this language could be. Today, of course, you can't open a physics paper or textbook without seeing vectors. In the early 1890s, however, vector analysis was still either the cheeky offspring of the venerable component method or the cocky younger sibling of the quaternion—depending on whom you spoke to. Thomson was in the first camp, Tait in the second.

Maxwell and Clifford were the earliest to head away from quaternions toward vector analysis. Yet although they were now gone, Heaviside was not alone in his battle for vectors. Across the Atlantic, Josiah Willard Gibbs had also been inspired by Maxwell's *Treatise*. And like Heaviside, Gibbs was interested in Maxwell's *applications* of vectors—unlike Tait, Gibbs wasn't interested in vectors for their mathematical beauty. Indeed, he had begun his research career in a very practical way, working on gears and brakes for railway carriages, and governors for steam engines—this was in the 1860s, when the building of railway lines in America was gathering pace. The first transcontinental line, a massive infrastructure project, was completed in 1869—but by then Gibbs had already turned his attention from engineering to physics. In fact, he'd just returned from three years in Europe, auditing lectures from experts in the new science of thermodynamics, and over the next few years he made major contributions to this new science. But when he read Maxwell's *Treatise* in 1873, he became increasingly drawn to the physics of electromagnetism—and, therefore, to the language of vectors.

Like Heaviside, Gibbs noticed that Maxwell had made use of scalar and vector products, rather than the full quaternion product. Which is why, in the 1880s, Gibbs and Heaviside both branched out on their own exploration of vectors sans quaternions—without yet knowing of each other's work.

While Heaviside was an academic outsider, Gibbs was a young professor on the other side of the world from where the scientific action was taking place—he'd been appointed professor of mathematical physics at Yale in 1871, but his geographical distance from the scientific mainstream made him something of an outsider, too. In fact, Maxwell had been just about the only one, at the time, to recognize the importance of Gibbs's early work on thermodynamics. But while Heaviside railed against academic snobbery, Gibbs quietly got on with his own work—and in 1881 and 1884, he self-published parts 1 and 2 of his little book *Elements of Vector Analysis*.

He wasn't the first American to work on what he called "multiple algebras," algebras whose symbols represent more than one number, so that "double algebras" were complex numbers, vectors were "triple algebras" (in 3-D space), quaternions were "quadruple algebras" because they included a scalar plus the three spatial vector components—and so on. Inspired by Hamilton's discovery of a new algebra without the commutative law, the Harvard mathematics professor Benjamin Peirce had long been working on creating dozens of other new algebras. Later, his son Charles proved that the only algebras allowing division or the existence of inverses—and hence ways of solving equations and of reversing rotations—are those of real numbers (that is, traditional school algebra), complex numbers, and quaternions. Benjamin, and later another of his sons, James, gave lectures on quaternions at Harvard—the Peirces were certainly a mathematically talented family.

Gibbs, on the other hand, taught classes on vector analysis, not quaternions. He also gave us the dot and cross notation for scalar and vector products, as I've mentioned—much to Charles Peirce's displeasure. Gibbs replied to Peirce's criticism, saying he felt unsure himself about which notation to use. But he followed Tait and Hamilton in using Greek letters for vectors; Heaviside, on the other hand, used boldface type for vectors but kept Hamilton's *S* and *V* for scalar and vector products. So, you can see how the modern notation—as in the modern form of the four Maxwell equations above—is a synthesis of both Heaviside's and Gibbs's symbols. It's another example of the way new notations for new concepts take a while to find their best form—or, at least, their most popular one.

Gibbs got his start in vectors from Maxwell's *Treatise*, but a few years later he discovered Grassmann's work. Like everyone else, it seems, he never did manage to read all of *Ausdehnungslehre*, but he was mightily impressed with the generality of Grassmann's products—especially the fact that they were generalizable to more than three dimensions. The scalar product is also generalizable: if two vectors $\boldsymbol{a}$ and $\boldsymbol{b}$ each have n components, labeled $a_1, \ldots, a_n$ and $b_1, \ldots, b_n$, then their scalar product is just:

$$\boldsymbol{a} \cdot \boldsymbol{b} = a_1 b_1 + a_2 b_2 + a_3 b_3 + \ldots + a_n b_n.$$

Today this general form is often called an "inner product," following Grassmann. By contrast, I've already mentioned that the vector (or cross) product is only defined in three dimensions, whereas Grassmann's analogous "outer product" was general, and we'll see this in more detail in chapter 11.

Gibbs was so impressed with Grassmann that he wrote to his son, Hermann Jr., urging him to publish his father's earliest work—Gibbs seemed very keen that Grassmann should have priority over Hamilton. So much so that in the preface to his *Elements of Vector Analysis* he acknowledged Grassmann and Clifford, but made no mention of Maxwell, let alone Tait and Hamilton. We know that his primary debt was to Maxwell's *Treatise*, though, because of a letter he wrote to a German colleague.[17] So Gibbs seems rather ungenerous in denying public credit where it was due. Perhaps he wanted to present vector analysis as a new subject quite separate from quaternions, for we'll see shortly how determinedly he defended this agenda. Still, Grassmann deserved Gibbs's efforts, which ultimately led to the publication of his long-neglected collected works.

THE VECTOR "WARS"

In 1888, Gibbs sent a copy of his *Vector Analysis* to everyone he could think of—including Thomson, Tait, and Heaviside. Heaviside was pleased to find he was not the only one working on vectors—but Tait was infuriated! And by the 1890s, a none-too-genteel debate had erupted in the pages of *Nature* and other leading scientific magazines. The question was: Should whole-vectorial systems be used at all—the likes of Thomson and Arthur Cayley said no, components were sufficient; but if they *were* to be used, then was it Hamilton's quaternions or Grassmann's algebra? Or should it be Gibbs's and Heaviside's utilitarian vector analysis, which Tait saw as little more than a rip-off of Hamilton?

Gibbs, in particular, roused Tait's ire: "a sort of hermaphrodite monster," he called Gibbs's *Vector Analysis*, "compounded of the notations of Hamilton and Grassmann." In an 1891 *Nature* article, Gibbs fought back, very politely, against Tait's slur. (*Nature* wasn't the glossy magazine it is today—the technology wasn't up to it—but by the 1890s it had become a very popular place for scientists to present their research and discuss their views.) Then Heaviside joined in, suggesting that round one had gone to Gibbs. Later, in a show of even-handedness, Heaviside said the resolution of the debate depended on your point of view. From the quaternionic rather than the physics viewpoint, he said, "Prof. Tait is right, thoroughly right, and Quaternions furnishes a uniquely simple and natural way of treating *quaternions*." Just to make sure his readers got the joke he added, "Observe the emphasis." But he did go on to speak of the mathematical beauty of quaternions, as opposed to what he saw as their "uselessness" in physics.[18]

Tait lost no time replying to Gibbs, through the pages of *Nature*. I've mentioned already that since the cross product is only defined in three dimensions, then so is the quaternion product. So, determined not to admit any limitations on his beloved quaternions, Tait challenged Gibbs's emphasis on the importance of extending vectors to more than three dimensions, asking rhetorically, "What have students of physics, as such, to do with space of more than three dimensions?" Which goes to show how hindsight can make history even more interesting—for while Tait was a leading scientist in 1891, just fourteen years later young Einstein will show that the special theory of relativity needs four dimensions, (t, x, y, z)—the usual three of space plus a time-coordinate. Back in 1891, though, Tait was plumping for three dimensions—but he did score a point for quaternions in terms of their compactness compared even with vector expressions, let alone in comparison with component equations.[19]

As we saw in chapter 7, Thomson had won out over Tait in T and T', their joint cutting-edge physics textbook from which quaternions were banished. Later Thomson recalled their fights on the topic, saying, "We have had a thirty-eight years war over quaternions." He did acknowledge that Tait had been "captivated by the originality and extraordinary beauty of Hamilton's genius in this respect"—and yet, he said, Tait could never give an example of how physical problems could actually be *solved* more easily with the use of quaternions. Of course, Tait, like Maxwell, maintained that if you wrote a problem in quaternion or whole-vector form it was easier to see the physics—but as Thomson repeatedly pointed out, you still have to solve the equations in terms of their components. For instance, you might remember from school trying to find the trajectory of a ball thrown obliquely into the air. You have to use Newton's second law, $\boldsymbol{F} = m\boldsymbol{a}$, and first you "resolve" the vectors into their components (as in fig. 8.1), and then you apply Newton's law to each force component—just as Thomson said![20]

Meantime, Gibbs and Tait continued their argument via *Nature*—Gibbs never losing his cool—with Heaviside adding his piece in *The Electrician*. Then Tait's former student Alexander Macfarlane joined in—he was now based in Texas, and in an address to the American Association for the Advancement of Science, he suggested that a more complete algebra will likely evolve, unifying quaternions and Grassmannian algebra. (He was right, as we saw with Clifford's geometric algebra, and as we'll see with tensor analysis.) He even attempted his own form of vector analysis, making i, j, and k real rather than imaginary, but it wasn't as neat as Gibbs's and Heaviside's, and it never caught on.[21]

By 1892, the debate reached as far as Australia, where Cambridge graduate Alexander McAulay was teaching mathematics and physics at the University of Melbourne. He'd given a paper on quaternions at the January meeting of the Australian Association for the Advancement of Science, which was published in the *Philosophical Magazine* in June, and he also added his voice to the debates in *Nature*. He was firmly in the quaternionists' camp, and he blamed Maxwell for the physicists' lack of appreciation of the power of quaternion analysis, because Maxwell had used vector and scalar products rather than the full quaternion product. Macfarlane shot

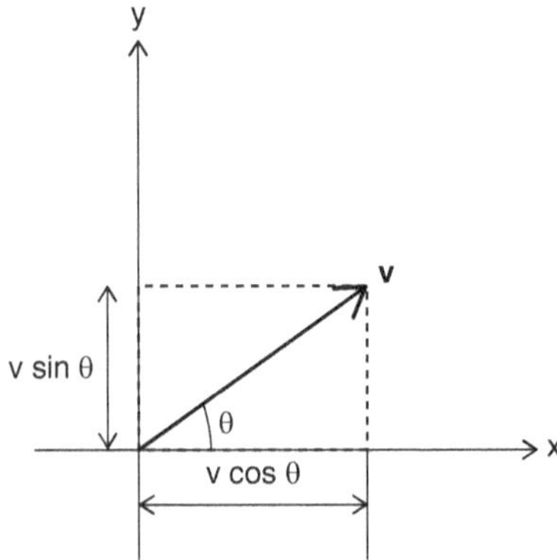

FIGURE 8.1. Finding the trajectory of a projectile. Once the ball is thrown, with initial velocity v, the only force acting on it is the downward force of gravity. So, there's no force component acting in the horizontal direction. Taking these components separately, first find the value of the horizontal component of motion: canceling the mass term in Newton's second law, and using Newton's dot notation for time derivatives, we have:

$$\ddot{x} = 0$$
$$\Rightarrow \dot{x} = C = v\cos\theta$$
$$\Rightarrow x = vt\cos\theta.$$

(C is the magnitude of the horizontal component of the initial velocity; the second constant of integration is zero for projection from the origin.)

Now find the vertical component of motion:

$$\ddot{y} = -g$$
$$\Rightarrow \dot{y} = -gt + C_1 = -gt + v\sin\theta$$
$$\Rightarrow y = -\frac{gt^2}{2} + vt\sin\theta + C_2$$

(but $C_2 = 0$ for projection from the origin). You can then find the parabolic trajectory by expressing t in terms of x and substituting into the equation for y.

back at this, saying it wasn't Maxwell who was to blame but the limitation of quaternions themselves—especially that annoying minus sign in front of the scalar product. And so it went, with quaternionists versus vector analysts arguing their cases with varying degrees of perspicacity and temperance.

By 1893, McAulay had taken a position as the first lecturer in mathematics and physics at the newly established University of Tasmania.

In the same year he published his book *Utility of Quaternions in Physics*, which he'd initially written in 1887 as an ambitious essay for Cambridge's prestigious Smith's Prize. Maxwell had shared the prize back in 1854, but McAulay had known he was blighting his chances with an essay on quaternions, which few at Cambridge understood. As he explained in the preface of his book, those Cambridge professors who did teach the subject treated it "*as an algebra*, but this [is] not Hamiltonian. . . . Hamilton looked upon Quaternions as a *geometrical* method." As I mentioned in chapter 4, while De Morgan and others were interested in algebra for algebra's sake, Hamilton's goal had been to find an algebraic language for geometry—in particular, for describing rotations in three-dimensional space. That's how he'd hit upon his new algebra of quaternions in the first place.

Perhaps McAulay hadn't realized that in his *Treatise* Maxwell had expressed this same view about the importance of geometry—which relates closely to the physical meaning in question. Quaternion analysis, which Maxwell called "the Doctrine of Vectors," is, at base, a language of geometrical arrows—so it focuses on the actual points and lines in space, rather than on their coordinates. This geometrical approach allows physical quantities to be modeled independently of the choice of coordinates—so it brings out their invariant, *physical* properties. We'll see more about this in the next chapter, but it's why Maxwell took so many pains to point out the importance of whole vectors rather than coordinate-based components. He was interested in modeling the electromagnetic field, using the geometry of the lines of force—like the lines of iron filings—around magnets, electric charges, and electric currents. Similarly, in 1915 Einstein will famously use the tensor extension of this "doctrine of vectors" to model the geometry of curved space-time. In hindsight it's easy to take for granted the development of mathematical and scientific ideas, and to assume an easy path from Maxwell to Einstein and from vectors to tensors. It can seem hard to believe that barely a century ago, mathematics, of all things, could have been so controversial. But in 1893 McAulay and his peers were far from done.

In his preface McAulay went so far as to advocate that students should study quaternions rather than Cartesian geometry—adding that the only way to encourage students to study the subject would be to make it "pay," as he tellingly put it, by including more exam questions on the subject. Not much has changed: most students, it seems, want to learn how to pass exams rather than how to think. Such are the pressures on our students, and the limitations of our education and employment policies.[22] Like Maxwell before him, McAulay clearly understood this problematic choice between teaching students to think and training them to pass exams and job interviews, and he sought to inspire future students with his book, appealing directly to them and encouraging them to immerse themselves timelessly in "the delirious pleasures" of quaternions. "When you wake," he continued, "you will have forgotten [exams] and in the fullness of time will develop into a financial wreck." But instead of wealth, McAulay's siren song continued, you will possess "the memory of that heaven-sent dream" of being "steeped" in quaternions, a memory that will make you "a far happier and richer man than the millionest millionaire." It's noble if bizarre and impractical advice! (Like all his colleagues McAulay used "man" and "he" when addressing students, for it was the norm at a time when women generally weren't allowed to take degrees. McAulay did have a feminist wife, though—he married Ida Butler in 1895, and she continued to fight for women's rights to higher education and the vote, and for sex education and family planning.)[23]

Heaviside privately told Gibbs that McAulay "seems to be a very clever fellow, and he knows it and shows it a little too much sometimes." As for Tait, he had lost none of his acerbic exuberance, and his review of McAulay's pro-quaternion book proved perfect clickbait—to use a pointed anachronism—for a cover story in *Nature*. "It is positively exhilarating," he wrote, to read McAulay after "toiling through the arid wastes" in the work of Gibbs and Heaviside. Then, with stunning blindness to his own stylistic excesses and personal attacks, but perhaps with some reason, Tait took issue with McAulay's rhetorical flourishes—he thought they would put people off quaternions rather than convert them. Still, he saw in McAulay a man of "genuine power and originality," who "snatches up the magnificent weapon which Hamilton tenders to all, and at once dashes off to the jungle on the quest of big game."[24] Such a macho, colonialist metaphor would not go down well today!

Gibbs, by contrast, countered McAulay's defense of quaternions by suggesting that Hamilton himself had obscured the simplicity and power of the vectorial approach—especially when he'd joined scalar and vector products into a single quaternion product. And in contrast to McAulay, Gibbs felt that Hamilton had actually sidelined the geometrical aspect of vectors. Who would have thought, when using vectors today, that there'd been so many issues to think through—geometry, economy, notation, physical interpretation, ease of computation—and with so much passion!

The debate continued, with new players joining in—notably Tait's former student and future biographer, Cargill Knott. Then Heaviside came back into the fray, deeming, in the pages of *Nature*, Tait's and Knott's defenses of quaternions irrelevant, and urging McAulay to give up quaternions. "The quaternionic calm and peace have been disturbed," he added triumphantly. "There is confusion in the quaternionic citadel; alarms and excursions, and hurling of stones and pouring of boiling water upon the invading host." He was right, and the invading vector analysts eventually won, as any undergrad math or physics book testifies. And McAulay did give up quaternions: in 1905 he went off to spearhead Tasmania's first hydroelectricity system—one of the first in the world. It was a fitting finale, for earlier he had applied quaternions to both electromagnetism and hydrodynamics.[25]

• • •

Gibbs stands out in all this fury as a man of calm and reason. In his *Nature* article "Quaternions and the Algebra of Vectors," he gave a brilliant summary of what the fuss was *really* about, cutting through questions of ego and priority to the age-old purpose of mathematical physics—that of clarifying the "relations and operations" between physical quantities. And although he was firmly on the side of vector analysis, he saw it as nothing so terribly new—it was just the simplest and most useful way of expressing the kinds of mathematical operations people had been developing for centuries. We owe it to history, he said, and to our students, to acknowledge this long line of contributors—and here he did include Tait, whom he praised for developing and applying Hamilton's work *beyond* Hamilton, no matter how zealously and loyally he proclaimed otherwise. After all, said Gibbs, "the current of modern thought is too broad to be confined by the [methods] even of a Hamilton." It's a view of history that is worth remembering—that the dazzling initiators of new ideas rarely leave those ideas in perfect form, and that many other brilliant minds set to work clarifying, applying, and extending them.[26]

BACK TO THE BEGINNING . . .

In fact, the history relevant to the "vector wars" goes right back to the beginning of recorded mathematics. For at its core, this fiery fin-de-siècle debate was about the best way to *represent information*—and the best way to calculate with it. Representation is what the Mesopotamians and Egyptians had been working on with their tables and arrays for recording information about the size of land parcels, or the volume of earth to be dug for a canal and the cost in money and labor of the digging. And they calculated this information by consulting multiplication tables, summaries of algorithms, and lists of Pythagorean triples such as that in Plimpton 322, the Mesopotamian tablet shown in figure 0.1 in the prologue.

A thousand years later, the ancient Greeks had experimented with the idea of representing not only commercial and sociological data but information about *locations in space*—and as we've seen, their idea of coordinates developed into our familiar mathematical Cartesian and polar grids. Fast-forward to the 1890s, and you can see that the vector/quaternion/components debate was about the best way to represent *and* utilize information about physical quantities located in space.

As Tait and the other quaternionists argued in the 1890s—quaternions beat vectors for compact *representation* of information, while both quaternions and whole vectors beat lists of components. But as Thomson maintained till the very end, since you have to *calculate* with components anyway, why bother with special notation for the entities as a whole? He was not alone—Cayley, for one, had the same view. But they would surely have changed their minds had they lived in the digital age, where representational economy matters in terms of cost, time, and energy.

Of course, Tait and Maxwell had long been arguing for quaternions and whole vectors on physical grounds, but the new vector analysis approach—with its real basis vectors ***i***, ***j***, ***k*** rather than the quaternionic imaginary numbers *i*, *j*, *k*—proved more natural to physicists. And so, by the opening decade of the twentieth century, it was clear the vector analysts had, indeed, won—at least for the time being. Quaternions would have to wait for the computer age before they could make a comeback—and today new research papers appear regularly, extending and applying quaternions to aerospace navigation, biometric recognition, robotics, molecular dynamics, control systems, color image processing, and more.

BEYOND VECTORS

It was physics, especially Maxwell's theory, that brought vector methods into the mainstream—and it was the physicist Gibbs and the electrician-physicist Heaviside who, in the 1890s, had led the charge for vectors over quaternions. But as we'll see in the next chapter, it was mathematicians who identified the key qualities—in both quaternions and whole vectors—that led first to *four*-dimensional vector analysis, then to *n*-dimensional vectors, and ultimately to tensor analysis. First, though, we'll check in with Cayley and Tait and their mathematicians' view of the vector wars.

(9)

FROM SPACE TO SPACE-TIME

A New Twist for Vectors

Since 1863 Arthur Cayley had held the Sadleirian professorship of mathematics at Cambridge—in fact, he was the first to hold this prestigious chair, which had been endowed by Lady Mary Sadleir. Not much is known about her, except that she was devoted to good causes—the math chair at Cambridge was far from her only beneficence. It is fitting, given such a benefactor, that Cayley was an early supporter of women's higher education. Although they couldn't take formal degrees at Cambridge, women were allowed to study lectures similar to those offered the men, and in 1869 and 1871 the first women's residential colleges were established—Girton and Newnham. One of these young Girton women was Grace Chisholm, who found Cayley very welcoming, yet stiflingly old-fashioned in his approach to mathematics. She recalled the "flapping sleeves" of his academic gown "as he stood with his back to the listeners chalking and talking at the same time at the blackboard."[1] But this was a one-off comment: she'd needed special permission from the headmistress of Girton, as well as from Cayley, even to attend his class; with few exceptions, women studied with tutors in their own colleges, rarely attending lectures with the men.

Chisholm qualified for the equivalent of a first-class degree in 1893. A motion to formalize degrees for women was defeated, several times, over the next few decades—including in 1897, when some male undergrads were so chuffed at retaining their privilege they went on a wild celebratory rampage through town, causing the equivalent of more than a hundred thousand dollars' worth of damage. Cayley must have been disgusted, for he'd been chairman of Newnham's council through the 1880s, and he had also taught at Girton. Virginia Woolf would give her famous "Room of One's Own" lectures at these colleges in 1928.[2]

In the years just before that wild student rampage, the vector wars had been playing out, and as we saw, the physicists' argument in favor of modern vector analysis was gaining ground. In the background, though, Cayley and Peter Tait had been quietly debating the issue from a mathematical point of view. Back in 1888 Cayley had exclaimed to Tait, "we are irreconcilable and shall remain so," but in the summer of 1894—when Cayley was seventy-three and Tait ten years younger—these two elder statesmen of British mathematics went public with their mathematicians' view of the vector debate. And at Tait's suggestion, they did it together, each reading a paper before the Royal Society of Edinburgh.[3]

THE BEAUTIFUL CONCEPT OF INVARIANCE

Cayley began his presentation diplomatically, quoting Tait's view of the advantage of quaternions: "They give the solution of the most general statement of the problem they are applied to, quite independent of any limitations as to the choice of particular coordinate axes." What Tait meant was that if you change your frame of reference—by rotating the axes, say—your points in space will have different coordinates, and your vectors (and quaternions) will have different components. But as you can see in figure 9.1, the *length* of a vector doesn't change when the frame is rotated through an angle of θ—it's just the same vector—and so the *scalar and vector products* of any two vectors will remain the same, too.

This remarkable property, where certain things stay the same even when the components are measured from different axes, is called "invariance." The relationship between the two sets of coordinates is called

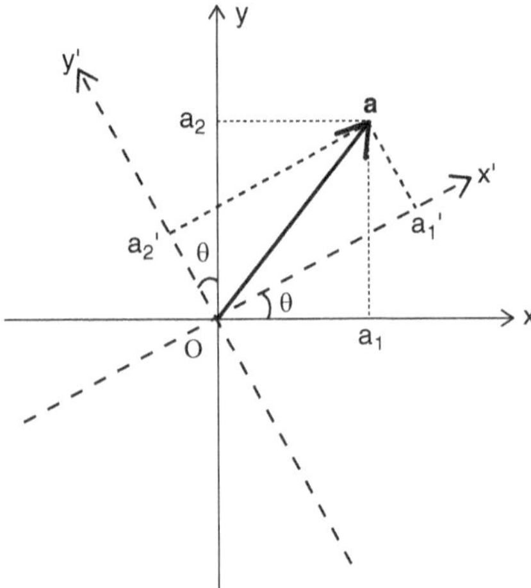

FIGURE 9.1. The vector ***a*** has components (a_1, a_2) in the usual *x-y* frame, and $(a_{1'}, a_{2'})$ in the rotated *x'-y'* frame. It is the same vector, with two different coordinate representations. The length of the vector is the same in both frames, as you can see from the geometry of the figure. Mathematically speaking, it is invariant under rotations. The same will be true for a second vector ***b***.

So the scalar and vector products are invariant under rotations, too, as you can see by using the geometric definitions of these products: $\boldsymbol{a} \cdot \boldsymbol{b} = ab \cos \Phi$, where a and b are the lengths of the two vectors, and Φ is the angle between them; because the vectors don't change, the angle between them isn't affected by the coordinate change. Similarly, $\boldsymbol{a} \times \boldsymbol{b}$ has a magnitude (length) of $ab \sin \Phi$, and a direction perpendicular to the plane of the two vectors, and this plane—the plane of the page in the diagram here—doesn't change when you rotate the axes in this way. When the axes are changed like this, the "coordinate transformation equations" are:

$$x' = x \cos \theta + y \sin \theta, y' = -x \sin \theta + y \cos \theta.$$

We saw an example of this for the robot arm in figure 4.2 (where we rotated the arm rather than the axes). But the key point is not the details but the fact that when you change your coordinate frame, the two sets of coordinates are related by specific equations. This is key to the idea of tensors.

a "coordinate transformation." There are two basic types of coordinate transformation: one is simply a "change of variables," such as transforming from Cartesian to polar coordinates as in figure 2.3, where your axes stay the same; the other is a transformation between *frames*—that is, where the axes themselves are changed. (More technically, a frame is a ruler, a clock, and a coordinate system, which allow the observer to coordinatize events in time and space.) It's this second type of coordinate transformation that is important in many practical problems.

Because whole-vector expressions such as scalar products are invariant under certain coordinate transformations, "coordinate-free"—or "coordinate-independent"—is a more mathematical way of saying "whole" when speaking of representing vectors. The 1905 special theory of relativity will show in spades just why (and how) invariance matters in physics, but in 1894, when Tait and Cayley were debating the issue, they were focusing on math rather than physics.

In fact, the idea of invariance is both mathematically fascinating and more broadly applicable than in physics alone. To take a modern digital application first, neural networks handle complex data by passing information from one node or "neuron" to another, just as the neurons in our brains do. At each node, a model (analogous to linear regression) assigns weights to the input data, weighting each piece of data according to its importance in the desired output. We saw something similar in chapter 4, with search engine ranking algorithms. In neural networks, though, each layer of nodes adds more complexity to the model, so when working out the math for mapping one layer of neurons onto the next, programmers must make sure that key features of the information are invariant under these "maps" or coordinate transformations.

For instance, in 2022 the DeepMind AI group's "AlphaFold" neural network succeeded in predicting the structures of virtually all the known proteins—two hundred million of them, which had been identified over the years through genetic sequencing of various species. Proteins are chains of amino acids that fold into three-dimensional shapes—and it's the shape that governs their function. So, it's the shape that scientists want to understand in order to create new types of drugs, or new enzymes for agriculture or pollution control, or to detect new variants of concern in the SARS-CoV-2 virus via its associated proteins, and so on—except that there are so many possible ways these chains of amino acids can fold that it had long been impossible to figure out the actual structures. The AlphaFold algorithm used the linear (1-D) sequence of each protein's amino acids, along with training data about known structures of related proteins, to predict the 3-D coordinates of all the key atoms in the folded protein.[4] As it progresses from one layer of nodes to the next, such an algorithm "learns" more about the protein's possible shape from this input data, so programmers have to make sure that key attributes, such as the distance between the atoms, stay the same when the information is transferred to the coordinate frame of the next node.

As with other molecular modeling, and also with computer vision applications, the protein algorithm also had to learn to recognize the correct shape even if it is rotated—which meant that the mathematical representation of the 3-D shape had to be invariant under rotations. The concept of invariance is important in many neural network and other technological applications, from satellite imagery and biomedical microscopy imagery to keeping the James Webb Space Telescope in place.

Cayley, and the other pioneers of the math of invariance, would be stunned at these sophisticated modern applications. On the other hand, they knew that all sorts of things can be invariant. For example, in chapter 4 we saw that if you rotated a book horizontally through 90° and then flipped it over vertically through 180°, it would have a different orientation when you performed these operations in reverse—whereas if you rotated a featureless box or ball in this way, it would look the same each time. That's because the box and ball are symmetrical. Similarly, a snowflake generally has six points or corners, and it is beautifully symmetrical—so it looks just the same when you rotate it through multiples of 60°. (To be pedantic, in nature not all snowflakes are *perfectly* symmetrical—but you can see the point.) In other words, it is invariant under these rotations—and also under 180° rotations, or reflections, about its axes of symmetry.

So, invariance is related to symmetry—and in math these two words are often used interchangeably.

As for Tait, he was interested in the invariance of vector and quaternion quantities, and he gave the example of $\boldsymbol{a} \cdot \boldsymbol{b} = 0$, an equation that stays

FIGURE 9.2. The symmetry of snowflakes. Plate 18 from Wilson A. Bentley, "Studies among the Snow Crystals during the Winter of 1901–2, with Additional Data Collected during Previous Winters," *Monthly Weather Review* 30, no. 13 (1903): 607–16, https://doi.org/10.1175/1520-0493-30.13.607.

true even if you measure the vectors' components in different coordinate frames, such as those shown in figure 9.1. That's because the scalar product itself is invariant under these coordinate transformations. You may remember from school that the equation $\boldsymbol{a} \cdot \boldsymbol{b} = 0$ can be interpreted as saying the vectors $\boldsymbol{a}$ and $\boldsymbol{b}$ are perpendicular to each other. Being perpendicular is a geometric property, and a vector is (part of) a quaternion—so Tait argued that whole vectors and quaternions give clear and immediate geometrical interpretations.

Cayley, on the other hand, was interested in using invariance to help solve equations in pure math. A simple example of an algebraic invariant is the discriminant, which, in the quadratic case we learn at school, is $b^2 - 4ac$. It's the expression under the square root sign in the formula for the solution of the general quadratic equation, $ax^2 + bx + c = 0$, and as you can see in the next endnote, it stays the same if you change the equation by changing the coordinates in certain ways—for example, by replacing x with $x' = x + h$. In other words, the discriminant is invariant under the coordinate transformation $x' = x + h$. (As I indicated earlier, a coordinate transformation is just a set of equations showing how to relate the coordinates in the original frame with those in the new one—in this case, when the axes are horizontally translated by a distance h.)[5]

What this all means for the final round of the vector wars is that while both men were exploring the idea of invariance, Cayley was interested in coordinates and coordinate transformations, whereas Tait was interested in whole quaternions and vectors. You can see why they remained "irreconcilable" on the best way of representing vectorial information!

Tait didn't know it, but the coordinate-free, invariant way of writing equations that he championed, such as $\boldsymbol{a} \cdot \boldsymbol{b} = 0$, is the key link between vector analysis and tensor analysis. Cayley, however, responded just like William Thomson, saying you still need coordinates to do the calculations. So, he proclaimed—in his 1894 address to the Royal Society of Edinburgh—that just as the full moon is more beautiful than a dimmer moonlit view, "so I regard the notion of a quaternion as far more beautiful than any of its applications." To emphasize the point, he added that a quaternion formula was like a pocket map, incredibly useful once you unfold it—once you translate the formula into its coordinate-based components. When Chisholm had first met Cayley just a couple of years earlier, she felt that "The fire was gone that they say had once gleamed from the eyes of the great mathematician." But there was still fire enough in Cayley's engagement with the vector wars.[6]

Peter Guthrie Tait in his study, demonstrating the physics of electricity; used with generous permission from the James Clerk Maxwell Foundation, Edinburgh.

Not to be outdone via literary analogies, Tait responded that rather than an unfolded pocket map, coordinate-based geometry was like a steam hammer, requiring expert manipulation so it would be useful rather than destructive. Quaternions, on the other hand, were so general they were "like an elephant's trunk, ready at *any* moment for *anything*," large or small—picking up a breadcrumb or strangling a tiger, as he put it. We don't usually associate hyperbole with mathematicians—but Tait, with his hearty laugh and twinkling eyes, was a bit of a prankster.[7]

• • •

Sparring over the relative advantages of coordinate-free vector and quaternion equations versus coordinate-based component ones continued for the next decade. It was an English-speaking debate, partly because Grassmann's vectorial system hadn't really taken off at that time, while quaternions were still virtually unknown outside Britain—and the young vector analysis mavericks Gibbs and Heaviside were English-speakers, too. But historian Michael Crowe suggests another reason for these long-running debates over the best notation to use for vectorial quantities. British mathematicians were aware that Leibniz's symbolism for calculus was much better suited to calculations than was Newton's, and that by the early nineteenth century this notational advantage had helped continental mathematics to leap ahead of that in Britain. So, they did not want the same thing to happen with Hamilton's quaternions and vectors.[8]

Whatever the reason for it, I'm emphasizing this seemingly arcane dispute because it also shows how difficult it was for even the best mathematicians to appreciate the value of vectors in their whole, coordinate-free form, as opposed to their component forms. Yet it is mathematicians who will turn vectors into tensors (or rather, who will identify and generalize the invariance and algebraic structure underlying vectors). So, although the physicists were ahead of the game in the creation of vector analysis, it is mathematicians who will pave the way for Einstein's masterpiece of physics, the (tensor) theory of general relativity.

Meantime, the next step in the story of vectors begins with the younger Einstein and his math professor Hermann Minkowski, at the Federal Polytechnic School in Zurich. In 1900 Einstein completed his degree at the "Poly," as it was affectionately known—but he was the only one of his small class of graduates not to be offered a job there, just as Maxwell hadn't been offered a fellowship at Cambridge when he graduated. Maxwell had been deemed too careless, while young Einstein was far too sure of himself for his professors' liking. (He was also Jewish, so anti-Semitism may have played a role in his lack of employment: Einstein himself thought so.)[9] And so, in desperate financial straits and with a pregnant fiancée to care for, he took that legendary job at the Swiss patent office. It was during those years that he developed his special theory of relativity, in his spare time.

Before I talk more about this theory, and how it led to the next step in the story of vectors, let me acknowledge the ongoing controversy over Einstein's relationship with his first wife, who'd been his fellow student at the Poly.

REMEMBERING MILEVA MARIĆ

Much has been written about this tragic saga, in which the idealistic young students Einstein and Marić fell in love, secretly had (and gave up) a baby out of wedlock, married against parental disapproval, and finally separated, torn apart by Einstein's growing fame and workload—and the fact that Marić, having repeatedly failed her exams, lost her confidence and her academic dreams, and increasingly took on all the day-to-day responsibilities for their two other children. I won't detail the sad drama further—except to say that it was bigger than the two of them. Einstein famously behaved very badly during the breakdown of the relationship, but the seeds of that breakdown, sown when they were still students, also have to do with the patriarchal culture in which they were both trapped. This includes what I believe is the sexism of the examiners who failed Marić twice: she was the only girl in her class, frightened and pregnant at the time of her final attempt at the exams, yet she had excelled in her earlier school studies.

As for the popular belief that Marić "did Einstein's math for him," or coauthored his 1905 relativity paper, as far as I know there is no confirmed evidence for it—and she herself never claimed it. She was certainly one of the first to believe in Einstein, as he struggled against the "old philistines" who kept refusing him an academic job—and her extensive study at the Poly made her, at the very least, a worthy and important sounding board for him. But the surviving letters between them suggest she didn't have Einstein's driving, creative scientific curiosity. Still, I can't help thinking that if they'd been at university today, with our more open society and with contraception readily available, she would have gotten her diploma and the doctorate she was planning, and gone on to make her mark in science—and things between them might have turned out very differently.[10]

Mileva Marić-Einstein and Albert Einstein in Prague, 1912. ETH-Bibliothek Zürich, Bildarchiv. Photographer: Jan F. Langhans/Portr_03106. Public domain.

THE SPECIAL THEORY OF RELATIVITY

Maxwell's critics had complained he had no model for the hypothetical ether, the medium they assumed must surely be necessary for the transmission of light waves, just as sound waves need air. Instead, he'd focused on describing concrete, measurable electromagnetic effects—and it was an astute move, because in 1887, the famous Michelson-Morley experiment "failed" to detect the ether via a state-of-the-art interferometer. The concept for the experiment was actually Maxwell's, but it was Albert Michelson who designed the Nobel Prize–winning equipment needed to put it into practice. Michelson was the first American to win a Nobel, in 1907, and his experiment helped set the stage for the special theory of relativity. So, there's a nice symmetry in the fact that the 2017 Nobel Prize for physics went to the founders of the Laser Interferometer Gravitational-wave Observatory (LIGO), which detected gravitational waves in 2015, as predicted by the general theory of relativity.

The idea behind the 1887 experiment was that if the ether existed, when Earth moved through it there would be an "ether wind"—just as you feel a breeze on your face when riding a bike on a still day. And just as you move faster swimming downstream in a river than upstream or across it, Michelson and his collaborator Edward Morley expected the speed of light would be fastest in the downstream direction—that is, "with" the ether wind. Maxwell had proposed using the timing of eclipses of Jupiter's moons when the giant planet was seen from Earth at two nearly opposite positions in its orbit. Earth would then be moving toward Jupiter at one position and away from it at the other, so that the speed of the light from the Jovian system would be measured both "with" and "against" the ether wind.[11] But Michelson and Morley used interference patterns—the same kind of patterns Thomas Young had used to show the wave-like nature of light in the first place. Specifically, they sent a beam of light downstream and another beam across-stream—that is, parallel to the direction of Earth's motion and perpendicular to it; any difference in light speed would cause a phase difference between the two beams, which would show up in the interference pattern. But the researchers found no such difference. Which meant that the ether wind had no discernible effect on the speed of light. (Experiments have been ongoing ever since, to see if tiny speed differences can be found with better equipment—and some physicists speak of the possibility of a "quantum ether." But the old mechanical idea of ether is out.)

By 1895, both George FitzGerald and Hendrik Lorentz had independently and ingeniously "explained" Michelson and Morley's result by suggesting that objects—including measuring rulers—*physically shrank* when they were traveling parallel to the direction of the ether wind. Such a physical "length contraction" would shrink the *measurement* of light's downstream speed, masking the supposed fact that it was "actually" faster.

In 1904, Lorentz made a brilliant start at applying this conjecture to Maxwell's equations of electromagnetism, devising a set of coordinate transformations between the frame of a stationary observer and that of an observer moving with constant speed relative to the stationary one, like the swimmer in the river moving relative to an observer on the riverbank. These transformations were designed to show that the rulers in the moving frame shrank in just the right way to account for the Michelson-Morley result—and Henri Poincaré dubbed them the "Lorentz transformations" when he developed Lorentz's ideas more fully in 1905. (We'll see these transformations in fig. 9.3 below.) Years later, during a visit to the University of Leiden, Einstein met Lorentz and immediately fell under his spell. They developed a friendship, for Einstein thought Lorentz was both an extraordinarily lucid thinker—he'd won a Nobel Prize in 1902 for his work extending Maxwell's theory in light of the discovery of electrons—and a person of perfect character. "The greatest and noblest man of our times," Einstein called him, "a marvel of intelligence and exquisite tact."[12]

Einstein and Poincaré never warmed to each other, although they admired each other's work. Poincaré was a superb mathematician—and a groundbreaking scientific philosopher to boot. In 1905 he was a famous fifty-one-year-old professor of mathematical astronomy and celestial mechanics at the Sorbonne in Paris. His 1905 paper on the Lorentz transformations was, in fact, a detailed theory of relativity, in the "special" case where the relative motion between observers (and therefore between their coordinate frames!) is constant. At the same time, twenty-six-year-old patent officer Einstein independently completed his own "special" theory of relativity. Marić had been captivated when she read his final draft, telling her husband, "It's a very beautiful piece of work!" More than a century later, physicists still describe it as being beautiful: compared with the brilliant but convoluted complexities of Poincaré's and Lorentz's papers, Einstein's is simpler and more intuitive. It is also the only one that ditched the old notion of a stationary ethereal medium for transmitting light through "empty" space, so it's the only one that is fully relativistic, as we'll see over the next couple of pages.[13]

Since Poincaré accepted the idea that the ether existed—and that the Lorentz transformations provided the necessary "length contraction" to support the ether hypothesis—he'd taken this as his starting point. Einstein, on the other hand, began his analysis of relative motion by invoking two simple principles. First, the relativity principle, which says that if a stationary observer deduces a particular law of physics, then another observer, moving relative to the first one with constant speed, should deduce the same law. Otherwise, there wouldn't be much point to physics, if the laws changed every time you changed your speed. Poincaré understood this, too—at least in the Lorentzian sense that Earth's motion through the ether had no effect on light, and therefore no effect on Maxwell's equations. What this means, as Lorentz, Poincaré, and Einstein all showed, is that the Lorentz transformations are, in fact, just the right coordinate transformations so that Maxwell's equations have the *same form*, regardless of how the components of the electromagnetic field vectors are measured—from a fixed coordinate frame, or one that is moving relative to it with constant speed.

A simpler example of what "keeping the same form" means is Newton's second law. Under the simple, so-called Galilean coordinate transformation shown in figure 9.3, the stationary observer deduces the horizontal

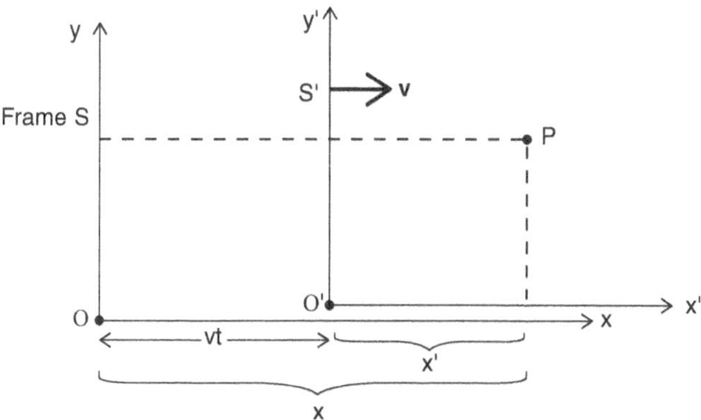

FIGURE 9.3. Two frames S and S' are moving relative to each other in such a way that S' is moving to the right with constant speed v relative to S. So, let's take S to be you standing on the sidewalk and S' to be someone driving by.

If the axes initially coincided—if you and the car were level with each other at time $t = 0$—then after a time t the S' frame (the car) will have moved vt units to the right. In this case the relative motion is horizontal, so for any point P the y and y' coordinates (and in 3-D the z and z' coordinates too) will be the same; during the time t, however, S' has moved closer to P, so a measurement made *at that moment* will give P's horizontal coordinate as x when measured in the S frame and x' $(= x - vt)$ when measured with respect to the moving S' frame.

This is the Newtonian/Galilean perspective, where time ticks away at the same rate for both observers. Even so, you can see that measurements of distance must be different in the different, relatively moving frames, and therefore speeds must be measured differently, too. You can see the mathematical consequences of this, from both the Newtonian and Einsteinian perspectives, in the related (and entirely optional!) box.

COORDINATE TRANSFORMATIONS AND INVARIANCE

CALCULATIONS FOR FIGURE 9.3

The coordinate transformations for the horizontal Galilean translation in figure 9.3 are:

$$x' = x - vt, y' = y, z' = z, t' = t. \tag{1}$$

It works well when v is much less than the speed of light. In such "everyday" situations, using (1) you find that Newton's laws keep their form, regardless of whether they are measured in S or S': for instance, if you deduce $F = m\ddot{x}$, then the person in the moving car deduces $F' = m\ddot{x}'$. That's because $x' = x - vt$, and when you differentiate this twice with respect to t' $(= t)$, you get $\ddot{x} - \dot{v}$ (using Newton's dot notation for the time derivatives); but since the speed is constant, the second term is zero and you have $\ddot{x}' = \ddot{x}$. So Newton's second law applied to this component of force has the *same form* in both frames: $F = m\ddot{x} = m\ddot{x}' = F'$. In fact, the force components have the *same value* in each frame, so they are *invariant* under Galilean transformations; and since there's no relative motion in the y and z directions, both observers will measure the same force components in those directions, too.

Although you both deduce *the same values of* F *and* $\ddot{x}$, you *disagree* on the horizontal speed component u of a moving object such as a ball being thrown, because in S, $u = \dot{x}$, and in S', $u' = \dot{x}' = \dot{x} - v$. So, the force and acceleration are invariant under Galilean transformations, but the speed is not.

Maxwell's equations do not keep their form under the coordinate transformations (1). Instead, the *Lorentz transformations* are the right ones for the laws of electromagnetism. (In this case, the individual components and indeed the electric and magnetic field vectors are not invariant: e.g., if a charge is at rest in one frame, a relatively moving observer will see a magnetic field, but an observer in the charge's frame will not. But the Maxwell equations linking the vectors *do* have the same form in each frame.) These transformations show that not just spatial measurements but also time measurements are relative—so you can no longer assume that $t' = t$. The Lorentz transformations for the set-up in the diagram are:

$$x' = \beta(x - vt), y' = y, z' = z, t' = \beta(t - \frac{vx}{c^2}), \qquad (2)$$

where c is the speed of light (measurement units are often chosen so that $c = 1$), and

$$\beta = 1/\sqrt{1 - v^2/c^2}. \qquad (3)$$

These equations represent a so-called boost (of S' relative to S) in the x-direction, but the full Lorentz transformations can describe boosts in any direction, and rotations, too. And note that if c is infinite as implied in action-at-a-distance, then equations (2) are just equations (1).

If you were to measure the *length* of a moving car, at a given time you would have to measure simultaneously both its endpoints x_a and x_b, say, and compute $x_b - x_a$; using the Lorentz transformations (2) to compare your result with the actual ("rest") length of the car in its own frame, $x_b' - x_a'$, you get

$$x_b' - x_a' = \beta(x_b - x_a)$$

(because $t_b - t_a = 0$ for a simultaneous measurement). Since $\beta > 1$ (because the denominator is less than 1 for nonzero v), you see that the actual car length is greater than the one you measured. In other words, your measurement suggests the car had "shrunk." Unlike Lorentz's idea, there is no *physical*, molecular shrinking of the car itself, but a shrinking in *measurements* used in calculations. But these measurements do have real, testable consequences in the stationary observer's physics.

Similarly, the relativity of the time measurement is why Earth-based observers deduce that time slows down for fast-moving airplane or spaceship travelers—or GPS satellites—just as distances shrink. (Both special and general relativity are needed to account for time in GPS measurements, as we'll see.)

As Einstein realized, however, you can also interpret figure 9.3 as saying that S is moving relative to S'—so it is moving to the left, and its speed is $-v$. So, the Lorentz transformations from this point of view are:

$$x=\beta(x'+vt'), y=y', z=z', t=\beta(t'+\frac{vx'}{c^2}). \quad (4)$$

And this time, it is measured lengths in *S*'s frame that contract.

GROUPS

Groups are important tools for studying symmetries, such as invariance. In this case, the "elements" of the group are coordinate transformations, and they form a group if they all obey a simple set of rules, which I'll illustrate for the Lorentz transformations (although we won't need these details in this story).

The fact that the Lorentz transformations (2) have an "inverse," (4), and also an "identity" element (that is, an element that is unchanged by the transformation—in this case, the transformation when $v = 0$), is key to knowing that they form a mathematical "group." Closure (the idea that all possible transformations in the group are of the same type) and associativity (of the group "product"—in this case, the composition of two transformations) are the other key features.

force component to be $F = m\ddot{x}$, and the relatively moving observer deduces $F' = m\ddot{x}'$. The equation has the same form in each frame, and both observers agree that force equals mass times acceleration. You can see the calculations for this in the box above. (The box also shows calculations with the Lorentz transformations, showing the math of "length contraction" in special relativity. We don't really need these calculations for our story—we just need the *ideas* and conclusions—so feel free to skim or skip them.)

Similarly, if you deduce $\nabla \times \boldsymbol{E} = -\frac{\partial \boldsymbol{B}}{\partial t}$, one of Maxwell's equations that we saw earlier, then using the Lorentz transformations you'd see that a moving observer deduces an equation with exactly the same form. So, you'd both agree that the changing magnetic field on the right-hand side of the equation gives rise to the curl of the electromagnetic field on the left. The same goes for the other whole-vector Maxwell equations.

In other words, if an equation representing a law of physics has the same form even when its components are measured in different frames, then all observers will deduce the same physics (and the same math, as we saw with the invariant equation $\boldsymbol{a} \cdot \boldsymbol{b} = 0$). That's the principle of relativity in action.

It's also another example of invariance—in this case, the invariance of the *form* of the equations. (Such form-invariant equations are also called "covariant.")[14]

Einstein's *second* principle was that the speed of light in empty space is independent of the motion of the source (which is what physicists seemed to have found experimentally). Together these two principles imply that the vacuum speed of light, c, is a universal constant.

From these two principles alone, Einstein had *derived* the Lorentz transformations from scratch, in a completely general way. By contrast, Poincaré had *assumed* Lorentz's "length contraction" was a real physical effect, and then rederived the Lorentz transformations by showing that they left the form of Maxwell's equations invariant. (Woldemar Voigt found something similar back in 1887, so these ideas were "in the air." We'll hear more about Voigt in chap. 11.) But Einstein had no need of hypotheses about the ether and no need to suppose any objective *physical* shrinking of rulers took place—rather, you get different measurements from different, relatively moving frames of reference. I showed a simple example of this in the box of calculations for figure 9.3, but the key fact, as Einstein made clear, is that the effect is reciprocal: *each* observer can consider themselves at rest and the other one to be moving. In other words, each observer would measure the other's ruler contracting, because they're each moving relative to the other. Lorentz and Poincaré, by contrast, believed that only one observer was "really" moving (relative to the all-pervading ether)—the one whose ruler "really" shrinks. That's why Einstein's was the only fully relativistic theory.

FOUR-DIMENSIONAL SPACE-TIME NEEDS FOUR-DIMENSIONAL VECTOR ANALYSIS

The Lorentz transformations include time as well as the three spatial directions (as you can see in equation (2) in the box above with fig. 9.3). So, when the term "relativity" is mentioned today, one of the first things many people think of is the four-dimensional nature of space-time. Yet the idea of a "fourth dimension" had been tantalizing the public since the 1880s. True, Hamilton had created a mathematical four-dimensional space when he defined quaternions in 1843—but as we saw with the vector wars, not even mathematicians could agree on its worth. So, while Tait, Cayley, Heaviside, and the others debated the academic merit of components versus whole vectors versus quaternions in the 1880s and 1890s, the possibility of a literal four-dimensional space was being tackled in popular books—such as mathematician Charles Howard Hinton's *Scientific Romances* and *A New Era of Thought*.

Hinton—who was George Boole's son-in-law—was also a spiritualist, and spiritualists loved the idea of a mysterious other dimension. These ideas had so penetrated popular culture that when Grace Chisholm stood for the oral exam related to her doctoral dissertation (which included n-dimensional spaces), one of her examiners mentioned the spirit idea, and asked what *she* meant by "higher dimensions." She replied that for her, it was simply a way of speaking about certain abstract relations in mathematics. She got her doctorate with highest honors, although obviously she'd fielded much harder questions than this![15]

As for Hinton, he hadn't lost his mind completely to the spirits, for he also focused on trying to visualize four-dimensional geometric objects, such as "hypercubes." You can think of a cube as a "hypersquare," a sequence of two-dimensional squares that forms a 3-D shape—that is, a cube—so a hypercube would be an arrangement of cubes in 4-D space. In 1954, Salvador Dali famously drew on this idea in his crucifixion painting "Corpus Hypercubus," in which the cross is made of cubes—but seventy years earlier, Hinton had inspired the schoolmaster Edwin Abbott to write his legendary 1884 *Flatland: A Romance of Many Dimensions*. Hinton also inspired his wife's sister Alicia Boole Stott, Boole's youngest daughter. Stott discovered several mind-bending results in four-dimensional geometry, including a collection of physical models she made of various sections of an imagined 4-D structure made of six *hundred* tetrahedrons, which together can be thought of as the three-dimensional surface of the four-dimensional analog of an icosahedron. The mind boggles at such visual geometric imagination.[16]

Then, in 1895—the same year that Lorentz found the Lorentz transformations to explain the Michelson-Morley experiment—H. G. Wells published his famous novel *The Time Machine*, in which he claimed that *time* was the fourth dimension. Brilliantly imaginative as Wells was, however, words alone were not enough to make this idea stick—to make it real. It took Lorentz, Poincaré, and Einstein to uncover the mathematical properties of such a four-dimensional construct—and it took Einstein to make it "real" by suggesting testable predictions to check his theory. (One of those predictions led him, completely unexpectedly, to deduce that $E = mc^2$.) So, when people hear the words "fourth dimension" today, they tend to think of Einstein, not Wells—and Einstein did it with the language of math.

As far as *vector* language is concerned, Poincaré worked entirely in components, but Lorentz also used whole-vector notation and vector calculus. In his special relativity paper Einstein, like Poincaré, wrote all his equations in terms of components, and although his coordinates (t, x, y, z) represented both time and space, he did not yet speak of space-time.

In principle Einstein could have used quaternions, which have four components, to represent quantities in his four-dimensional coordinate system.[17] He'd likely never heard of quaternions, though—his math professor, Minkowski, had said no one outside Britain used them—and perhaps he didn't even know much vector algebra, although Minkowski did teach that. As a student Einstein had figured that he needed a single focus if he was to succeed in unraveling nature's secrets, and this focus was firmly on physics. Besides, he felt there were so many topics to choose from in math that he was overwhelmed. So, although he admired Minkowski, he'd cut quite a few of his classes so he could study on his own the physics he wasn't being taught in lectures—including Maxwell's theory. Minkowski, on the other hand, thought Einstein was simply "a lazy dog who never bothered about mathematics at all." By contrast, when Minkowski himself had been a seventeen-year-old student, he'd won a prestigious award for his mathematical work—and had secretly given the prize money to an impoverished classmate.[18]

Minkowski certainly ate his words in 1905: "Oh that Einstein, always missing lectures—I really would not have believed him capable of it!" But for all its conceptual elegance, compared with Poincaré's paper young Einstein's *was* a little rough around the edges, mathematically speaking. In fact, Minkowski, now professor of pure mathematics at Germany's prestigious Göttingen University, told his students, "Einstein's presentation of his deep theory is mathematically awkward—I can say that because he got his mathematical education in Zurich from me."[19]

For instance, Poincaré was fluent in the mathematical language of invariance and group theory, and used it to show something surprising: the four-dimensional expression

$$x^2 + y^2 + z^2 - (ct)^2$$

doesn't change if you transform the coordinates via Lorentz transformations. In other words, it is *invariant* under this "group" of transformations. (Actually, Poincaré used $c = 1$ here, and today it is common to choose units so that $c = 1$ in the Lorentz transformations and other equations of relativity.) Einstein found the same result, although he didn't express it in such a sophisticated way. Formal "group theory" was pioneered by Évariste Galois—the impetuous prorevolutionary agitator and foolhardy lover who famously died in a duel in 1832, when he was just twenty years old. Cayley was one of the many others who made important early contributions to group theory, whose key features I listed at the end of the boxed caption to figure 9.3. But it was Minkowski who really made sense of this unusual expression.

It was unusual because, as Poincaré had pointed out, physicists were used to quadratic expressions where all the terms were added. For example, the length of a position vector is found from its components via Pythagoras's theorem (as you can see by extending fig. 0.2 to 3-D and representing the components simply in terms of the coordinates):

$$\boldsymbol{a} = x\boldsymbol{i} + y\boldsymbol{j} + z\boldsymbol{k} \Rightarrow |\boldsymbol{a}| = \sqrt{x^2 + y^2 + z^2}.$$

Hermann Minkowski, ca. 1896 (when Einstein was his student). ETH-Bibliothek Zürich, Bildarchiv/Photographer unknown/Portr_02711. Public domain.

This is the formula for measuring lengths and distances in flat, 3-D Euclidean space, and it is invariant under coordinate transformations such as the rotations in figure 9.1. Minkowski realized that the analogous concept in special relativity is the expression:

$$\sqrt{x^2 + y^2 + z^2 - (ct)^2},$$

which Poincaré and Einstein had shown is invariant under Lorentz transformations. This expression looks rather like the Euclidean distance formula, but it is not the measure of distance in ordinary space. Rather, it is the "distance" in what Minkowski called "space-time." In other words, it is the interval between "events" in space-time, rather than the distance between points in space. You can see what it means physically in the endnote, but here I'm focusing on its mathematical analogies with the Euclidean distance formula.[20]

The space-time used in special relativity is now called "Minkowski space-time"; it is a four-dimensional extension of flat Euclidean space, so it is also called "flat" space-time, as opposed to the curved space-times of general relativity. A distance or interval measure is called a "metric," and a measure with the form $\sqrt{x^2 + y^2 + z^2 - (ct)^2}$ is called the Minkowski metric in his honor. (We'll see a more precise definition later.) The concept of "world lines" is due to Minkowski, too. In Euclidean geometry, objects are located at a *point* in space; but even if the object is stationary in space, it is moving in time, so its location in space-time is represented by a *line* through the point—a line that is parallel to the time axis, and which gets longer as each second ticks by.

Minkowski first put forward his new concept of space-time in a lecture he gave in November 1907. A year later, he developed this more fully in his famous talk, "Space and Time," which opened with the memorable proclamation, "Henceforth space by itself, and time by itself, are doomed to fade away into mere shadows, and only a kind of union of the two will preserve an independent reality." He gave Einstein the credit for recognizing the true, two-way principle of relativity, politely dismissing Lorentz's "fantastical" hypothesis of a literal, physical contraction of moving objects. The confident tone of this address—on paper, at least—belies the fact that the gentle Minkowski used to turn deep red and stammer in front of an audience.[21]

The key thing for our story, though, is that Minkowski then made a start at developing *four-dimensional vector analysis*. At one point he even tried quaternions, because you can divide with quaternions but not vectors, as we saw earlier. In the end, he found the ordinary, Heaviside-Gibbs–style vector analysis more flexible.

In ordinary vector analysis, which takes place in the flat Euclidean space used in diagrams such as figure 9.1, the length of a vector can also be written in terms of its scalar product with itself:

$$\boldsymbol{a} = x\boldsymbol{i} + y\boldsymbol{j} + z\boldsymbol{k} \Rightarrow |\boldsymbol{a}| = \sqrt{\boldsymbol{a} \cdot \boldsymbol{a}} = \sqrt{x^2 + y^2 + z^2}.$$

In (flat) space-time, Minkowski defined the scalar product by analogy, using the interval metric: if a 4-D vector $\boldsymbol{a}$ has components (x, y, z, t), then the scalar product is

$$\boldsymbol{a} \cdot \boldsymbol{a} = x^2 + y^2 + z^2 - (ct)^2.$$

More generally, and using modern notation (and choosing units so that $c = 1$), while the scalar (or dot) product of two 3-D vectors is

$$\boldsymbol{a} \cdot \boldsymbol{b} = a_1 b_1 + a_2 b_2 + a_3 b_3,$$

the 4-D dot product in Minkowski space-time is

$$\boldsymbol{a} \cdot \boldsymbol{b} = a_1 b_1 + a_2 b_2 + a_3 b_3 - a_4 b_4.$$

(Today, it's also common to denote the time component with a suffix 0 instead of 4.) Later mathematicians will generalize this to any kind of space—curved or flat, 3-D, 4-D, or n-D—defining the scalar product via the metric, in a beautiful example of the power of mathematical analogies.

In another paper, Minkowski came close not just to 4-D vector analysis; he dipped his toe into tensor analysis, too. Tragically, he never got the chance to develop his work fully, or to know that today his name lives on in the "Minkowski metric." Not long after he gave space-time to the world, he died suddenly, after surgery for appendicitis. He was only forty-four. When his old college friend and Göttingen colleague, David Hilbert, stood in front of his students to tell them the sad news, he wept.[22]

• • •

At first, Einstein wasn't impressed with what his former professor had done to his theory—at that stage he preferred coordinates and components to whole, coordinate-free vectors. So, it was Minkowski's friend Arnold Sommerfeld who took up his ingenious space-time invention and began to explore 4-D analogs of scalar and vector products, and of the vector calculus operations of divergence, curl, and grad. Sommerfeld, who was then a professor of theoretical physics at the University of Munich, opened his 1910 paper, "On the Theory of Relativity I: Four-Dimensional Vector Algebra," with a tribute to Minkowski, "the friend who suddenly passed away."

Sommerfeld had been a member of Germany's "Vector Commission," which was established in 1903. In the wake of Britain's vector wars, the commission's goal was to settle on a standard vector notation—although its founder, Göttingen math professor Felix Klein, thought that all it achieved was a proliferation of notations! Klein was a fan of Heaviside (and Maxwell), but he had also been impressed with Grassmann's *Ausdehnungslehre*. He'd heard about it soon after Grassmann's son Justus enrolled as a math student at Göttingen in 1869—Justus had proudly brought along copies of his father's book, earmarked for two professors who had expressed interest in Grassmann's work. One of these professors was Alfred Clebsch, whose enthusiasm for Grassmann also inspired his colleague Klein. Two decades later, Klein—along with Gibbs—was instrumental in the publication of Grassmann's collected works.[23] At the same time, Sommerfeld had become first Klein's assistant at Göttingen and then his collaborator. He moved to Munich in 1906 and had taken with him the interest in vectors—and invariants—that Klein had nurtured.

In his two 1910 papers on relativity, Sommerfeld introduced the term "four-vector" for 4-D vectors in space-time. He emphasized the significance of the invariance of the space-time interval, for it expresses the constancy of the speed of light (as you can see in the endnote).[24] Poincaré, Lorentz, and Einstein knew this, too, of course. But Sommerfeld also pointed out the importance of invariant symbolism in adapting Maxwell's equations to space-time, noting "the complicated calculations" that Lorentz and Einstein had used in order to show that the ordinary component form of Maxwell's equations stayed the same under these transformations. Instead, Sommerfeld aimed to build on Minkowski's work, showing how to write these equations in invariant four-dimensional form.

He didn't find a 4-D analog of the 3-D "Heaviside" vector form of the equations shown in chapter 8; rather, he showed that what we now call "tensors" are needed to express the form-invariance of the 4-D Maxwell equations.

YOU CAN SAY EVEN MORE WITH TENSORS!

Tensors were a new concept then, whose intriguing origin we'll explore soon—but to get the flavor, this is how Sommerfeld explained them. With vectors you are exploring the geometry of *lines* (or arrows); for example, in three dimensions, perpendicular lines are characterized by the vector equation $\boldsymbol{a} \cdot \boldsymbol{b} = 0$, and parallel lines by $\boldsymbol{a} \times \boldsymbol{b} = 0$ (as you can see from the geometric definitions of the dot and cross products given in the caption to fig. 9.1). But sometimes you also need to describe *planes*, and they have an additional feature: *orientation* in space. This is defined by the direction of the plane's "normal," which is represented by a unit vector perpendicular (or normal) to the plane. But to get a clearer idea of what planes have to do with tensors, we can go back to a surprising source: Maxwell's 1873 *Treatise*.

Imagine a box submerged in water—an idealized version of the hull of a ship, perhaps. It keeps its shape under the pressure of the water because of the balance of forces acting on each of its faces. All sorts of rigid bodies, from bridges and airplanes to tiny crystals, balance these "stress" forces. ("Stress" is the force per unit area.) There are also many situations where the forces may not balance—in deformable or elastic materials, of course, but engineers also need to account for potential additional stress impacts, such as heavy traffic on a bridge or heavy seas battering a ship. Then there are situations where you want to move or rotate the body, so again you want a net *imbalance* of forces. Maxwell's friend Thomson was one of the pioneers in the mathematical study of these situations, which he analyzed back in 1856 in a paper on elasticity. Even earlier, Augustin Cauchy had published his foundational 1828 paper on stress in bodies in equilibrium.[25] Maxwell, however, wanted to analyze the forces on a magnetized object immersed in an electromagnetic field. By analogy with the immersed box, he considered the forces acting on each face of a little cube of the magnetized body. (In other words, he considered a "volume element" that can be integrated through the whole body.) Thomson and Cauchy had used a similar approach. But Maxwell did something special, not only by extending these analyses to electromagnetism but also by representing stress with a symbol with *two indices*.

In math an "index" refers in this context to a label, not a power. For example, the components of vectors are generally written with one index, as in $\boldsymbol{a} = (a_1, a_2, a_3)$; the indices refer to the three axes from which the

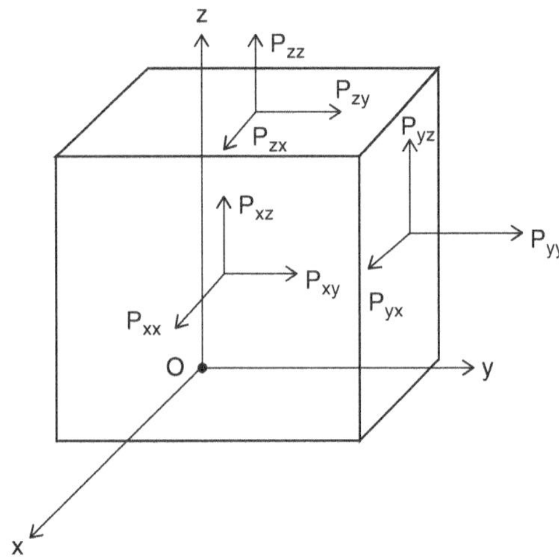

FIGURE 9.4. The fundamental stress components acting on the faces of a cube. Maxwell noted that if $P_{hk} = P_{kh}$, the stresses won't produce a rotation (he was interested in the rotation produced by magnetism—as expressed in his curl equation). Indeed, the diagram suggests that if, say, $P_{yx} > P_{xy}$, the right-hand face will be pulled forward, and the cube will begin to rotate.

components are measured. Maxwell recognized that stress was an example of "physical quantities of another kind which are related to directions in space, but which are not vectors." He said that's because in 3-D space a vector has three components, but a stress needs nine of them (as you can see in fig. 9.4). So, he wrote the components of stress as P_{hk}, explaining that the first label, *h*, indicates the surface on which the stress is acting—it is the one whose normal is parallel to the *h*-axis—and the second label shows the direction of the force producing the stress. There are three choices for *h*—one for each dimension in space, that is, one for each coordinate axis—and three choices for *k*. That's nine possible combinations, one for each of the nine components.[26]

What Maxwell had done here was to give a concise definition of the components of a single quantity he referred to simply as "stress," but which is now called the "stress tensor"—just as the components (a_1, a_2, a_3) form a single *vector*, $\mathbf{a}$. The fascinating thing about this is that tensors hadn't yet been invented—at least, not as mathematical objects on a par with vectors.

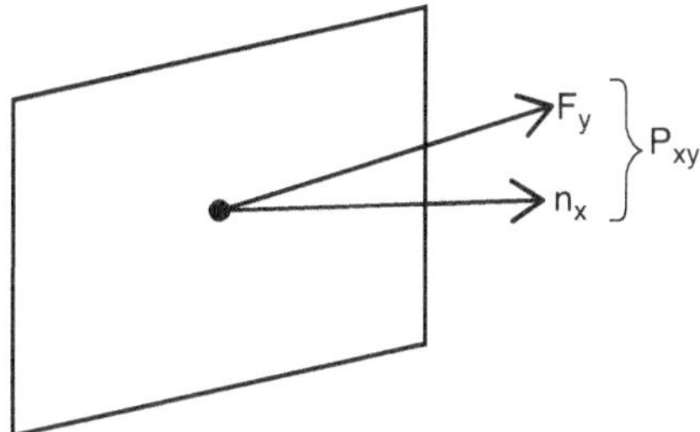

FIGURE 9.5. You need two vectors to make a stress tensor. First, the vector ***n*** giving the direction of the surface on which the stress is acting—I've labeled it with an x to indicate that here the relevant component of the normal is parallel to the x-axis. Second, the vector giving the strength and direction of the force: the label y indicates that the relevant component for the force acting on this face is in the direction parallel to the y-axis.

Yet Cauchy and Thomson had come close to the idea, too, although they didn't have Maxwell's concise and general notation.

From a modern point of view, the key idea here is that you need two vectors to form a stress tensor: one to represent the direction of the surface on which the stress is acting, and one to represent the force itself. You can see this in figure 9.5, which is another way of expressing Maxwell's definition of his components P_{hk}. When the mathematicians enter the field, they'll turn the essence of this concept into a much more precise and more general definition of a tensor. But already you can see that with nine components, a tensor such as stress can store much more information than a 3-D vector can.

By 1910 when Sommerfeld was writing, tensors hadn't yet found their way into the mainstream, but he understood that they are far more versatile than just ways of representing stresses. And they don't necessarily have anything to do with planes, for they can be adapted to any number of dimensions and applications. Sommerfeld referred in passing to Grassmann, but his focus was on his friend Minkowski, and on rewriting Maxwell's equations in the language of space-time. To do this, he followed Minkowski in rewriting the components of ***E*** and ***B*** as two-index quantities—giving essentially the modern tensor form of Maxwell's equations.[27]

We'll see these beautiful equations later. First, though, we need to delve further into the evolution of tensors. In particular, we'll find out how they encode and extend the idea of invariance—and why Einstein needed them for his masterpiece. But we'll also go back in time a little, to find out what non-Euclidean geometry has to do with our story. For the road to tensors is a long one—and, as we've seen with vectors, too, it was built with creative insights from many surprising directions.

(10)

CURVING SPACES AND INVARIANT DISTANCES

On the Way to Tensors

When Einstein began wrestling with the problem of how to extend the special theory of relativity to a general one, where the relative motion didn't have to be constant, he needed a whole new mathematical toolkit. And who better to call on than his old friend from the Swiss "Poly," Marcel Grossmann. Grossmann had never cut Minkowski's math classes! He was also a first-rate mathematician and a loyal friend. In fact, it was through Grossmann's family that the impoverished Einstein had landed that life-saving job at the patent office. Grossmann himself had been given a place at the Poly—as a PhD student and an assistant to his supervisor—as soon as he graduated, and seven years later he'd become professor of mathematics there.

Einstein had tried for years to find someone to take him on as a doctoral student, but all he'd got for his trouble were rejections—partly because of his sassy reputation among the "old philistines" of mainstream academia. Still, he persevered, and in the summer of 1905, he'd finally had some luck with a landmark paper on measuring molecules: it earned him

Marcel Grossmann, 1909. ETH-Bibliothek Zürich, Bildarchiv/Photographer unknown/Portr_01239. Public domain.

a doctorate from the University of Zurich. He dedicated it to Grossmann, in appreciation of his generosity in their student days—from lending him notes from lectures he'd skipped to helping him find a job. Unlike his professors, Grossmann had seen from the outset that Einstein was destined for greatness.[1]

In 1911, Grossmann began seeking ways to bring his now-famous friend back to their old school—which had recently been renamed the Swiss Federal Institute of Technology (still known as ETH for its German acronym). Marie Curie and Henri Poincaré were among those who wrote glowing and perceptive testimonials for Einstein, and in 1912 he was appointed to the newly established chair of theoretical physics at the ETH. Perhaps he savored such a glorious return. He certainly relished the chance to work once again with his old classmate, the way they used to study and explore ideas together as students, smoking their pipes and drinking coffee at the nearby Café Metropole.

For the next two years, the two friends collaborated closely on the herculean task of building mathematical foundations for the general theory of relativity. Einstein had already realized that in generalizing the special theory, he was working on nothing less than a whole new theory of gravity. After all, an example where the relative motion isn't constant is when one observer is falling because of gravity—in a free-falling elevator, say—and the other is fixed to the ground: the falling observer is accelerating relative to the ground-based one, so their relative speed isn't constant. In 1912, however, when Einstein and Grossmann began their collaboration, there were still two major mathematical problems to solve before Einstein could find his general theory.

First, how to transfer the laws of physics from the special theory to the general one; and second, how to find the appropriate "distance" or interval measure in curved space-time (we'll see *why* it's curved in chap. 12), analogous to the flat Minkowski metric we saw in the previous chapter. As Einstein later recalled, "We found that the mathematical methods for solving problem 1 lay ready in our hands in the absolute differential calculus of Ricci and Levi-Civita"—that is, in tensor calculus. "As for problem 2," he continued, the answer lay in Bernhard Riemann's work on curved surfaces.[2]

"Ready-made" these tools may have been, but first Einstein had to master them. As he told Arnold Sommerfeld, "In all my life I have never before labored [so] hard." He went on to say that with Grossmann's help, he had "become imbued with a great respect for mathematics, the subtle parts of which, in my innocence, I had till now regarded as pure luxury." Minkowski would have been thrilled.[3]

As we've seen, Sommerfeld had already dabbled with tensors as a way of writing Maxwell's equations in flat 4-D space-time—and Maxwell himself had used two-index tensorial quantities to describe stresses in ordinary 3-D space. What Gregorio Ricci and Tullio Levi-Civita did was to develop a calculus that could handle curved spaces, too—and it was Grossmann and Einstein who first applied tensor calculus to curved *space-time*. But they all built on Riemann's work—and he'd built on the ideas of his legendary teacher, Gauss.

THE MATH OF CURVED SURFACES: CARL FRIEDRICH GAUSS

Grossmann had done his PhD on non-Euclidean geometry, which had been pioneered by Gauss, Janos Bolyai, and Nicolai Lobachevsky in the 1820s. With the geometry of curved surfaces, Euclid's axiom about parallel lines never meeting no longer holds, and you can see this most readily with lines of longitude on the globe: they're parallel at the equator but they meet at the poles. So, can you ever speak definitively about "parallel" lines on a curved surface? This is a crucial question for vector analysis because the whole concept of addition of vectors is based on the parallelogram rule, where you slide vectors to the correct position by keeping them parallel (as in fig. 3.1).

The answer to this question came much later, as we'll see when we meet Levi-Civita. But he was indebted to Ricci, who, in turn, was indebted to Gauss's work on a related problem. In everyday life, we measure the distance between two points with a straight ruler laid along the straight line between the points, and Pythagoras's theorem gives the distance measure or "metric." But if there are no straight lines on a curved surface, what is the distance formula? The solution, Gauss decided, was to look at points that are very close together.

Gauss knew all about measuring distances firsthand. Ever since he was a teenager he'd been interested in geodesy—the mathematical analysis of the shape of Earth and the area of its surface—and in triangulation, the method used for surveying. Since then, he'd worked on various government and military surveying projects—war was still seemingly never-ending. In fact, it was because of the Napoleonic Wars, which had raged during the years 1803–15, that Sophie Germain had finally revealed her true identity to Gauss. As we saw in chapter 3, she'd sought mathematical advice by writing to him under the pseudonym Monsieur Le Blanc. But when French forces took a Prussian town not far from Gauss's home, she was so afraid for him that she persuaded a family friend—who was also a military general—to check on his safety. Gauss was most surprised by a courteous visit from an enemy soldier—and even more perplexed when the soldier said he'd been sent on behalf of Mademoiselle Germain!

From 1821 to 1825, Gauss, now in his forties, led a series of surveying expeditions mapping the entire kingdom of Hanover. These expeditions required not just mathematical know-how and expertise with instruments but also traveling through difficult terrain and living rough. Then, at night, Gauss would have to process by hand the data he and his team had collected—and to help himself he used the method of least squares that he'd already invented. This is a method for fitting a line to data points, or for finding the most accurate estimate from a bunch of repeated measurements—and among its many modern applications are the linear regression models I mentioned in connection with machine learning. Gauss was almost as reluctant to publish as that enigmatic Elizabethan Thomas Harriot had been two centuries earlier, for both men were perfectionists—and so the least squares method, and countless other of Gauss's discoveries, were reinvented later by others.[4]

Anyway, after all this hands-on experience Gauss was ready to present his seminal 1828 paper, *Disquisitiones Generales circa Superfices Curvas* (*General Investigations of Curved Surfaces*). In fact, it is Gauss who created the concept of the metric, the distance measure we met in the previous chapter, where I wrote the flat 3-D Euclidean metric as

$$\sqrt{x^2 + y^2 + z^2}.$$

I assumed that x, y, z were the components of the position vector giving the distance along the straight line from the origin to the point (x, y, z), but the distance between two arbitrary points (x_1, y_1, z_1) and (x_2, y_2, z_2) is given by

$$\sqrt{(x_2 - x_1)^2 + (y_2 - y_1)^2 + (z_2 - z_1)^2}.$$

Gauss's idea was that if two points on a *curved* surface are close enough together, so that $x_2 - x_1$ is very small, and similarly for the other coordinate differences, the little piece of surface between them is approximately flat. You can see this intuitively by imagining two nearby dots on an orange or a ball—and on a larger scale it's why we can speak of a patch of "flat" land on the curved Earth. If your two points are *infinitesimally* close together, the surface between them will be flat to all intents and purposes, and the line between them will be straight. In modern terms, the surface is "locally"

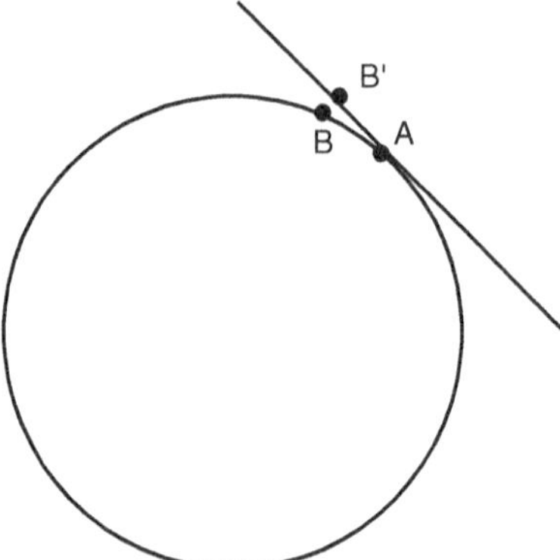

FIGURE 10.1. When two points on a curved line, or a curved surface, are very close together, the distance between them, AB, is approximately the straight-line distance AB'. In computer graphics, for example, curved lines can be built from tiny tangential segments.

flat. It's the same idea as when a small section of a curved line is approximated by a tangent line. You can use this straight tangent line to approximate the distance to a nearby point on the curve, as in figure 10.1—except that with the curved *surface*, you have a tangent *plane* rather than a tangent line.

What this means is that you can use the Euclidean distance measure on a curved surface, except that instead of $x_2 - x_1, y_2 - y_1, z_2 - z_1$, you need the kind of infinitesimal distances used in differential calculus: dx, dy, dz. In other words, instead of

$$\sqrt{x^2 + y^2 + z^2} \text{ or } \sqrt{(x_2 - x_1)^2 + (y_2 - y_1)^2 + (z_2 - z_1)^2},$$

Gauss showed that the general distance measure or metric is

$$ds = \sqrt{(dx)^2 + (dy)^2 + (dz)^2},$$

where the length of a line on the surface is denoted by s. Often this expression is squared, and, taking liberties with notation, the brackets are left out to make it easier to write—so the Euclidean metric for the surface is generally written as

$$ds^2 = dx^2 + dy^2 + dz^2.$$

(Writing eighty years after Gauss, Minkowski knew to use this kind of differential notation for his 4-D space-time metric:[5]

$$ds^2 = dx^2 + dy^2 + dz^2 - c^2\,dt^2.)$$

If all this isn't familiar to you, you might be thinking that it's all very well to talk about distance measures as the length of tiny straight lines, but what do you do when you want to measure longer curved distances? On the face of it, you'd have to lay down your infinitesimal rulers end-to-end along the curved line across the surface. Fortunately, the Leibnizian differential notation makes it beautifully clear how to answer this question much more simply: just integrate ds to find the distance s. I did this in figure 2.3b, using the 2-D version of this metric. If you *are* familiar with this, I hope you agree that it's interesting to look back and see where the math we learn today comes from—in this case, the mighty Gauss. He didn't use the term "metric," though; instead, he called this distance measure the "linear element" of the surface—and it's still often called the "line element."

The curved surface of a ball, say, is only two-dimensional, so you might also be wondering why the metric looks the same for flat 3-D space and a small patch on a 2-D curved surface. When we look at the ball's surface, we are looking at it from the outside, so we see the points on the surface as points in 3-D space. But as you can see in figure 10.2, and analogously in figure 10.1, the infinitesimal distance between two of those points, measured *on the surface*, is virtually the same as the straight-line distance measured in space.

So, to highlight the *intrinsic*, 2-D nature of the surface—the way that a super-intelligent ant or 2-D alien would see it—Gauss followed Leonhard Euler's lead by parameterizing surfaces in terms of two "curvilinear" coordinates, which he called p and q. On the surface of Earth, for example, the p-axis could be the line of latitude around the equator, and the q-axis the meridian of longitude through Greenwich. An intelligent ant-alien crawling over this surface would take measurements from these two axes—it wouldn't be aware there was a third dimension.

Gauss showed that when you do this, the metric on the 2-D surface becomes

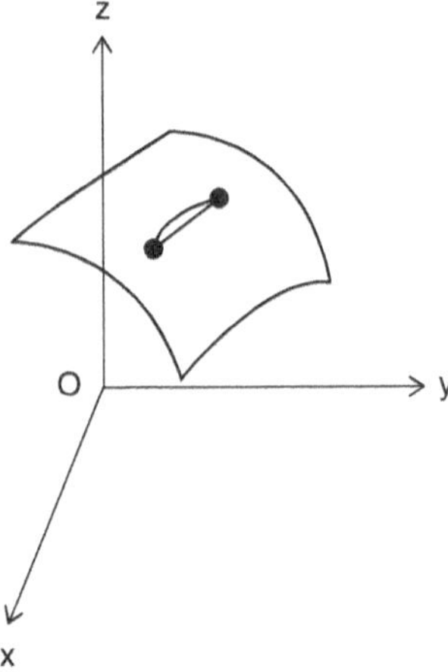

FIGURE 10.2. The curved distance between two points that are separated by an infinitesimal distance is almost the same as the straight-line distance between them, measured in 3-D space as if the surface weren't there.

$$dx^2 + dy^2 + dz^2 = Edp^2 + 2Fdpdq + Gdq^2,$$

where the E, F, G are functions of p and q. (You can see how he did it, in the next endnote.) In a virtuoso feat he also showed that the coefficients E, F, G in this expression, along with their derivatives, contain all that our 2-D creature would need to know to figure out the intrinsic geometry of the surface.

For example, the ant-alien would be able to tell if it were crawling over a flat space, where the angles in a triangle add up to 180°, a positively curved space such as a sphere, where they add up to more than 180°, or a negatively curved surface such as that of a saddle, where they add to less than 180° (as in fig. 10.3).

That's because Gauss defined the "intrinsic curvature" in terms of the *area* of the curved triangle and the *difference* between 180° and the triangle's angle sum. He didn't know it, but for the case of a sphere Thomas Harriot had already discovered this formula, during his own work on map-making and navigating the surface of the earth:

$$\frac{\alpha + \beta + \gamma - \pi}{\text{Area of triangle}} = \frac{1}{r^2},$$

where the triangle's angles are designated by α, β, γ radians, π radians equal 180°, r is the radius of the sphere, and $\frac{1}{r^2}$ is its "Gaussian" or intrinsic curvature. Because Harriot didn't publish his result, Albert Girard rediscovered and published it several decades later—but Gauss's version was more sophisticated, and not just restricted to spheres. What's more, Gauss showed that both the area and the angles in this formula could be found from the metric alone. This is quite astonishing at first sight. In fact, Gauss was so excited he called this discovery his *theorema egregium*, his "remarkable theorem."

So let me unpack this remarkable result a little more. We've already seen that the metric tells you the length of a line—such as the magnitude of a vector, as we saw in chapter 9, or the circumference of a circle as in figure 2.3. And long ago Archimedes had worked out the surface area of a sphere ($4\pi r^2$), and had even shown how to find the area of cylindrical segments of the sphere. But Gauss showed that the area of *any* curved surface could, in fact, be found from the metric coefficients, E, F, G, via the surface integral of the expression $\sqrt{EG - F^2}$. Without calculus and the concept of a metric, when Harriot derived the curved area of a triangle on a sphere—he was the first known person to do this, and he needed it to prove the intrinsic curvature formula above—he had to use many pages of ingenious geometrical arguments.

Mind you, it took Gauss many pages, too, to establish these fundamental relationships between curvature and the metric, but whereas Harriot had solved only the spherical triangle problem, Gauss showed the way for areas of any shape on any surface. As for the angles needed in Gauss's and Harriot's curvature formula, amazingly it turns out that E, F, G are *scalar products* of the unit vectors tangent to the coordinate lines p and q shown

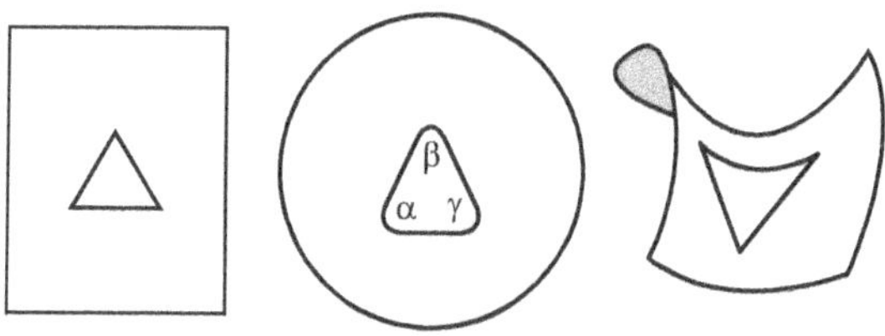

FIGURE 10.3. The angles α, β, γ in a triangle on a flat piece of paper add up to 180° or π radians, but on a sphere, they add to more than π, and on a saddle they add to less.

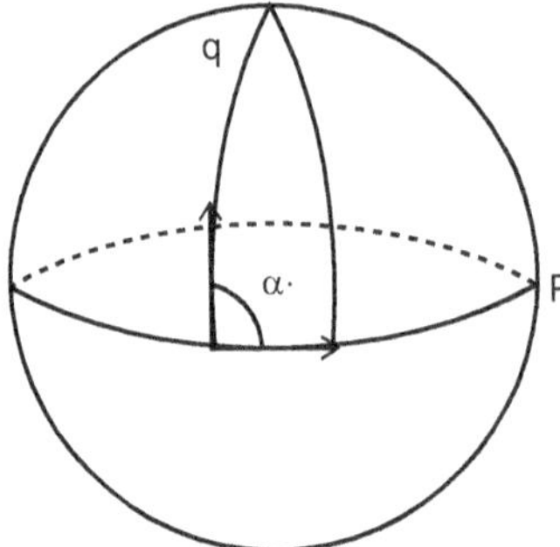

FIGURE 10.4 A spherical triangle bounded by three great circle coordinate lines. The angle between two such lines is the angle between their tangent vectors, as illustrated by the arrows. On a sphere all three angles are right angles.

in figure 10.4—and the geometric formula for scalar products gives the cosine of the angle between these vectors. Of course, in 1828 Gauss didn't know about vectors and scalar products—at least not in the formal sense—but he had the equivalent coordinate form, nonetheless. I've explained how he did it in the endnote.[6]

It's telling that the impetus for both Gauss and Harriot to make their breakthroughs on curvature was practical—mapmaking and navigation. So, all this isn't just for ants and aliens. We, too, can use the metric to determine the curvature of the surface we live on—we don't have to travel into space to see Earth from the outside. It really is remarkable, as if one tiny equation, $ds^2 = Edp^2 + 2Fdpdq + Gdq^2$, encodes an all-seeing god's-eye view of the whole surface. And as we'll see when we meet Gregorio Ricci, the coefficients of the differentials in a metric—such as the E, F, G here—turn out to be the components of a *tensor*, the next step on from a vector. In Einstein's hands, it will be a key to the whole cosmos.

FROM MAPMAKING TO BLACK HOLES: INVARIANCE, TOPOLOGY, AND "STRAIGHT" LINES

Einstein will also draw on something else that Gauss highlighted: the concept of invariance. In fact, invariance is critical to the economy and power of tensor equations, as we saw for vectors in the previous chapter. In particular, it's important to have an invariant distance measure, in the sense that the Euclidean distance, say, or the Minkowski interval between events in space-time, would remain the same if it were being measured from a different point of view—from a rotated frame, for example, or a Lorentz-transformed one. We saw examples of this earlier, and they illustrate the fact that the Euclidean and Minkowski metrics we met above are invariant under rotations and Lorentz transformations, respectively.

In 1828 Lorentz transformations were more than half a century into the future, but Gauss did show that his 2-D curvilinear metric—or what he called the "linear element" or "measure of curvature"—was invariant (or as he put it, "unchanged") when you bent the surface into a different shape. At least, this was true if you didn't tear or cut the surface: rolling up a piece of paper to form a cylinder, for example, or molding a soccer ball into a football. This latter kind of molding is what happens with the molten and fluid Earth, which bulges at the equator because of its rotation yet doesn't change its total volume.

The invariance of the metric under this kind of bending or squashing relates to the counterintuitive idea that the curved surface of a cylinder is intrinsically "flat." (It is only curved when viewed from the outside, so this kind of curvature is "extrinsic.") To see the invariance, imagine a flat sheet of paper with a straight line drawn on it; as far as our 2-D ant can tell, it is just the same line—with the same length, and therefore the same metric—as when the paper is rolled up into a cylinder. But a sphere is intrinsically curved, as you may have noticed after juicing a half-orange: you can't flatten out the hemispherical shell without tearing it. It's only flat locally, as in figures 10.1 and 10.2. So, we're talking not only coordinate transformations here but also topology—which deals with properties of surfaces and shapes that can be molded without tearing—although Gauss didn't use this term. We saw the idea of topology briefly in chapter 5, with the work of Gauss's students Möbius and Listing in the 1840s, which was two decades after Gauss's paper. Nonetheless, what Gauss had shown about the topological invariance of the metric explains why the bulging

Earth can be treated as a sphere when it comes to measuring distances and angles.

An even more remarkable connection between curvature and topology is expressed in what's called the global Gauss-Bonnet theorem. The "local" version of the theorem is just Gauss's definition of curvature in terms of the angles and area in a curved triangle on a patch of the surface—with Harriot's formula as a special case for spherical triangles—and, when needed, Pierre Ossian Bonnet's 1848 extension of Gauss's theorem to open surfaces such as disks. When you apply this not simply to a patch but to the whole ("global") surface, such as a whole sphere, topology comes into it. In 1972, Stephen Hawking used the global Gauss-Bonnet theorem and Einstein's equations to prove that the boundary, or event horizon, of a stationary black hole is topologically equivalent to a sphere. It's another example of the way little human "ants" can sit with their pens and paper and use the math of curvature to discover vast and mysterious new territory. The year before, Hawking had proved mathematically that the area of this boundary could never decrease—a result that wasn't confirmed experimentally until 2021. But it was Roger Penrose who, in 1965, had used topology to prove that if general relativity is correct, then black holes really should exist in nature—they were not just mathematical artifacts. And then, in 2019, thanks to the international collaboration behind the Event Horizon Telescope, we were all treated to that extraordinary first direct image of a black hole—or rather, its roughly (topologically!) circular shadow.[7]

Penrose shared the 2020 physics Nobel Prize after this spectacular confirmation of the existence of black holes. The other two recipients, astronomers Rienhard Genzel and Andrea Ghez, discovered the "supermassive compact object"—assumed to be a black hole—at the center of our galaxy. Ghez is only the fourth woman to have won a Nobel Prize for Physics. But let me come back to Gaussian curvature, for it isn't only useful in geodesy, surveying, and cosmology: today it has a wide range of down-to-earth applications, including cutting-edge materials science.[8]

• • •

Gauss had another brilliant insight that paved the way for the sophisticated math behind the headlines today. Just as Hamilton realized that quaternions formed an algebraic system in their own right, with different rules from ordinary algebra, Gauss realized that a curved 2-D surface was a "space" in its own right, just like the 3-D space that we live in. Such a space, with its two curvilinear coordinate axes and its own distance measure, has its own intrinsic 2-D geometry, and we've already seen that the metric is the key to this geometry, for it is needed to calculate distances, angles, and curvature. But there's even more to it than this. In ordinary (flat) space, we have Euclid's geometry. It is the geometry of straight lines, and the angles they make with each other, and there are countless theorems about these lines and angles, which have served us well for more than two thousand years. So, the question for Gauss was this: How do you find geometrical rules in a space that has no straight lines? His answer is ingenious. The thing about a straight line is that it is the shortest distance between two points in ordinary Euclidean space—so the question becomes, what is the shortest distance between two points on a curved surface?

On the surface of a sphere, like Earth, mathematical astronomers, mapmakers, and navigators had known for millennia about "great circles"—circles that have the same center and radius as the sphere, such as lines of longitude, and the equator. But it was Johann Bernoulli and his brother who'd figured out that the shortest distance between two points on this surface is along the great circle that runs through them. Even today, airplane pilots follow great circle lines where possible, to make their journeys more efficient. In the 1720s, Bernoulli's former student Euler invented what is known as "the calculus of variations" to find the equation of this shortest line; it's a sophisticated extension of school calculus, where you set derivatives of a function equal to zero to find its maxima and minima. A century later, Gauss showed how to apply the calculus of variations to the metric, to find the shortest distance between any two points on a 2-D curved surface.

This "shortest distance" lies on a line called a "geodesic"—a name that reminds us of non-Euclidean geometry's connection with ancient navigation and Earth's great circles. Einstein will make stunning use of geodesics when he rewrites the laws of motion for curved spaces. But as he recalled later, first he and Grossmann needed to understand the work of Gauss's remarkable student Bernhard Riemann, for he's the one who took Gauss's analysis into higher dimensions.

BERNHARD RIEMANN TAKES UP GAUSS'S BATON

In the late 1840s and early 1850s, Riemann studied at the University of Göttingen where Gauss was a professor, but it wasn't the vibrant center of intellectual activity it would become later in the century. The professors were formal and remote, and their lectures were old-fashioned—even Gauss taught only elementary classes. So, Riemann transferred to the University of Berlin for a couple of years—the professors there were also excellent mathematicians, and they gave much more cutting-edge lectures. Still, he chose to do his PhD with Gauss back at Göttingen.

Three years later, Riemann set himself quite a challenge in a paper first translated into English by William Kingdon Clifford—Maxwell's brilliant young colleague and George Eliot's friend, who developed a synthesis of Hamilton's and Grassmann's vectorial ideas as we saw in chapter 7. As I mentioned in the prologue, if you generalize x, y, z, the familiar Cartesian coordinates, and write x_1, x_2, x_3, then it's easy to imagine as many dimensions as you like, with coordinate axes $x_1, x_2, x_3, \ldots, x_n$. Vector components—of velocity, say—in such an n-D space would be denoted by, say, $v_1, v_2, v_3, \ldots, v_n$. Riemann's challenge was to adapt Gauss's work on curved 2-D space and find the rules of geometry for curved n-dimensional space.

First, Riemann had to define a way of measuring distances on an n-dimensional curved surface. He called this surface, this curved space, a "manifold." (It's a topological concept, for Riemann followed Gauss in exploring *intrinsic* curvature.)

Before we see how he did it, I should mention that he wasn't the only one investigating n-dimensional spaces in the 1850s. Arthur Cayley, for example, wasn't just the inventor of matrix theory, and a spirited opponent of vector analysis; he was also a pioneer in both invariant theory *and* n-D geometry—specifically, the geometry of a curved surface when it is projected onto flat Euclidean space, like a shadow on the ground.

Speaking of Cayley, twenty years later, in 1874, he was to be honored with a portrait; it still hangs in the Trinity College dining room, next to Maxwell's portrait and across from Newton's (Maxwell had returned to Cambridge in 1871, as the first director of the university's first science lab, the Cavendish). The occasion inspired another of Maxwell's famous poems, this time addressed to the Cayley portrait fund committee. For those who are "to space confined," he began, what honor could you pay to one whose mind has penetrated beyond these bounds? Then, after poetically listing Cayley's achievements, he hoped viewers would pause before the portrait's two-dimensional form and reflect on the man whose "soul, too large for vulgar space, in *n*-dimensions flourished unrestricted."[9] It's a lovely evocation of the sense of freedom mathematicians can feel when they imagine spaces untethered to our ordinary world.

Cayley was ultimately working in Euclidean geometry, though, whereas Riemann was looking at the intrinsic geometry of curved spaces, or manifolds. In modern terminology, a manifold looks "locally" like a flat *n*-D Euclidean space—so it is an extension of Gauss's idea that curved surfaces look flat if you zoom in close enough, and that flat spaces can be handled with simple generalizations of school geometry, especially Pythagoras's theorem. So, Riemann defined the distance measure—the metric, or what he called the "line element"—of a flat *n*-dimensional manifold by analogy with the Euclidean metric:

$$ds^2 = dx_1^{\,2} + dx_2^{\,2} + \ldots + dx_n^{\,2}.$$

In fact, it is Riemann who first used the term "flat" for a surface whose line element is the sum of squares of differentials like this.

On a flat piece of paper, the Euclidean metric $ds^2 = dx^2 + dy^2$ holds regardless of the size of the sheet—and in ordinary Euclidean space, $ds^2 = dx^2 + dy^2 + dz^2$ holds everywhere. On a curved manifold, however, Riemann followed Gauss, noting that you could only write the metric as a sum of differentials like this if you focused your attention locally, in the flat "neighborhood" around a *point*. Information about the intrinsic curvature of the *surface* is contained in the *coefficients* of the differentials in the surface's *intrinsic* metric, as we saw earlier with Gauss's 2-D metric,

$$ds^2 = Edp^2 + 2Fdpdq + Gdq^2.$$

In general, though, the coefficients in a metric depend not just on the curvature but also on the choice of coordinates: the same metric will look different when expressed in different coordinates, just as the equation of a circle looks different in Cartesian coordinates than it does in polar ones: $x^2 + y^2 = a^2$ and $r = a$, respectively, for a circle with radius a and centered at the origin. So, the simple fact of having coefficients in the metric is *not* enough to tell you if the surface is curved. What Riemann hinted at, however, was that there *is* a way to decipher the manifold's curvature from these metric coefficients. We'll see what he meant later, for he didn't go into detail in this paper, which was designed for a general audience.

He presented it when he was applying to become a *Privatdozent* at Göttingen, in 1854. *Privatdozents* were lecturers who were paid by their students, not by the university—so it was pretty tough if you didn't get many students. But it was the first step on the academic ladder—and part of the grueling application process included a "habilitation" thesis and lecture. Einstein's first academic job, in 1908, was as a *Privatdozent* at Bern University; he'd failed the year before with an application offering his 1905 special relativity paper as a habilitation thesis: it was rejected as "incomprehensible"! Which shows just how radical his theory was at the time. Riemann's lecture, too, was beyond most of his listeners' comprehension, although seventy-seven-year-old Gauss, who was in the audience, certainly appreciated its significance. Gauss was a good person to have on your side, and he thought Riemann was a true mathematician "of a gloriously fertile originality." High praise indeed, for Gauss wasn't one to hand out compliments, as Grassmann and many others had found. Perhaps Gauss's difficult temperament had something to do with the death in childbirth of his beloved first wife, for it seems he never recovered from his grief.[10]

Three years later, Riemann became an assistant professor, and eventually a professor, at Göttingen. He made many brilliant contributions to mathematics, including his PhD thesis, which laid the foundations for complex analysis, taking the idea of complex numbers into deeper territory. He also pioneered the topological idea of classifying a surface according to its "genus" (as it is now called)—essentially the number of "holes" in it, like a teacup with one hole as opposed to a sphere with none. But what concerns our story now is a paper he wrote in 1861, for it is at the heart of both the algebraic theory of curvature and the concept of tensors.

RIEMANN'S LANDMARK ESSAY: HOMING IN ON TENSORS

Riemann had written this essay for a competition sponsored by the Parisian Academy of Sciences, on the topic of heat conduction for a specific type of heat distribution. As we saw with Grassmann's and Germain's prize-winning papers, such competitions were an important way to stimulate research on cutting-edge topics. And since it was *so* cutting-edge, here things become a little more complex, not least because of Riemann's Greek notation, so feel free to skim this section for the key take-away points. In particular, note that his Greek symbols have more than one index or subscript—a hallmark of tensors—and that metrics are examples of "quadratic differential forms"; also, the notation Σ means a sum.

Riemann began with the heat equation that Joseph Fourier had derived in 1822. As I mentioned in chapter 6, this equation shows how temperature changes over time as heat diffuses through a body. You can imagine, say, holding a metal fire poker where the other end is heated by the fire, and gradually you feel the heat flowing up and through the handle—so the temperature is changing in all three dimensions of space. For the competition, however, the Academy had specified that as the heat flowed, the temperature should change only in two dimensions—say, when the interior of the poker is insulated so the heat only flows along and across its surface. Riemann's innovative approach to this problem was to transform the coordinates in the heat equation from the usual Cartesian coordinates $x \equiv x_1$, $y \equiv x_2$, $z \equiv x_3$ to new coordinates s_1, s_2, s_3, which he defined to be functions of only two dimensions—say x and y but not z. This is what Gauss had done when he wrote his 2-D curved metric in terms of p and q.

I've been talking a lot about coordinate transformations here and in chapter 9, and how quantities such as scalar and vector products, and the distance/interval measures given by the Euclidean and Minkowski metrics, remain invariant when expressed in terms of the new coordinates. Similarly, Riemann ended up with the equation

$$\Sigma\alpha_{\iota,\iota'}dx_\iota dx_{\iota'} = \Sigma\beta_{\iota,\iota'}ds_\iota ds_{\iota'}.$$

You can see that the form of the expression on the left-hand side of the equation is the same as that on the right, and so the expression is invariant under Riemann's coordinate transformation from x_1, x_2, x_3 to s_1, s_2, s_3.

Riemann's two-index symbol $\alpha_{\iota,\iota'}$ is related to the "conductivity coefficients," which appear in the heat equation—and his $\beta_{\iota,\iota'}$ is related to the conductivity coefficients in the heat equation written in terms of the new coordinates s_1, s_2, s_3. (He wrote the conductivity coefficients themselves as $a_{\iota,\iota'}$ and $b_{\iota,\iota'}$, respectively.) But the technical details[11] are not important: it's the way Riemann represented his coefficients that matters here, and the fact that he intuited—forty years before tensors were formally defined—that they were what we now call the "components" of a tensor.

The Σ in Riemann's equation stands for sum. (It's the upper-case Greek letter sigma, analogous to the Latin S, used as shorthand for "sum.") The indices ι, ι' in the sum $\Sigma\alpha_{\iota,\iota'}dx_\iota dx_{\iota'}$ each refer to the three dimensions of space, represented by the three coordinates x_1, x_2, x_3. All the possible combinations of ι, ι' in this case are (1,1), (1,2), (1,3), (2,1), (2,2), (2,3), (3,1), (3,2), (3,3). So $\Sigma\alpha_{\iota,\iota'}dx_\iota dx_{\iota'}$ is a shorthand way of writing

$$\alpha_{1,1}dx_1dx_1 + \alpha_{1,2}dx_1dx_2 + \alpha_{1,3}dx_1dx_3 + \alpha_{2,1}dx_2dx_1 + \ldots + \alpha_{3,3}dx_3dx_3.$$

(And similarly for $\Sigma\beta_{\iota,\iota'}$.) This looks rather like the metrics we've been discussing, although this kind of "differential form" had also been studied purely for its algebraic properties—and Riemann was talking about heat conduction, not metrics. Differential forms are expressions with differentials such as dx_1, dx_2, . . . ; Riemann's expression is a "quadratic differential form," because it is made from products of *two* "differentials"—dx_1dx_1, dx_1dx_2, and so on—analogous to the squares in quadratic equations.

To see the similarity of Riemann's differential form with a metric, in the Euclidean metric $dx_1^2 + dx_2^2 + dx_3^2$ the coefficients analogous to Riemann's $\alpha_{\iota,\iota'}$ would be:

$$\alpha_{11} = \alpha_{22} = \alpha_{33} = 1,$$

with all the other α_{ij}'s equal to zero. (Today the indices in a tensor component aren't separated by commas as Riemann did, because commas now refer to partial derivatives. And I've used the subscript ij because Riemann's use of ι and ι' is confusing: as we've seen in figure 9.3, dashes often denote transformed coordinates. But I'll keep to Riemann's notation for the rest of this chapter.) Riemann was well aware of this similarity with metrics. But although he said that he was using similar methods to those Gauss had used for his work on curved surfaces, his focus was on the algebra that came out of his analysis of the heat equation.

The first hint that this algebraic analysis involved quantities that we now call tensors is in the *two-index notation* he used to represent the conductivity coefficients ($a_{\iota,\iota'}$ and $b_{\iota,\iota'}$) in the heat equation, and the associated coefficients $\alpha_{\iota,\iota'}$ and $\beta_{\iota,\iota'}$ that we saw in the quadratic form above. At the end of chapter 9, I showed how Maxwell, too, had intuited the idea of a tensor, when he wrote the components of stress with two indices, P_{hk}. I also showed—with the help of figures 9.4 and 9.5—why he needed two indices, rather than the one you need for the components of a vector (as in fig. 9.1, for example). Riemann didn't discuss why he used two indices when he represented his conductivity coefficients as $a_{\iota,\iota'}$. But to visualize them, you can take a small volume of the body conducting the heat, just as in figure 9.4—so the coefficients $a_{\iota,\iota'}$ behave like the P_{hk}, except that they are measuring conductivity rather than stress. For example, when $\iota \equiv y$ and $\iota' \equiv x$, the conduction of heat is going in the y-x direction like the arrow P_{yx}.

Riemann did specify that he was considering only the case where the conduction was the same in both directions—for example, when the conduction is the same from y to x as from x to y, so that $a_{y,x} = a_{x,y}$, or more generally, $a_{\iota,\iota'} = a_{\iota',\iota}$. This is another example of invariance: interchanging the indices on $a_{\iota,\iota'}$ doesn't change the value of the coefficient. Another name for this invariance is "symmetry," because interchanging the indices ι, ι' to get ι', ι is like reflecting them, just like reflecting your image in a mirror, or reflecting a snowflake about an axis of symmetry.

But there's more to Riemann's "tensors" than just two indices.

THERE'S MORE TO TENSORS THAN NOTATION!

In fact, two indices alone are not necessarily the marker of a tensor. For instance, two decades after Riemann's paper, Maxwell, too, represented conductivity with two indices—in his case, K_{pq}, where he explained that the conduction was flowing from a point p to q; but he also denoted a current flowing in the same direction as C_{pq}.[12] It is true that a current has a direction and a magnitude, but in a circuit it adds like an ordinary number or scalar, so it is neither a vector nor a tensor.

The question of notation *is* important, and the index notation is brilliant for doing computations with tensors. But it is only a way of *representing* a mathematical quantity, not of *defining* it. What matters is the properties that mathematical quantities have to have if they are to be called vectors and tensors—including not just their rules of addition and multiplication but also the ability to express invariant expressions, such as the scalar product $\boldsymbol{a} \cdot \boldsymbol{b}$ and Riemann's $\Sigma\alpha_{\iota,\iota'}dx_\iota dx_{\iota'} = \Sigma\beta_{\iota,\iota'}ds_\iota ds_{\iota'}$. Later we'll see in more detail what it takes to define a tensor. But it does turn out that Riemann's $\alpha_{\iota,\iota'}$ and $\beta_{\iota,\iota'}$ are tensor components—and so are the conductivity coefficients, although Riemann's working doesn't show it.[13] After all, tensors haven't yet been invented! Still, Riemann's paper—like Maxwell's, Cauchy's, and Thomson's papers on stress—shows the kinds of problems that made it necessary to invent them.

There's more, though, for tensors can have more than two indices, and in his search for the transformation that kept $\Sigma\beta_{\iota,\iota'}ds_\iota ds_{\iota'}$ invariant—that is, the same as $\Sigma\alpha_{\iota,\iota'}dx_\iota dx_{\iota'}$—Riemann came up with three- and four-index quantities, too. Then he did something really special. First, he tossed off, as an aside, the fact that "the expression $\sqrt{\Sigma\beta_{\iota,\iota'}ds_\iota ds_{\iota'}}$ can be regarded as a line element in a more general space of n dimensions extending beyond the bounds of our intuition." Then he said that if you imagine a surface in this space, then his three- and four-index quantities are key to measuring the curvature of the surface. They are built from combinations of the coefficients $\beta_{\iota,\iota'}$ and their derivatives—so this is what Riemann meant in his habilitation lecture, when he suggested there *was* a way to tease out the curvature information from the metric coefficients, regardless of the coordinate system.[14]

Riemann didn't give a name to these three- and four-index quantities. They are essentially the components of what are now—thanks to Ricci and Levi-Civita—called the Christoffel symbols and the Riemann tensor, respectively. The name "Christoffel *symbols*" indicates that these three-index quantities *aren't* components of a tensor—at least, not of a single tensor. The details don't matter here, for the point once again is simply that if something has indices, it isn't necessarily a tensor; as I indicated, it needs to have other properties—notably invariance under linear coordinate transformations.

Christoffel symbols are built from derivatives of the coefficients in a quadratic differential form such as the metric, and the Riemann tensor is built from the Christoffel symbols and their derivatives. So, in a space with a metric, the key idea is that if the Riemann tensor equals zero, then the space is flat. What's more, if the space is flat, then the Riemann tensor is zero. Which means the Riemann tensor is *the* thing that tells you if your space is curved or flat. So it's often just called the "curvature tensor."

Since the Riemann tensor is built from the coefficients in the metric, which are functions of the coordinates, it has a value at each point in space. So technically it is a tensor field rather than a tensor, just as the electric and magnetic vectors $\boldsymbol{E}$ and $\boldsymbol{B}$ are vector fields. In practice, though, researchers tend to refer simply to the *tensors* and *vectors* that define gravitational and electromagnetic *fields*. (Some, however, demand more mathematically precise language, but such rigorous definitions came long after Maxwell's mathematical development of Faraday's idea of physical fields, and Riemann's pioneering work on curvature.)

The Riemann tensor is sometimes called the Riemann-Christoffel tensor, and the symbols comprising it are called the Christoffel symbols because, surprisingly, Riemann wasn't the only one to come up with these quantities. So did Elwin Christoffel, a math professor at Einstein's old school, the Zurich Polytechnic—although Christoffel published his work in 1869, three decades before Einstein was a student there. Christoffel probably didn't know about Riemann's 1861 essay, but he *was* inspired by Riemann's 1854 habilitation lecture, which led him to explore conditions for the invariance of quadratic differential forms. Unlike Riemann, though, Christoffel didn't connect his three- and four-index symbols with curvature, for his focus was pure algebra.[15]

Riemann didn't develop his curvature idea much further—that would be up to Ricci, Einstein, and Grossmann. For like his English translator Clifford—and like Maxwell, Minkowski, and so many others—Riemann died too young to fully develop his potential. In 1862, he'd contracted tuberculosis. Over the next few years, he and his new wife and baby daughter spent time in Italy, desperately hoping he'd recuperate in the warmer climate. But in 1866, he lost his battle with this dreadful disease, just like his mother and three sisters. He wasn't yet forty.

• • •

Riemann's habilitation lecture wasn't published until 1867. Clifford's English version, *On the Hypotheses Which Lie at the Bases of Geometry*, appeared in 1873, in *Nature*. Riemann's essay on heat was published only in his collected works of 1876—it hadn't won the essay competition! The judges were looking for something more specific—they hadn't appreciated Riemann's extraordinarily general approach to their heat problem, let alone the fact that it contained the seeds of both tensor analysis and the theory of curvature. So, his paper had languished for years, unappreciated and unknown, and Riemann never knew how important it would become. He never knew that his name would live on in the Riemann tensor, the bedrock of general relativity.

In 1912, however, Einstein and Grossmann seized upon Riemann's work in order to build the geometry of curved space-time. We saw earlier that Einstein had identified two problems he and Grossmann needed to solve: how to transfer the laws of physics from the special theory to the general one; and how to find the appropriate metric in curved space-time.

Riemann gave them the answer to the second problem: the line element will look like Riemann's $\sqrt{\Sigma \beta_{\iota,\iota'} ds_{\iota} ds_{\iota'}}$, and the coefficients $\beta_{\iota,\iota'}$ will be functions of the coordinates. They won't be constants, like the 1s in the Euclidean metric, or the 1, 1, 1, $-c^2$ in Minkowski's metric,

$$ds^2 = dx^2 + dy^2 + dz^2 - c^2\, dt^2,$$

for then their derivatives would be zero and therefore so would the Riemann tensor, so the space-time would be flat.[16]

To solve the first problem, however, Einstein and Grossmann will need a rigorous theory incorporating all the tensorial ideas I've been discussing—ideas intuited over many decades by many mathematicians, from Cauchy and Christoffel to Maxwell, Riemann, and more. So, it's time to meet Gregorio Ricci!

(11)

INVENTING TENSORS— AND WHY THEY MATTER

In 1861, when Gregorio Ricci was eight years old, a long series of political and military machinations culminated in the proclamation of the Kingdom of Italy. This meant that most of the disparate Italian states, duchies, and kingdoms were formally, if not always willingly, united at last. But it was religion that dominated Ricci's childhood. His father—an aristocratic landowner, businessman, and engineer—was a devout Roman Catholic, who expressed his faith not just through hefty donations to the church but also by feeding the hungry. And his mother would take her four children with her as she walked the streets seeking out the poor—especially the women. She seemed to act as a counselor, hearing the women's troubles and offering comfort. Young Ricci grumbled at all the periodic stopping and waiting while his mother listened to each tale of woe, yet her dignity in the situation left a lasting impression on him, and he, too, was a lifelong and devout Catholic.[1]

Incidentally, Ricci's family name was Ricci Curbastro, but in his landmark paper on tensor calculus he signed himself Ricci, so that's the name he's known by today. (I should mention that similarly, James Clerk Maxwell's family name was Clerk Maxwell, but his friends referred to him as Maxwell.) Young Ricci was a keen student, with an unusually "penetrating mind" and a "lively ingenuity," as one of his teachers put it.[2] Then in 1869 he began to study mathematics at the papal university in Rome—but in the summer of 1870, he had to return home because of yet another war. The French had been allied with the papal troops, and when the Prussians prevailed, the new king of Italy seized the chance to take over Rome and bring it into the new Italian kingdom.

With Rome now the secular capital of the new nation, and the former papal university building taken over by the state, Ricci set his sights on the venerable University of Bologna, which was much closer to his hometown of Lugo di Romagno in northeast Italy. He had to do some extra study to qualify for admission—two years' worth, so all that political upheaval cost him dearly. Still, he was a supporter of the unification, and being at Bologna also meant, he told a friend, "I can follow more attentively and with more zeal the political reforms that our countrymen who have created a united Italy are going to carry out [in Lugo]."[3]

After a stellar year at Bologna—he scored full marks for his exams in calculus, chemistry, and geometry—he decided to move once again, this time to Pisa for its vibrant mathematical school. He got his doctorate in 1875, and his teaching certificate the following year—not that he could get a job at the time, for university positions were scarce. So, he stayed on at Pisa as an independent scholar, reading up on the latest math and physics. Like Einstein, he was particularly excited when he discovered Maxwell's theory of electromagnetism—and in 1877, this really was cutting-edge research you couldn't learn in school. Maxwell himself was still alive, and there was still a decade to go before Heinrich Hertz generated the radio waves that so spectacularly confirmed the theory.

After a couple of unsuccessful applications for teaching positions, Ricci obtained a fellowship from the ministry of public instruction—which is how he came to study under the famous Felix Klein. Klein, who was then based in Munich, was already celebrated for finding the relationship between invariant theory and groups of coordinate transformations—and three decades later, Henri Poincaré and Einstein would show that the Lorentz transformations form a group. They're the coordinate changes under which the Minkowski metric and Maxwell's equations remain invariant, as we saw earlier—and I briefly explained why they form a group in the boxed caption to figure 9.3. Klein also developed Arthur Cayley's work on projections of curved surfaces, and all in all he had such wide-ranging mathematical interests that when Ricci arrived in Munich in the autumn of 1878, he was overwhelmed by the study and research program Klein offered him. Still, he found Klein to be a "kind" advisor who gave him "vigorous help" in his studies.[4]

Even more important, working with Klein helped Ricci develop confidence in his own ability—a wonderful gift from a teacher to a student. When Chisholm did her PhD with Klein in the 1890s, she, too, would be struck not just by his brilliant mind, but also by the way he would encourage his students to have the confidence to "never be dull!" (Not long after she took her doctorate in 1895—the first official doctorate awarded to a woman in Germany—she married her former Girton tutor William Young, so she is better known today as Grace Chisholm Young. Her doctoral thesis was on algebraic groups, Klein's specialty, applied to spherical geometry.)[5]

After his year with Klein, Ricci spent a couple more years fruitlessly seeking a full-time university position—until finally, in the winter of 1880, he was appointed associate professor of mathematical physics at the University of Padua. Galileo, Copernicus, and Cardano were just three of his illustrious predecessors at this ancient and progressive institution, and Ricci would remain there for the next forty-five years.

His first published papers as a Padua academic were on electromagnetism and differential equations, but he soon became intrigued by the math of invariance. He also decided it was time to find a wife. He'd first fallen in love when he was a student at Pisa, but the girl had disdained his awkward infatuation. Then his older brother's "unsuitable" love interest had aroused fury from his conservative Catholic parents—she came from a poor and unconventional family, but what really upset Ricci senior was that she was a relative, and he was sure God would not approve of such an incestuous union. So, thirty-year-old Ricci decided the best way to avoid both heartache and parental displeasure was to seek matchmaking advice from the local priest. The priest was happy to oblige, and he arranged for Ricci to meet a lively, intelligent, and eminently suitable young woman named Bianca. It was a happy courtship, and she became his bride and lifelong companion.[6]

Ricci had been doing more than courting, though. In the same year as his wedding, 1884, he published his first paper on the road to tensor analysis. He'd been inspired by the works of Riemann and Gauss, and his paper was on the transformation properties of "quadratic differential forms"—that is, sums of the squares or other products of pairs of differentials, such as we've seen in the distance metrics based on Pythagoras's theorem. We saw some of these properties with Gauss's metric, which is invariant under bending transformations—like the rolled-up piece of paper that shows the surface of a cylinder is just as intrinsically flat as the unrolled paper—and with the Lorentz-invariance of Minkowski's metric. After the 1876 publication of Riemann's 1861 essay on heat, which extended Gauss's work to n-dimensions, several mathematicians had taken up his ideas. But Ricci's aim was to "avoid lazy discussions about the existence and nature of spaces of more than three dimensions," and to look beyond his colleagues' focus on *applications* of differential forms. Rather, he wanted to provide a clear understanding of the mathematical *theory*.[7]

We saw this emphasis on theory with Hamilton's development of the rules of vector algebra and vector calculus, forty years earlier. He'd begun with the rules for multiplying his i, j, k, which he'd triumphantly carved on Broome Bridge, and they enabled him to define new kinds of multiplication—quaternion, scalar, and vector products. Once he had these in hand, he'd created the differential calculus vector operator nabla, ∇. It was only after Maxwell had learned all these rules from Tait—and had given names to the nabla operations grad, divergence, and curl—that he'd felt comfortable using vector calculus in his theory of electromagnetism. So now, establishing similar rules for tensors is just what Ricci set out to do.

Not that he called them tensors: rather, he simply referred to them as "systems of functions." For example, the coefficients $\beta_{\iota,\iota'}$ in Riemann's "line element" (or metric) $\Sigma\beta_{\iota,\iota'} ds_\iota ds_{\iota'}$ are now called the components of a tensor, and, as we saw in chapter 10, on curved surfaces they are functions of the coordinates—so, in fact, they're a set (or "system"!) of functions. (So technically they are "tensor fields," but as I mentioned in the previous chapter, less rigorously "tensors" will do nicely.) We'll see mathematically why the metric is a tensor later—and Riemann's four-index quantity, too, which Ricci called "the system of Riemann," so it's now called the Riemann tensor. And instead of saying "tensor calculus," Ricci named the calculus of his systems the "absolute differential calculus." By "absolute" he meant unchanging, because the interesting thing about tensors is that they encode the idea of invariance.

Before I get into Ricci's math, though, I'll sketch out the basic idea of tensors as a way of representing, and calculating with, information—just like vectors. But first I'll tell you how tensors got their modern name.

HOW TENSORS GOT THEIR NAME— AND HOW THEY REPRESENT DATA

The term "tensor" comes to us from Hamilton via Göttingen math professor Woldemar Voigt. But it is Einstein who would firmly establish this name, once he and Marcel Grossmann got their heads around Ricci's absolute calculus. Hamilton had used the term in a different context—as the magnitude of a quaternion, which he defined by analogy with the modulus of a complex number. It's the square root of the sum of the squares of the components of the quaternion—just as the magnitude of a vector is found from the sum of the squares of *its* components. But Voigt is the one who first used the term "tensor" in its modern context, in his 1898 book on crystallography—which, incidentally, Chisholm Young and her husband favorably reviewed in *Nature*.[8]

Voigt was referring specifically to the stresses and tensions in crystals—

and "tension" comes from the Latin "tensio," which in turn comes from "tendere," meaning "to stretch." But Voigt said he was simply extending Hamilton's use of "tensor." For example, the vector or cross product, $\boldsymbol{a} \times \boldsymbol{b}$, produces a third vector c whose magnitude (or Hamiltonian "tensor") is related to the magnitudes ("tensors") of $\boldsymbol{a}$ and $\boldsymbol{b}$. In other words, the cross product gives a new magnitude ("tensor") from two others—and as we're about to see, one of the novel things about Ricci's tensor analysis is that tensor multiplication produces new, "higher-order" tensors from old ones.

Grassmann had had this idea, too. As we saw in chapter 5, instead of the term "vector," he'd used the German word *strecke*, which can be translated as "line or stretch"—and his basic geometric objects were "lines" that could be "stretched" or "extended" to form planes, just as two vectors form a plane in the parallelogram rule. Grassmann defined the outer product of two 3-D vectors as the *oriented area* of the parallelogram bounded by the two vectors. The usual vector or cross product does this, too, in effect, but only for parallelograms, whereas Grassmann's definition of an outer product implied that you could then add a third vector to extend the parallelogram to a box, and so on, adding as many new dimensions as you like.

In the 1880s, Gibbs developed Grassmann's idea further, essentially hitting on a similar definition of tensor multiplication as Ricci was creating around the same time. Today it is called a "tensor product" or, following Grassmann, an "outer product," although Gibbs and Ricci didn't use those names. It is a way of combining information from two tensors into one—a kind of multiplicative version of Roman numerals, where you add more symbols to increase the magnitude of a number: I, II, III, V, VI, VII, VIII, and so on. This analogy also illustrates the fact that outer products are generally not commutative: VI is a different number from IV.

You can see the idea of tensor (or outer) products in figure 11.1—which also highlights the fact that vectors and matrices can be thought of as tensors. There's a sophisticated mathematical reason for this, but for now it's enough to notice that like tensors, their components are represented with indices—one index for vectors, two for matrices, as the caption

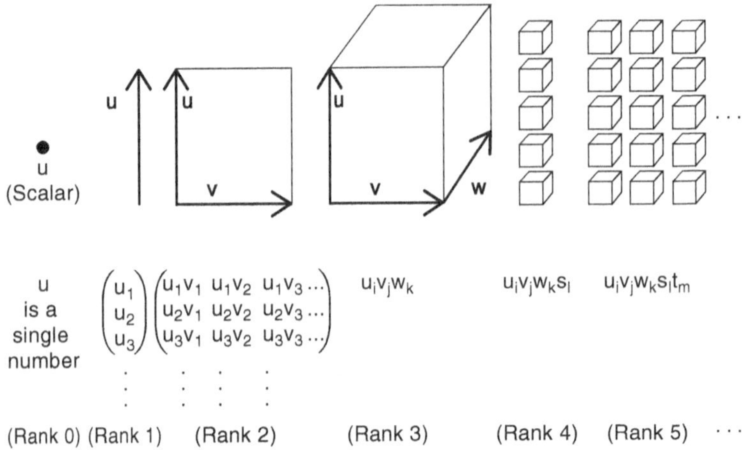

FIGURE 11.1. Tensor (or outer) products combine information from the tensors being multiplied. Ordinary numbers are represented by a symbol such as a. If they represent quantities that don't depend on coordinates—such as temperature—then they are scalars, and since scalars don't change under coordinate changes, they are tensors. We'll see that vectors are tensors, too, and we've already seen that their components are denoted with an index to represent the axis from which the component is measured.

The tensor product of a column vector $\boldsymbol{u}$ and a row vector $\boldsymbol{v}$ can be represented as a matrix. You're likely familiar with the symbol a_{ij} for the element in the *i*th row and *j*th column of a matrix, so here $a_{ij} = u_i v_j$. I'll spell this out in the 2-D example in the narrative. In the same way, you can build up more tensor products. For example, the tensor product of the 1-index vector $\boldsymbol{u}$ and a matrix A has components with 3 indices, as you can also see in the narrative. Similarly, if two matrices A and B were each formed from the outer product of vectors, the components of their tensor product would be $c_{ijkl} \equiv u_i v_j w_k s_l$, and so on. This is not the only way to build new tensors, but notice that you can represent more information as the tensor "rank" increases.

The "rank" is also called the "order" of the tensor, for this is what Ricci called it when he introduced the idea. It has to do with the number of transformation matrices needed to transform from one coordinate system to another, but it essentially corresponds to the number of indices, each one representing a different type of information. (If you're familiar with matrix algebra, note that for matrices viewed as tensors, this is a different use of "rank" from that in linear algebra.) We'll see more about this as we go.

to figure 11.1 spells out. Of course, as I intimated in chapter 9, the mark of a tensor goes beyond the fact that its components are denoted with indices—but let's go with this for now. For the tensor product is a way of producing new tensors, with even more indices, each index telling you something specific about the data represented by the tensor component.

TENSORS AND DATA SCIENCE (AND A PEEK AT QUANTUM MECHANICS)

Figure 11.1 also illustrates how tensors enable data to be stored and combined in data science today. In chapter 4, I outlined how vectors and matrices are used in machine learning and search engines, but with tensors you can add in not just *more* data but different *kinds* of data. For instance, we saw that in a search engine, information can be stored as a matrix, with rows representing key words, and columns representing different documents containing these key words. Tensors allow you not just to add more words or more documents—you can do that by making the matrix larger—but *additional* information that you can't fit into the matrix. For example, if you want to add the date of publication of the document, and the author, you have to extend your original matrix to the 4-D shape shown in figure 11.1. The key thing is that each type of information has its own index.

Another modern application of tensors is signal processing, which is used in interpreting an electroencephalogram (EEG) or electrocardiogram (ECG), for example. In addition to the spatial components of the signal, you might want to know such things as its temporal and frequency components—and again, each type of information requires its own index.

So, there's a difference between the number of components and the number of indices. In ordinary vector analysis we've seen that in n-dimensional space (or more properly, in an n-dimensional "vector space"—an n-D space with group properties), a vector has n components, each measured from one of the n coordinate axes. The particular axis gives the particular label on the component—so a vector has components $v_1 \equiv v_x$, say, and so on up to v_n. In other words, you have n components, but each has only one index—and this one index takes one of n values, one for each dimension. We've also seen that each component (or element) of a matrix has two indices: one locating the row and the other locating the column. You need both locations to pinpoint the position of a particular element, so you need two indices. Similarly, we saw in figure 9.5 that the component of a stress tensor needs two indices, one to indicate the surface on which the stress is acting, and the other to indicate the force; in three-dimensional space, *each* index takes values from 1 to 3 (or from x to z), so that's $3 \times 3 = 9$ components, as we saw in figure 9.4.

And so it goes for higher-order tensors. For example, I referred in the previous chapter to the Riemann tensor, which has four indices. It's tempting at first glance to think that it needs four indices because it represents the curvature of 4-dimensional space-time—but the fact that we're working in four dimensions just means that each index on the Riemann tensor can take four values, one for each dimension. The indices themselves each represent a different attribute or building block of the Riemann tensor.[9] So, while the stress tensor in ordinary space has $3 \times 3 = 9$ components, a 4-index tensor in space-time will have $4 \times 4 \times 4 \times 4 = 256$ components: four coordinate choices for each of the four indices. That's a lot of information you can fit into one tensor! And you can go on to tensors with as many dimensions and orders as you like. (But if a tensor has "symmetries," as the Riemann tensor does—such as when interchanging or "reflecting" two indices doesn't change the value of the component—some of its components are the same, so the possible amount of data it can represent is reduced.)

To take another digital tensor application, in image processing the location of each pixel is represented in a matrix whose rows and columns represent the dimensions of the picture—but to produce color images *three layers* of matrices are needed, one for each of the colors red, green, and blue. (These three colors are a legacy of Maxwell's discovery of color slide photography: he and his assistant Thomas Sutton used red, green, and blue filters to take the first-ever permanent color photograph, the tartan ribbon I mentioned in chap. 6.)[10] So, the third index in the tensor encoding the relevant pixel information represents the color. Similarly, and taking a different example, medical diagnoses are more accurate if they combine data from a variety of different types of test, each represented by an index. These kinds of multi-index constructions—these tensors, and their products—are so important in data science that Google named one of its machine-learning platforms TensorFlow, and there are various other programs and tools, such as Tensorlab and Tensorly.

Since tensor products are so important, I'll spell out the idea of producing a matrix from the tensor product of a column vector $\boldsymbol{u}$ and a row vector $\boldsymbol{v}$, as in figure 11.1. For simplicity, I'll make the vectors 2-D here:

$$\begin{pmatrix} u_1 \\ u_2 \end{pmatrix} \begin{pmatrix} v_1 & v_2 \end{pmatrix} = \begin{pmatrix} u_1 v_1 & u_1 v_2 \\ u_2 v_1 & u_2 v_2 \end{pmatrix}.$$

This is a neat way of combining the information from two vectors, and it also follows the rules of ordinary matrix multiplication in this case. But you can see the difference between matrix multiplication and tensor products when you try to multiply $\boldsymbol{u}$ by a 2×2 matrix: the ordinary matrix rules don't allow you to multiply a 2×1 matrix by a 2×2 one at all, but the tensor product does:

$$\begin{pmatrix} u_1 \\ u_2 \end{pmatrix} \begin{pmatrix} a_{11} & a_{12} \\ a_{21} & a_{22} \end{pmatrix} = \begin{pmatrix} u_1 \begin{pmatrix} a_{11} & a_{12} \\ a_{21} & a_{22} \end{pmatrix} \\ u_2 \begin{pmatrix} a_{11} & a_{12} \\ a_{21} & a_{22} \end{pmatrix} \end{pmatrix} = \begin{pmatrix} u_1 a_{11} & u_1 a_{12} \\ u_1 a_{21} & u_1 a_{22} \\ u_2 a_{11} & u_2 a_{12} \\ u_2 a_{21} & u_2 a_{22} \end{pmatrix}.$$

So, the tensor product is a way of combining information from two (or more) systems or sets into a single large one. For instance, consider the natural language processing (NLP) programs behind such marvels as email spam filters, language translators, converting spoken words to text (for use by those with hearing problems, for example, such as in captioning TV programs), the helpful voice on your GPS, the polite text from a company's chatbot, the predictive text used in web searches, say, and apps such as email and Instagram, and the spectacularly human-like text generated by bots such as OpenAI's ChatGPT. Tensor products can offer a way to combine a set of words with a set of grammatical instructions—where words are represented as vectors by assigning them a position within a dictionary of words, and similarly for the grammar. I should add here that much of the training data used to develop sophisticated NLP (especially large language models or LLMs) is human-generated content scraped from the web without the content creators' knowledge or permission, and I'm heartened that writers and artists are attempting to fight back against this theft.[11] But since tensors themselves are not the problem, and since there are benefits as well as problems[12] with NLP and LLM programs and applications, I'll venture on with this brilliant application of tensor products.

To keep it simple, suppose the word dictionary contains the three words, "cats, love, mice," and each word is assigned a position number from 1 to 3. The dimension of the vector representations of these words will therefore be three, so if "cats" is in the first position, "love" is in the second, and "mice" is in the third, then these words would be represented as the vectors (1,0,0), (0,1,0), (0,0,1), which I'll label ***C, L, M***, respectively. To create the sentence "Cats love mice," just add the vectors: $\boldsymbol{C} + \boldsymbol{L} + \boldsymbol{M} = (1,1,1)$. But this is no different from the vector representation of "Mice love cats," which is definitely not the case. So it's here that tensor products can come to the rescue. Take a second set of vectors, representing key grammatical instructions for the roles these words will play: subject, object, verb, labeled as, say, ***S, O, V***, and represented as (1,0,0), (0,1,0), and (0,0,1), respectively. The tensor product of "cats" (represented as a column vector) and "subject" (a row vector) would be represented as we saw just before for column and row vectors:

$$\begin{pmatrix}1\\0\\0\end{pmatrix}\begin{pmatrix}1 & 0 & 0\end{pmatrix}=\begin{pmatrix}1 & 0 & 0\\0 & 0 & 0\\0 & 0 & 0\end{pmatrix},$$

and similarly for the tensor products of "mice" and "object," and "like" and "verb." Now you can construct an unambiguous sentence,

$$\boldsymbol{C} \otimes \boldsymbol{S} + \boldsymbol{L} \otimes \boldsymbol{V} + \boldsymbol{M} \otimes \boldsymbol{O} = \begin{pmatrix}1 & 0 & 0\\0 & 0 & 1\\0 & 1 & 0\end{pmatrix},$$

where $\otimes$ is the symbol for tensor products. This is different from "Mice love cats":

$$\boldsymbol{M} \otimes \boldsymbol{S} + \boldsymbol{L} \otimes \boldsymbol{V} + \boldsymbol{C} \otimes \boldsymbol{O} = \begin{pmatrix}0 & 1 & 0\\0 & 0 & 1\\1 & 0 & 0\end{pmatrix}.$$

Although this sentence is false, it *is* unambiguous.

This example is just one way of applying tensor products in NLP.[13] And one way of applying them in quantum mechanics is in representing the "quantum state" of several particles.

We saw in the prologue that the spin of an electron can be "up," represented by the vector (1, 0), or "down," represented by (0, 1); so a "superposition" of these two possibilities—an in-between state in which either outcome is possible—is (α, β), where α is the "weight" or probability amplitude of being in the "up" state and β the probability amplitude of being in the "down" state. ("Probability amplitude" just means that $|\alpha^2| + |\beta^2| = 1$. To make the probabilities work, α and β are complex numbers.) We also saw that spin can be used to represent the 0s and 1s in quantum computing, with the spin "up" state representing the binary digit 0, say, and "down" representing 1. I wrote these "up" and "down" states as row vectors for convenience, but in quantum mechanics state vectors are written as column vectors, and they are often called "kets." In chapter 4 I spoke about the unit vectors ***i***, ***j***, ***k*** being a "basis" for constructing a vector ***v***—and analogously, the basis for the spin state ψ of an electron, or a qubit, can be chosen so that $\begin{pmatrix}1\\0\end{pmatrix}$ represents the spin "up" state, and $\begin{pmatrix}0\\1\end{pmatrix}$ for spin "down." Using what is known as "Dirac notation," after quantum pioneer Paul Dirac, these basis vectors are denoted by the kets $|0\rangle$ and $|1\rangle$. So the state vector for a qubit is represented in component form as

$$|\psi\rangle = \alpha|0\rangle + \beta|1\rangle.$$

What this means is that until it is actually observed, the qubit is in a state of superposition between the two states $|0\rangle$ and $|1\rangle$, α and β representing the respective likelihoods of it being in each state. Tensor products come into it when qubits are *combined*, as of course they must be to make a usable quantum computer. To go gently, though, let's take just two qubits, and represent their states as

$$|\psi_1\rangle = \begin{pmatrix}\alpha\\\beta\end{pmatrix}, \ |\psi_2\rangle = \begin{pmatrix}\gamma\\\delta\end{pmatrix}.$$

To find the state of the combination of these two systems, take their tensor product:

$$|\psi\rangle = |\psi_1\rangle \otimes |\psi_2\rangle = \begin{pmatrix} \alpha\gamma \\ \alpha\delta \\ \beta\gamma \\ \beta\delta \end{pmatrix}.$$

It's a 4 × 1 vector giving the probability amplitudes for four possibilities: both qubits are in the up state (both represent zeroes), the first is up and the other is down, the second is up and the first is down, and both are in the down state. It's this ability for each qubit to represent a superposition of 0s and 1s that makes quantum computers so potentially powerful, for they can carry out multiple computations at the same time. You can sense this power from the fact that in a system of n qubits, the tensor product will have 2^n complex-number components. Even with a relatively modest number of qubits that's a lot of 0's and 1's being processed simultaneously.[14]

• • •

While column vectors representing quantum states are called "kets," represented as $|A\rangle$ for an arbitrary state, row vectors are called "bras," denoted by $\langle A|$. That's because when you put a bra and a ket together—such as when you take the scalar (or inner) product of two states $|A\rangle$ and $|B\rangle$—you complete the bracket ("bra-ket"):

$$\langle B|A\rangle.$$

Scalar products are needed in the "normalization" of state vectors that ensures weights such as α and β do relate to the probabilities of a measurement result—but the point here is that the content of the $B\rangle$ column vector is now acting as a bra or row vector, $\langle B|$, operating on the ket $|A\rangle$ to give the scalar product. This sounds complicated (and technically, a bra is a "dual" of a ket, and both reside in complex vector spaces)—but you can look at it as an application of a result from ordinary vector analysis, where vectors can play different roles, depending on how you write them.

For instance, going back to our column vector $\boldsymbol{u}$ and row vector $\boldsymbol{v}$, we saw a bit earlier that ordinary matrix multiplication of $\boldsymbol{u}$ times $\boldsymbol{v}$ gives a 2×2 matrix (which in this case is also their tensor product). But if you swap the order, ordinary matrix multiplication of $\boldsymbol{v}$ by $\boldsymbol{u}$ gives not a matrix but a number, $v_1u_1 + v_2u_2$. In fact, it is the scalar product of the two vectors. (At least, it is the scalar product in flat two-dimensional Euclidean space. As I mentioned in chapter 9, the scalar product depends on the metric. In Minkowski space-time, for example, the scalar product is

$$\boldsymbol{a} \cdot \boldsymbol{b} = a_1b_1 + a_2b_2 + a_3b_3 - a_4b_4, \text{ or } \boldsymbol{a} \cdot \boldsymbol{b} = -a_0b_0 + a_1b_1 + a_2b_2 + a_3b_3$$

if the time component is denoted by a 0 subscript rather than a 4). So, you can see that it makes a difference whether you write your vector, your data, as a row or column. This is the kind of thing that might make math seem bizarre and contradictory—but it's just this sort of detail that piques a creative mathematician's curiosity. And Ricci and his successors came up with a neat way around it.

First, represent the two types of vector with different notation. Following Ricci, write the components of column vectors with an "upstairs index" or superscript—so you no longer write u_1 and u_2 but u^1 and u^2. (Actually he put upstairs indices in brackets, presumably to make it clear that these are labels, not powers. As mathematicians and physicists such as Einstein and Grossmann became more adept at using index notation, they discarded the brackets.) Keep the downstairs indices, the subscripts, for the row vectors. So, in the case of a matrix formed from a column vector times a row vector, the elements can be represented as u^1v_1, u^1v_2, u^2v_1, u^2v_2, and so on. Straightaway you can see, just by looking at the notation, that you are multiplying two different kinds of vector. Decades later, Dirac would apply this distinction via his bra and ket notation.

Second, give these two entities different names, to avoid confusion. Today the word "vector" in this context refers to column vectors, while row vectors are called "one-forms" or "dual vectors." Early twentieth-century researchers coined these terms; the idea originates with Grassmann, who had used the term "complement" instead of "dual." Bras are examples of one-forms. Back in the 1880s, Ricci called vectors and one-forms "contravariant vectors" and "covariant vectors," respectively, and these names are also used today.

Actually, Ricci didn't talk specifically about row vectors and column vectors, for these are just examples of his two types of tensor. As we'll see in the next two sections, the general idea behind this distinction—and behind Ricci's choice of names—comes from a concept that goes beyond index notation and tensor products, which are often the main aspects of tensor math needed in data science.

In fact, even the position of indices is not so important in data science as it is in math and physics, for often the data are entered without any need for abstract symbols at all. Rather, it's the rank and "shape" that many programmers use to characterize each different tensor. For example, in TensorFlow (and other programming language libraries such as Python's Numpy), the shape refers to the dimension. A scalar has a shape 0, represented as an empty bracket: [ ]. A vector is programmed as a string of numbers, one for each component; its shape is the number of components, so a 3-D vector has shape [3]. Rank 2 tensors can be represented as matrices, and their shape is the number of rows and columns, so the 2 × 2 matrix above would have shape [2, 2]. A rank 3 tensor, such as a 2 × 3 × 5 array, has shape [2, 3, 5], and so on.

An advantage of emphasizing shape is that programmers can include what TensorFlow calls "ragged tensors," arrays with strings of different sizes—a string of words or sentences, perhaps, where the number of letters or words has nothing to do with the dimension of space. As I've mentioned, in n-dimensional space, each index on a tensor must take a value between 1 and n, but "ragged tensors" are allowed to have variable sizes. This is a nice example of the way data scientists have adapted a mathematical concept for their own needs.

All this is a long way from the way tensors were first used, albeit unwittingly—as stress and metric tensors in mathematical physics. Still, these earlier uses also had to do with representing and handling information. By "handling" I mean knowing how to combine information into new tensors and how to interpret and apply the results. So, to earn the title of tensor, it's not enough simply to put information into a list or array—the Mesopotamians were doing that sort of thing four thousand years ago. To be a tensor, the arrays have to obey certain rules, just as we saw with vectors and matrices in chapter 4. We've already met tensor products, and there are also rules for addition, of course. We saw how the parallelogram rule for adding vectors takes account of their magnitude and direction, but more generally, vector and tensor addition—and multiplication, too—must obey the laws of "linearity." (This means, for example, that

$$(2\boldsymbol{a}) \cdot \boldsymbol{b} = 2(\boldsymbol{a} \cdot \boldsymbol{b}) = \boldsymbol{a} \cdot (2\boldsymbol{b}), \text{ and } \boldsymbol{a} \cdot (\boldsymbol{u} + \boldsymbol{v}) = \boldsymbol{a} \cdot \boldsymbol{u} + \boldsymbol{a} \cdot \boldsymbol{v}$$

—analogous to the distributive law in arithmetic.)

But the most important thing in math and physics is the ability of tensors to represent information invariantly—"absolutely"—without spurious data coming from the choice of coordinates. This is not an issue for many data science applications, although it is important in some of them, such as the neural networks we saw earlier. Either way, there's much more to the idea of a tensor than figure 11.1 suggests.

INVARIANCE (AND AN ACADEMIC SCANDAL)

The idea of invariants—things that stay the same when you rotate or move or otherwise change your frame of reference—had been intriguing mathematicians at least since Cayley and Boole were working together in the 1840s. In chapter 9 we saw many examples of invariance, from the shape of snowflakes to the scalar product $\boldsymbol{a} \cdot \boldsymbol{b}$ to the Minkowski metric,

$$ds^2 = dx^2 + dy^2 + dz^2 - (cdt)^2$$

(where units are often chosen so that $c = 1$). But each of these shapes and expressions is invariant only with respect to a particular group of coordinate transformations, as we saw in figures 9.1 and 9.3.

In studying invariants, mathematicians such as Riemann had their eye on physical applications—in his case, the curvature of surfaces—while others, including Cayley and Klein, were interested in the purely mathematical structure of these groups of coordinate transformations. Klein had been the managing editor of the journal *Mathematische Annalen*, a position he took up when his mentor Alfred Clebsch suddenly died of diphtheria. Clebsch is the professor to whom Justus Grassmann had given a copy of his father's *Ausdehnungslehre* when he arrived as a student at Göttingen in 1869. Clebsch was impressed, and he went on both to extend Grassman's ideas and to cofound the *Mathematische Annalen* as an outlet for research on invariant theory.

There are many other names I could add to the list of mathematicians working on invariants and/or differential forms during the latter half of the century—including Ricci's former math professor at Pisa, Enrico Betti. There's an overlapping thread here, for one of Betti's achievements was his generalization of Stokes's theorem to *n* dimensions. We saw earlier that the original 3-D version of this theorem had been first published in the 1854 Smith's Prize exam, which Maxwell sat—and that it relates a surface integral to a line integral. The key thing is that it does this in an *invariant* way—you should get the *same* surface area no matter the coordinates you use. So Betti's work on this is an example of the diverse reasons mathematicians were interested in coordinate transformations and invariance.

Betti also reminds us of the revolutionary upheavals taking place in Europe in the nineteenth century. As a student in 1848 he'd fought in two of the first battles for Italian independence—his thesis advisor had led the Tuscany university battalion. It lost, to the Austrians, but luckily Betti survived, and went on to become a significant mathematician and teacher—and it was on his advice that Ricci had gone to Berlin to study with Klein.[15] Betti also contributed to the new Italian journal for pure and applied math, *Annali di Matematica pura e applicata*. Specialist journals, such as *Annali* and Clebsch and Klein's *Annalen*, were important places for mathematicians to publish their work—and had they existed in Thomas Harriot's day, perhaps his work would not have been lost for so long: one of the world's first modern scientific journals was the *Philosophical Transactions* of the Royal Society of London, which began in the 1660s, nearly half a century after Harriot's death.

Countries with established universities, scientific societies, and journals tended to be at the center of mathematical progress, and in 1884 Ricci published some of his first results on invariance in the *Annali*. He didn't know of Riemann's work when he first set out on this journey. Instead, he took his initial inspiration from the purely mathematical approach of Elwin Christoffel's 1869 paper. As I've mentioned, Christoffel and Riemann both independently discovered what are now called the Christoffel symbols and the Riemann (or Riemann-Christoffel) tensor, which, as Riemann showed in his 1861 paper, gives the invariant condition for knowing whether a surface is flat or not. Christoffel didn't refer to curvature at all—other than in a note at the end of his paper, saying that in his 1854 habilitation thesis Riemann had applied quadratic differential forms to the line element. But that was enough for Ricci, who went searching for Riemann's papers.

At the same time, he had his teaching—not always an easy task for a reserved, diffident person like Ricci. But he was passionate about his subjects, and if his lectures lacked color, they were nonetheless clear and rigorous—Ricci was a great one for proof. As he told a colleague, "I don't deny that the proofs [in my lectures cause] some difficulty when they are presented to students who unfortunately don't take their education seriously." Still, he went on, that wasn't going to stop him from presenting math the way it should be presented. After all, "if my own judgment doesn't deceive me, these proofs are beautiful." Besides, the best students were inspired by such rigor, and the rest, he thought, might benefit from it, for they had come from secondary school with insufficient grounding in mathematical fundamentals.[16] I'm sure many lecturers today can empathize with him.

Ricci was also busy applying for promotion to a full professorship. On his first attempt, in 1884, he and his young rival, Guiseppe Veronese, missed out to a more senior candidate. Ricci was happy to defer to seniority, but Veronese appealed. He also pointed out that he was from a working-class background (unlike Ricci) and needed the increased salary to "help my poor parents and my brothers." The plea was treated kindly by the authorities, and he was given a permanent teaching position with an increase in pay. Unruffled, Ricci was hopeful when he tried again for a full professorship in 1887. He was now thirty-four and had a reputable publishing record—but so did Veronese, who applied for the same position. The ensuing battle made front-page news, with rumors that faculty skullduggery ultimately deprived Ricci of his rightful promotion. The saga dragged on, and it would take another three years for Ricci finally to gain his professorship.[17]

Meantime, what else could he do, brilliant mathematician that he was, but throw himself into the work that would, ultimately, immortalize his name?

HOW TO HANDLE ALL THOSE INDICES

Over the next few pages, I'm going to spend time showing why Ricci had two different kinds of vector—or in modern terms, a vector and a one-form—because they are prototypes for all tensors. (Actually, he started with general tensors and gave vectors as an example: perhaps because vector analysis wasn't so well established then as it is today, but also because his theory arose primarily from the study of invariant differential forms rather than from vector analysis.) These two types of vector relate to Ricci's index notation, which is a marvelous example of the way mathematicians use symbolism to bring out the underlying structure of mathematical concepts. We saw something of this with the rise of algebra in chapter 1 and the war over vector notation in chapter 8. Here, though, there's detail and an approach that may be unfamiliar to you, although the only math tools you need are vector and matrix multiplication.

As always, though, if you get to the point where you just want to take my word for it, skip down to the next section. And if, once you get there, you decide you just want to see how Einstein and Grossmann used tensors in formulating the framework for general relativity, then go ahead to the next chapter. Ricci would understand: it always takes effort to learn a new skill, he wrote in the introduction to his seminal overview of his "absolute differential calculus."[18] Tait had said something similar when he was trying to convince people of the advantages of quaternions. But like Tait, Ricci was sure that after "surmounting the difficulties of initiation," readers would soon convince themselves of the "elegance and clarity" of these methods—methods that hinge on the right *notation* for the idea of invariance. It's a notation that relies on spotting patterns, so it's also rather fun.

• • •

If tensors were to encode the idea of invariance under coordinate changes, Ricci had to find the specific relationship between the components of a tensor in one frame and those of the same tensor in the new frame. We saw an implicit example of this for the position vectors in figure 9.1, where I showed geometrically that the magnitude of $\boldsymbol{a}$, and also the scalar product $\boldsymbol{a} \cdot \boldsymbol{b}$, are invariant when you rotate the coordinate axes. What I didn't show then was how each vector component changed, but this is the kind of question that Ricci was trying to answer.

He began by generalizing the way coordinate transformations work. But I'll start with a specific example, the rotation in figure 9.1, where the transformation equations from the usual x-y coordinates to the rotated x'-y' ones are:

$$x' = x\cos\theta + y\sin\theta,\ y' = -x\sin\theta + y\cos\theta.$$

You might have noticed already that these are linear equations (they contain just x and y, with no powers or other products), and that you can write them as a matrix equation:

$$\begin{pmatrix} x' \\ y' \end{pmatrix} = \begin{pmatrix} \cos\theta & \sin\theta \\ -\sin\theta & \cos\theta \end{pmatrix} \begin{pmatrix} x \\ y \end{pmatrix}.$$

This is another instance of the way vectors and matrices pop up again and again when mathematicians want to represent and handle information. (You might also have noticed that the rotation matrix here is similar to the one in fig. 4.2, but there we were rotating the robot arm, whereas here, as in fig. 9.1, the vector stays the same, but the axes themselves rotate.)

If you let $X = \begin{pmatrix} x \\ y \end{pmatrix}$ and let A represent the "transformation matrix," you can write this coordinate transformation equation more economically as:

$$X' = AX.$$

As we saw in chapter 1, when algebraists began to use symbols instead of words or specific numerical examples, they were able to generalize their results—so this equation can be generalized, and the symbol A can stand for *any* 2-D linear homogeneous coordinate transformation. ("Homogeneous" here just means that the transformation maps the origin O to the origin O' of the new frame, and you need this for tensors, because then tensor equations such as $\boldsymbol{a} \cdot \boldsymbol{b} = 0$—or the condition for flatness, Riemann tensor = 0—will remain invariant.) And from what we've already seen of vectors and matrices, this equation also suggests we can easily move from our original 2-D rotation to transformations in any number of dimensions.

Different authors use different symbols, but in teasing out and generalizing the way coordinate transformations behave, I'll adapt the notation widely used in textbooks on tensor analysis today. It differs only slightly from Ricci's; in particular, both contravariant vector components *and* coordinates are written with upstairs indices, but Ricci denoted coordinates as we usually meet them in college math: $x_1, x_2, \ldots$. So, to explore a bit further the structure of a coordinate rotation—and to get a handle on how coordinate transformations work in general—first up, I'll use x^1, x^2 for the original coordinates (x, y), and $x^{1'}, x^{2'}$ for the new ones (x', y'). Then the original rotation transformation equation $x' = x\cos\theta + y\sin\theta$ can be generalized like this:

$$x^{1'} = A_1^{1'} x^1 + A_2^{1'} x^2,$$

where $A_1^{1'} = \cos\theta, A_2^{1'} = \sin\theta$ in our specific rotation case. (For the elements of the matrix representing the coordinate transformation, I've used $A_1^{1'}$, and so on, rather than the a_{ij} notation of linear algebra. You'll see why shortly.) Similarly, the transformation equation $y' = -x\sin\theta + y\cos\theta$ can be generalized as

$$x^{2'} = A_1^{2'} x^1 + A_2^{2'} x^2.$$

There's a pattern in the indices here: in each of these general equations, the same dashed upstairs index appears throughout. This means you can represent these two equations in one, where the general upstairs index μ'

(pronounced "mu-dash") is presumed to take the values 1 and 2 (each in turn), since there are two independent coordinates in 2-D space:

$$x^{\mu'} = A_1^{\mu'} x^1 + A_2^{\mu'} x^2.$$

Notice, too, that in each term in the sum on the right-hand side, the downstairs index on the matrix component is matched by an upstairs one on the original coordinate; so, using the Greek letter σ (sigma) and the summation notation I flagged at the end of chapter 10, you can simplify the above expression to:

$$x^{\mu'} = \sum_{\sigma=1}^{2} A_\sigma^{\mu'} x^\sigma.$$

Once Einstein gets on top of all this, he'll make this notation even simpler. He'll say, look at the pattern and notice that whenever you have the same up and down index—in this case, a σ appearing twice like this—you add all those terms. And since you also know what dimension you're working in, why not leave out the summation sign altogether, and let the repeated indices tell you that this is a sum:

$$x^{\mu'} = A_\sigma^{\mu'} x^\sigma.$$

Today this notation is called the Einstein summation convention.

You can see where I'm going with this: when you add more variables—so that you're working with an n-dimensional space (Riemann's manifold!)—you can use the *same* symbolic equation, except that now your indices μ' and σ will run from 1 to n, and your sum implicitly has not two terms but n of them. That's n equations (one for each value of μ'), each one a sum of n terms (one term for each value of σ), all encapsulated in just one small equation. It's brilliantly economical!

I've chosen μ' and σ for the index labels here, but they are meant to represent general indices, so there's nothing specific about the letters μ and σ themselves—just as the letter x for the unknown in school algebra is an arbitrary choice (as was my choice of $A_\sigma^{\mu'}$ for the transformation matrix components). So don't focus on the specific letters here, but rather on the patterns of the indices. As we'll see, it's this economical symbolism that makes the equations of physics so beautifully elegant when written in tensor form.

INVARIANCE AND TENSORS

Now we're getting to what all this has to do with invariance. The components of a vector $\boldsymbol{a}$ are measured from the coordinate axes, so they transform in the same way, $a^{\mu'} = A^{\mu'}_{\sigma} a^{\sigma}$. For example, for the rotation in figure 9.1, the components will transform just like the coordinates in the equations above, so we have—using Ricci's upstairs indices for the (contravariant) vectors $\boldsymbol{a}$ and $\boldsymbol{b}$—

$$a^{1'} = a^1 \cos\theta + a^2 \sin\theta,\ a^{2'} = -a^1 \sin\theta + a^2 \cos\theta,$$

and similarly for $b^{1'}$ and $b^{2'}$. When you multiply these pairs of components to form the scalar product in the rotated frame, you find that

$$a^{1'}b^{1'} + a^{2'}b^{2'} = a^1 b^1 + a^2 b^2.$$

You get the same number—the same scalar product—in both frames, which is what it means to be invariant. So this is an algebraic version of the geometric argument in figure 9.1.

But what about the scalar product as we write it in college notation, $a_1 b_1 + a_2 b_2$?

In Ricci's notation this would be the scalar product of covariant or row vectors (or one-forms). As we'll see later, it turns out that in the usual Euclidean space and Cartesian coordinate system, there's no need to distinguish between upstairs and downstairs indices on vector components. But first we need to see more about what the downstairs indices mean for tensors.

The matrix of components $A^{\mu'}_{\sigma}$ for transformations of coordinates and (contravariant) vectors shows how to write the new coordinates or vector components in terms of the old ones. So, to write your original coordinates in terms of the new ones, the transformation goes the other way. We saw this for the Lorentz transformations in figure 9.3, but you can see how to do it for *any* coordinate transformation by going back to writing general transformation equations as $X' = AX$. Then matrix algebra tells you that to transform the other way, you'll have

$$X' = AX \Rightarrow A^{-1}X' = X.$$

For example, the inverse of the rotation matrix $\begin{pmatrix} \cos\theta & \sin\theta \\ -\sin\theta & \cos\theta \end{pmatrix}$ is $\begin{pmatrix} \cos\theta & -\sin\theta \\ \sin\theta & \cos\theta \end{pmatrix}$, since $AA^{-1} = I$. This suggests that the *first column* in the original matrix becomes the *first row* in the inverse, and similarly for the second row. In other words, the columns of the original matrix—the contravariant vectors, with their upstairs indices—become rows, or covariant vectors with downstairs indices, in the inverse matrix. Which means that all you have to do to represent the components of the inverse matrix is to interchange the indices on the original one: $A^{\mu'}_{\sigma} \to A^{\sigma}_{\mu'}$. This is why Ricci wrote covariant vectors (one-forms or dual vectors) with a downstairs index: the transformation rule for their components is $a_{\mu'} = A^{\sigma}_{\mu'}a_{\sigma}$. Using this rule, you can prove that $a_1b_1 + a_2b_2$ is invariant, just as $a^1b^1 + a^2b^2$ is invariant.

Generalizing vectors to tensors, Ricci said that if all the indices on the components of a tensor, of any rank, are upstairs, it is called "contravariant"; it will transform just like contravariant vector components, but with the appropriate number of transformation matrices $A^{\mu'}_{\sigma}$ (as you can see in the endnote).[19] If all the indices are downstairs, Ricci called it a "covariant" tensor. If some indices are up and some down, it's called "mixed." For example, we saw that the components of the tensor product of a column vector by a row vector were u^1v_1, u^1v_2, u^2v_1, u^2v_2, and so on, so they are the components of a mixed tensor.

This is why Ricci said that vectors are tensors: their components—like higher-order tensor components—transform in a *specific way* under a change of coordinates. And scalars are tensors because they are just numbers or numerical expressions that don't depend on the coordinates at all—so they are automatically invariant under coordinate transformations. (Just to dot i's, not all numbers are invariants or scalars—for instance, frequency depends on the relative motion of the observer, as exemplified in the Doppler effect that we'll see in the next chapter. And if the Unruh effect is finally detected, it may even turn out that temperature is not exactly the coordinate-independent scalar I said it was in chap. 7 and fig. 11.1—although you'd have to be traveling close to the speed of light to detect one degree of temperature change.)[20]

A MODERN VIEW

Index notation isn't all about coordinate transformations, though, as we'll see in the next section. So here I want to outline a modern view of tensors. Coordinate transformations are still at the heart of it, but rather than *defining* tensors through the way their *components* transform under these coordinate changes, modern mathematicians define "whole tensors" as *linear operators* that yield invariants.

We saw in chapter 6 that $\frac{d}{dx}$ is an operator and so is its vector extension nabla, $\nabla = \frac{\partial}{\partial x}i + \frac{\partial}{\partial y}j + \frac{\partial}{\partial z}k$. You have to "insert" a function into the operator $\frac{d}{dx}$ to get its derivative, while inserting a function f into nabla gives you "grad f," a vector whose components are partial derivatives of the function. But a tensor is more like the divergence operator, $\nabla\cdot$, which operates on a vector to give a scalar. And scalars, as we just saw, are always invariant. For example, in Maxwell's equations, $\nabla\cdot$ operates on the electric and magnetic field vectors, giving a scalar:

$$\nabla \cdot \boldsymbol{E} = 4\pi\rho$$
$$\nabla \cdot \boldsymbol{B} = 0.$$

Similarly, multiplying a row vector (a covariant vector or one-form or dual vector) by a column vector (a contravariant vector) gives a scalar—the scalar product in Euclidean space. So, you can think of a one-form as something that *operates on* a vector to give a scalar, an invariant. This definition gets right to the heart of what tensors are all about: not component and coordinate transformations per se but *invariance* under those transformations.

Going up an order (or rank), you can think of a matrix as a mixed second-order tensor that operates on a vector *and* a one-form to give a scalar. More specifically, it operates on a (column) vector to give another (column) vector—like the rotation matrix A in the transformation equation $\boldsymbol{X}' = A\boldsymbol{X}$, where A "operated on" $\boldsymbol{X} = \begin{pmatrix} x \\ y \end{pmatrix}$ to give a new vector, $\begin{pmatrix} x' \\ y' \end{pmatrix}$; then, as we just saw, when you operate on this new vector with a row vector (one-form), you get the scalar product. The higher the order of the tensor, the more vectors and/or one-forms it must operate on to give a scalar.

This conception of tensors as operators is fundamental in quantum theory, for example, while pure mathematicians take the idea into the more abstract territory of multilinear mappings. So there's much more to say about the notion of linear operators—and about "vector spaces," too, and other subtleties such as the difference between vectors and one-forms, and the significance of transforming "basis" vectors and one-forms rather than vector and tensor components—but that's beyond my scope here. Still, if you've stayed with me so far, I hope this section and the previous one have given you a feeling for the way mathematicians develop their ideas—how they sort out the rules that their mathematical constructs have to obey, and how they interpret these rules and constructs in terms of important ideas such as invariance. These interpretations evolve as mathematicians build on their forerunners' insights. This is a key theme of this book, where we've already seen this kind of mathematical development, from cuneiform tables and computational algorithms to symbolic algebra, vectors, and matrices, and from calculus to vector analysis. Later in this chapter we'll take the final step to tensor calculus.

THE AMAZING COMPUTATIONAL POWER OF TENSOR SYMBOLISM

The position of the indices on tensor components plays a vital role in tensor equations and computations. For example, we've seen that in Euclidean space, multiplying a row (covariant) vector $\boldsymbol{v}$ by a column (contravariant) vector $\boldsymbol{u}$ gives their scalar product. Using Ricci's index notation and Einstein's summation convention we have a beautifully economical representation of this:

$$v_1 u^1 + v_2 u^2 \equiv v_\mu u^\mu.$$

There's nothing special about my choice of the letter μ here—again, I could have chosen any letter, because what matters is that both indices are the same (so this is a sum).[21] The amazing thing about this representation is this: the fact that each pair of up and down indices is the same *actually tells us* that this scalar product is invariant under appropriate coordinate transformations. You can *prove* it's invariant, simply by inserting the transformation equations, as you can see in the next endnote. But once you understand how to do that, the index notation saves you the bother.[22]

It really is remarkable the way tensor notation makes things easier. Nonetheless, you might be grumbling that so far, we've had three different types of scalar product, with upstairs indices, downstairs indices, and now mixed indices. That's simply because in tensor analysis there are two types of vector, but I'll show shortly how it all comes together in one general expression. For now, I want to focus on the remarkable way the mixed form of the scalar product shows invariance *through its very symbolism*. This carries over for higher-order tensors, too, so whenever you see an expression where each downstairs index is matched by the same upstairs one—such as $T_{\mu\nu}h^{\mu\nu}$ (ν is pronounced "nu")—you know it is invariant. It's quite extraordinary, really—and this is just one example of why Ricci's index notation is such a brilliant innovation. He was right to say it is worth the struggle of initiation.

Tensor expressions such as $v_\mu u^\mu$ and $T_{\mu\nu}h^{\mu\nu}$ are examples of the tensor operation called "contraction"—because when you set a pair of upstairs and downstairs indices equal, you're reducing, or contracting, the rank of your tensor. For instance, $v_\mu u^\lambda$ (where λ is pronounced "lambda") is a general component of a mixed rank 2 (two-index) tensor, but when you set $\lambda = \mu$, you reduce the rank (or order) to 0, because $v_\mu u^\mu$ is a scalar (the scalar product).

You don't have to contract all the indices unless you want to find the invariants. For instance, $T_{\mu\nu}h^{\lambda\sigma}$ is a general component of a rank 4 (4-index) tensor, but if you set $\lambda = \mu$ you get a two-index tensor, with components $T_{\mu\nu}h^{\mu\sigma}$. It's a 2-index tensor because you're summing on the repeated index μ, leaving only the ν and σ indices free. It's as if contracting the μ indices "cancels" them, rather like the way you "cancel" terms in the chain rule, $\frac{dy}{dx} = \frac{dy}{du}\frac{du}{dx}$, although in this case you're summing terms rather than "deleting" them.

There's an especially important contraction now called the "inner product" in tribute to Grassmann. As we saw earlier, Hamilton's system, which morphed into our university-level vector analysis, was perfectly adapted for three-dimensional problems—remember those 3-D rotations that had set him on the path to discovering quaternions! Grassmann's was more abstract, so although it was harder to apply, it was more readily adaptable to the n-dimensional spaces that Riemann created, and in which Ricci's tensors operate. So, by the early twentieth century, Grassmann's ideas had begun filtering into the mainstream, giving added conceptual substance to the vector and tensor analysis that had descended from Hamilton. Ricci was in the Hamiltonian tradition, and in his 1900 overview of his calculus he doesn't use the term "inner" (*or* "outer") product; by 1916, however—and to take just one example—in his overview of general relativity theory Einstein will use these Grassmannian terms.

So, what is the inner product? It comes from contracting a pair of indices on a mixed tensor formed from the outer product of two other tensors. For example, suppose you form the outer product of a covariant tensor $\boldsymbol{T}$ with components $T_{\mu\nu}$ and a vector $\boldsymbol{u}$ with components u^{σ}. You get a new mixed tensor whose components are $T_{\mu\nu}u^{\sigma}$. Now contract the indices by setting $\sigma = \mu$, to give $T_{\mu\nu}u^{\mu}$; this is the general component of the inner product of $\boldsymbol{T}$ and $\boldsymbol{u}$. (In component form, it is $T_{1\nu}u^1 + T_{2\nu}u^2 + \dots$, the number of terms depending on the dimension of the space.)

But look what happens if you take the *outer* product of this tensor with another (contravariant) vector, $\boldsymbol{v}$: you get a new tensor, with components $T_{\mu\nu}u^{\mu}v^{\lambda}$. (So inner products reduce or contract the rank, and outer products increase it.) If you now set $\lambda = \nu$, you get yet another new tensor, with components $T_{\mu\nu}u^\mu v^\nu$. This is the inner product of $T_{\mu\nu}u^\mu$ and v^λ. Like the scalar product $v_\mu u^\mu$, this tensor is a scalar (an invariant number or function), because each pair of indices is the same.

In fact, if $\boldsymbol{T}$ is a metric tensor—which from now on, and cribbing from Einstein, I'll denote by $\boldsymbol{g}$, with components $g_{\mu\nu}$—then this particular inner product is, in fact, just what we've been used to calling the scalar product of $\boldsymbol{u}$ and $\boldsymbol{v}$. For, as we've seen, the metric actually defines the scalar product. For example, the 2-D Euclidean metric

$$ds^2 = dx^2 + dy^2 \equiv (dx^1)^2 + (dx^2)^2$$

has components $g_{11} = g_{22} = 1$, with the other components zero; so, writing out the sums indicated by the repeated indices, the inner product in this case is:

$$g_{\mu\nu}u^\mu v^\nu = g_{11}u^1v^1 + g_{12}u^1v^2 + g_{21}u^2v^1 + g_{22}u^2v^2 = u^1v^1 + u^2v^2.$$

This is, indeed, the usual vector analysis scalar product $\boldsymbol{u} \cdot \boldsymbol{v}$, except with Ricci's upstairs indices because here both vectors are contravariant.

A PEEK AT SYMMETRY, WHY METRICS ARE TENSORS, AND SORTING OUT THE INDICES ON SCALAR PRODUCTS

We saw in chapter 4 that the scalar product is commutative (it's only the vector product that isn't). And we just saw that $\boldsymbol{u} \cdot \boldsymbol{v} = g_{\mu\nu}u^\mu v^\nu$ (and by implication $\boldsymbol{v} \cdot \boldsymbol{u} = g_{\nu\mu}v^\nu u^\mu$). So, this commutativity, $\boldsymbol{u} \cdot \boldsymbol{v} = \boldsymbol{v} \cdot \boldsymbol{u}$, means that we must have $g_{\mu\nu} = g_{\nu\mu}$. The indices on the metric tensor components are, therefore, *symmetric*, like a reflection in a mirror. This symmetry is invariant under linear coordinate transformations because the scalar product is, so it's handy in computations. As we've seen, invariance in general is a mathematical "symmetry," because when something is invariant, it stays the same—just like the shape of a reflected image or a rotated snowflake.

In chapter 9 we met the Euclidean and Minkowski metrics, where the coefficients of the differentials are constant, signifying that they define flat spaces. In chapter 10, we saw that Gauss proved information about the curvature of a surface is contained in the *coefficients* of the differentials in the general 2-D metric, and that Riemann generalized this to curved n-dimensional spaces. So, with arbitrary coordinates x^μ, a metric in curved space can be expressed as

$$ds^2 = g_{\mu\nu} dx^\mu dx^\nu,$$

where now the coefficients $g_{\mu\nu}$ are not constants but functions of the coordinates.

From the repeated indices you can see straightaway that the "distance" or space-time interval measure, ds^2, is invariant, and you can prove it by using the coordinate transformation equations.[23] We've already seen examples of this general result: the Euclidean metric is invariant under transformations such as rotations, while the Minkowski metric is invariant under Lorentz transformations (fig. 9.3). But why is the metric a tensor? The clue is in the invariance; after all, representing invariance is the whole point of tensors.

To see this in more detail, we saw just before that $g_{\mu\nu} u^\mu v^\nu$ is the scalar product of the vectors $\boldsymbol{u}$ and $\boldsymbol{v}$. This suggests that the metric $\boldsymbol{g}$ "operates on" these two vectors to produce a scalar—the invariant scalar product. Which means the metric is a tensor according to the modern definition I gave two sections back.

It's also a tensor according to Ricci's definition, because we know the transformation equations for the contravariant vector components u^μ, v^ν, so the transformation equation of $g_{\mu\nu}$ must be that of a covariant rank 2 tensor if $g_{\mu\nu} u^\mu v^\nu$ is to be an invariant scalar. The previous endnote illustrates the calculations that show this, but you can see already that the modern view is more elegant.

• • •

There's one last thing I want to show you before I briefly outline Ricci's crowning achievement, tensor derivatives. When I spoke earlier of the scalar product in the various forms $v_1 u_1 + v_2 u_2$, $v^1 u^1 + v^2 u^2$, and $v_1 u^1 + v_2 u^2$,

I was assuming we were in 2-D Euclidean space, where the metric is $ds^2 = dx^2 + dy^2$. More generally, we've just seen that in a space with a metric whose components are $g_{\mu\nu}$, the scalar product of two (contravariant) vectors $\boldsymbol{v}$ and $\boldsymbol{u}$ is $g_{\mu\nu}u^\mu v^\nu$. Now look what happens if I swap the positions of the indices and write $g^{\mu\nu}v_\mu u_\nu$. This suggests the scalar product of two covariant vectors. But what is $g^{\mu\nu}$? Ricci defined $g^{\mu\nu}$ so that it has a very special property: it "raises the index" of a covariant tensor. We saw a little earlier that $T_{\mu\nu}h^{\mu\sigma}$, with repeated (contracted) μ indices, is a two-index tensor, as if we'd "canceled" the μ's when we summed them. So, Ricci defined the special contractions

$$g^{\mu\nu}g_{\mu\sigma} = g^{\nu}_{\sigma}, \text{ and } g^{\lambda\sigma}\, g^{\nu}_{\sigma} = g^{\lambda\nu}.$$

(Actually, Ricci used a^{rs} rather than $g^{\mu\nu}$, but otherwise I'm using his definitions.) In other words, $g^{\mu\nu}$ has taken $g_{\mu\sigma}$ to g^{ν}_{σ}. And $g^{\mu\nu}g^{\lambda\sigma}$ has taken $g_{\mu\sigma}$ to $g^{\lambda\nu}$. It works the other way, too: $g_{\mu\nu}$ can lower the indices. (That's because Ricci essentially defined the matrix representations of the metric components $g_{\mu\nu}$ and $g^{\mu\nu}$ to be inverses of each other. But the key thing is that these *are* definitions.)

Ricci defined this as a general property, so that the metric tensor can raise or lower indices when it is contracted with *any* tensor. This is important in tensor equations such as Einstein's, as we'll see. But it also brings all those different forms of the scalar product together in a rather brilliant way. The contraction $g_{\mu\nu}v^\mu$ lowers the index on the vector, giving v_ν. This means that in general, $g_{\mu\nu}v^\mu u^\nu = v_\nu u^\nu$, just as I had earlier for the particular scalar product of the row and column vector! Similarly, $g^{\mu\nu}v_\mu u_\nu = v^\nu u_\nu$.

But here's the really interesting thing. In Euclidean space, using Cartesian coordinates, we know that the metric is $ds^2 = dx^2 + dy^2$. I'm keeping to 2-D for simplicity, but of course you can add more dimensions and the result I'm heading to will be the same: namely, in this situation it doesn't matter if you write your vectors' indices with upstairs indices or downstairs ones. That's because (in 2-D)

$$v_\nu = g_{\mu\nu}v^\mu = g_{1\nu}v^1 + g_{2\nu}v^2,$$

but the only nonzero components of the Euclidean metric are $g_{11} = 1 = g_{22}$, so, you have

$$v_1 = g_{11}v^1 + g_{21}v^2 = 1 \times v^1 + 0 \times v^2 = v^1,$$

and similarly, $v_2 = g_{12}v^1 + g_{22}v^2 = v^2$. In other words, in Euclidean space with the usual Cartesian coordinates, there's *no difference* between components of a vector and a one-form (that is, between Ricci's contravariant and covariant vectors). That's why, in ordinary vector analysis, there's no need to worry about this distinction in terminology or in the position of indices.

TENSOR CALCULUS, VERY BRIEFLY

The biggest question for Ricci was, what happens if you differentiate a tensor? More specifically, is the derivative a tensor? If it's not, then tensors are not much use in physics, where physical phenomena are widely modeled by differential equations—as in Newton's laws of motion and Maxwell's equations of electromagnetism, for example. As we saw in chapter 9, the equations of physics must also keep their form invariant, even when you change from one reference frame to another. Otherwise, different observers would deduce different laws of physics, and we'd never be able to agree on the nature of physical reality.

Form-invariant equations are called "covariant," and Ricci called his invariant derivative a "covariant derivative." It's not the same as an ordinary or partial derivative, because it turns out that the partial derivative of a vector component u^μ with respect to one of the coordinates in that frame, say x^λ, doesn't, in general, transform like a tensor. Ricci found the right covariant form of the derivative by using an invariant expression discovered by Christoffel. It amounts to adding to the partial derivative a term involving the Christoffel symbols I mentioned in chapter 10. Ricci followed Christoffel and denoted these symbols with curly brackets, but today, following Einstein and others, they're usually denoted by $\Gamma^\mu{}_{\sigma\lambda}$. (The symbol Γ is the upper-case Greek letter gamma; the indices fit in with the relevant transformation equations, although today they are interpreted as the coefficients of the derivatives of basis vectors. For instance, in Cartesian coordinates, the basis vectors are ***i***, ***j***, ***k***. They are constant, so their derivatives are zero, and hence so are the Christoffel symbols in this case. So, in Euclidean space, you only need partial derivatives!)

Taking the partial derivative of a vector or tensor component gives it a second index—to show the variable with which the component is being differentiated. Ricci represented the partial derivative of a vector simply by adding another index, but today, the new index is indicated with a comma to show it's a derivative. So, the partial derivative of u^μ is represented as

$$\frac{\partial u^\mu}{\partial x^\lambda} \equiv u^\mu{}_{,\lambda}.$$

A semicolon is often used to designate the *covariant* derivative:

$$u^\mu{}_{;\lambda} = u^\mu{}_{,\lambda} + \Gamma^\mu{}_{\sigma\lambda} u^\sigma.$$

I've shown this equation just to give you a visual, so the details are not important—except to say that there's a natural extension to covariant derivatives of any tensor, not just vectors. The key thing is that it's a tensor—so its value is invariant under the relevant coordinate transformations. In other words, you get the same result when you transform to new coordinates and take the covariant derivative of the transformed component $u^{\mu'}$ with respect to $x^{\lambda'}$.

In Cartesian coordinates, the difference between partial and covariant derivatives disappears in flat spaces and space-times. This is another example of the way distinctions that are important in curved spaces disappear in flat space: in the Euclidean vector analysis we learn in college, there's no need to talk about covariant derivatives, just as there's no need to worry about the position of the indices on vectors.

NOBODY CARED!

People had been studying invariance and differential geometry for years when Ricci brought it all together in tensor analysis—his "absolute differential calculus." He drew on some of these earlier researchers—chiefly Gauss,

Riemann, and Christoffel, but also others such as Sophus Lie, whose work on groups and invariance is still important, and the renowned differential geometer, Eugenio Beltrami, who'd been Ricci's professor when he was a student at Bologna.

When Ricci entered some of his papers on tensors for Italy's Royal Mathematics Prize in the late 1880s, Beltrami was a judge. Speaking on behalf of the judging committee, he admired Ricci's mathematical virtuosity, but he wondered if the effort that had gone into creating this new calculus would ever be repaid by sufficiently fruitful applications—applications that could not be made using existing methods.[24] He rather seemed to doubt it—just as William Thomson and Arthur Cayley could see no benefit in whole-vector analysis over separate component calculations. (The vector wars were going on in Britain at the very same time as Beltrami's pronouncement about tensors in Italy.) But just as Maxwell had argued that the importance of whole vectors in physics was the physical insight they offered, so Ricci would later write that when it came to understanding curved n-dimensional surfaces and spaces, his calculus and its notation "contribute not only to the elegance, but also to the agility and clarity of the demonstrations and conclusions."[25]

Back in the 1880s, though, it must have seemed to Ricci—who still hadn't got his promotion to professor—that for all the promise of his brilliant student years, he would forever go unrecognized and unfulfilled.

(12)

EVERYTHING COMES TOGETHER

Tensors and the General Theory of Relativity

True trailblazer that he was, Ricci never stopped believing in the importance of his tensor calculus. Then, imperceptibly at first, the tide of its fortune began to turn. The first augury came when a remarkable student, Tullio Levi-Civita, enrolled in Ricci's class in 1890—the same year thirty-seven-year-old Ricci finally won his full professorship. Levi-Civita had graduated from his secondary school with an almost perfect score in all his subjects, and in 1894 he graduated from the University of Padua with an almost perfect score, too. Even his career began with a perfect trajectory, from Ricci's assistant to his fellow professor by the time he was twenty-nine.

The ebullient, secular Levi-Civita couldn't have been more different from his straitlaced, devout former professor—yet the two men would remain close friends and colleagues till the day Ricci died. They wrote many joint papers developing Ricci's "absolute differential calculus," and this led to tensor calculus's second change of fortune—when Felix Klein suggested they write an overview of it, to put it out into the mainstream. In 1900,

Klein published this seminal paper in his journal *Mathematische Annalen*—and twelve years later, Einstein and Grossmann read it, and were blown away.[1]

THE HAPPIEST THOUGHT: EINSTEIN'S THINKING ABOUT GRAVITY BEFORE HE DISCOVERED TENSORS

We saw in chapter 10 that in the autumn of 1912—soon after Einstein joined his old friend Marcel Grossmann on the faculty of the Swiss "Poly" or ETH—Grossmann suggested Ricci's "absolute" calculus as the way to transfer the laws of physics from the framework of special relativity to a general theory that included gravity. But to get to that point Einstein had already done a lot of thinking.

The difficulty in finding a theory of gravity in those days before space travel was that gravity is always with us here on Earth. With other forces, such as electromagnetism, you can build up a theory based on experiments that show how one charged particle moves in the field of another, for example. But you cannot readily isolate the gravitational effect of one body on another, because the force between everyday objects is negligible compared with the effect of Earth's gravitational field, which has the remarkable property that when there's no air resistance, all material bodies are drawn downward at the same rate. This of course is Galileo's famous result, which the Hungarian Baron Roland von Eötvös confirmed with unprecedented accuracy in 1909, and which was confirmed in 2022 to a spectacular accuracy of about one part in a thousand trillion, by observations made with the French satellite MICROSCOPE.[2]

Newton had based his theory of gravity on this law of falling motion—and on ingenious deductions from the observed motion of the moon and planets. It is still extremely accurate for most applications within the solar system, but in 1907 when Einstein began thinking about gravity, there *was* a well-known exception. Each planet revolves around the sun in an elliptical orbit, but because of the gravitational pull of the other planets, these elliptical orbits slowly precess—the way a tilted spinning top precesses around its axis because of the downward pull of Earth. This means that the line between the sun and the perihelion (the point in the planet's orbit that is closest to the sun) shifts slightly with each planetary revolution—but in the case of Mercury, the Newtonian calculation for the precession of the perihelion was out by a tiny 43 angular seconds a century. An angular second, or arc second, is 1/3600th of a degree, so it is amazing that nineteenth-century astronomers had managed to detect this miniscule difference.[3] So, how to create a more accurate theory of gravity?

Einstein was still at the patent office when he had a marvelous insight—"the happiest thought of my life."[4] He realized that we on Earth can, in fact, imaginatively "transform away" the effect of Earth's gravity, simply by changing our frame of reference. Our usual point of view is from a frame fixed firmly to the ground, but what if we are in free fall—falling off a roof, for example? Imagine that you are in this vertiginous situation, and then suppose you let go of a ball you were holding. A person watching from the ground will see that both you and the ball continue to fall at the same rate. That is how falling objects behave when we view them from terra firma. Relative to you, however, the dropped ball stays motionless—just as if there were no gravity acting on it at all.

In other words, you could "unmake" a gravitational field by using a free-falling observer like the poor person falling off the roof—Einstein didn't know that you could "free fall" more safely in a spaceship! But that's not all: Einstein realized that you could "make" a gravitational field, too—by putting yourself in an upwardly accelerating frame such as an elevator. We'll explore this idea in figure 12.1, but first, what Einstein took from his "happy thought" was that he could indeed devise a relativistic theory of gravity, since we can make or unmake a gravitational field simply by changing reference frames. (We can only do this "locally," though.)[5] This is why results such as those of the 2022 MICROSCOPE collaboration are so crucial: if all bodies didn't fall at the same rate under the same gravitational force, then a free-falling observer could not "unmake" gravity, and the whole conception of general relativity would fall apart.

Almost immediately Einstein saw that this equivalence between gravity and acceleration—what he called "the principle of equivalence"—

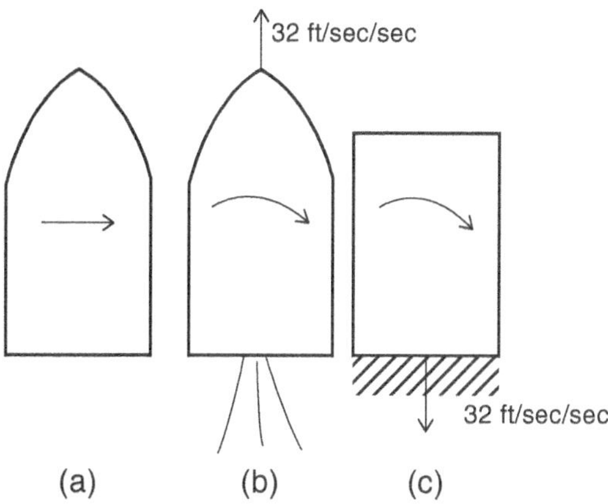

FIGURE 12.1. The principle of equivalence. The diagram shows three frames of reference: (a) an "inertial" frame, such as a gravity-free (free-floating) spaceship or a free-falling person or elevator, (b) a spaceship or elevator accelerating upward at 32ft/sec/sec (or 9.8m/sec/sec), and (c) a room here on Earth. In (b), with the floor of the spaceship or elevator pushing up on you as it accelerates upward, you would feel just as heavy as you do on Earth (frame (c)), which pushes up on us in response to our downward gravitational acceleration of 32ft/sec/sec or 9.8m/sec/sec. (These values are for the gravitational acceleration on the surface of Earth at sea level—at higher altitudes, of course, the acceleration decreases according to Newton's inverse square law.)

In the inertial frame (a), in the absence of forces, once launched a projectile will move in a straight line with constant speed. This is Newton's first law—the law of inertia, hence the term "inertial frame." In (c), a projectile has a downward parabolic path, as Harriot and Galileo showed—and the same happens in (b), which is equivalent to (c). (As the spaceship accelerates upward, the floor comes up to meet the projectile, so an outside observer sees the projectile move closer to the floor. In other words, its path appears bent.) Similarly, a light ray sent from one side of the rocket or elevator to the other will be straight in frame (a), but the floor in (b) will rise to meet it so that it appears bent to an outside observer. Since (b) is equivalent to (c), Einstein deduced that gravity bends light.

meant that gravity must bend light rays. Newton had suggested this, too, because he believed light was made of material particles that would be affected by gravity, so they'd follow a parabolic trajectory like Galileo's and Harriot's cannonballs. But Einstein had shown in 1905 that light particles—photons—have no rest mass, so his argument was very different from Newton's. You can see the nub of it in figure 12.1, which illustrates the "principle of equivalence." And there's more: since light takes longer to reach a distant observer when it travels on a curved rather than a straight path, gravity appears to *slow down* the speed of light. This is very different from special relativity, where the speed of light is constant, the same for all observers. (The *local* vacuum light speed, measured in a locally inertial frame such as figure 12.1(a), is still the constant *c*, but the observer measures a slower speed for light traveling on the curved path from a *distant* source.)

As early as 1911, when he was a professor at the German University in Prague,[6] Einstein published a mathematical derivation quantifying the amount of light bending near the sun. In this paper he also developed another idea he'd had back in 1907: the gravitational redshift, where gravity causes a shift in light frequency toward the longer-wavelength (or red) end of the spectrum. It's analogous to the Doppler effect that gives the sound of a wailing siren on an ambulance or police car its distinctive sound: higher as the vehicle accelerates toward you, lower as it moves away. That's because the sound waves are first compressed relative to you, making their wavelength shorter, and then expanded as the vehicle recedes—and the same happens with light waves. Imagine a pulse of light sent from the bottom of the spacecraft in figure 12.1(b) to an observer at the top. As the light travels upward, the observer is accelerating upward with the spacecraft, so they are moving away from the point where the light was emitted; this means the light's wavelength appears elongated or reddened. But since figures 12.1(b) and (c) are "equivalent," an observer at the top of a ladder in (c) should see redshifted light from a source on the ground, where gravity is stronger. This is a simple argument for the existence of *gravitational* redshift! (The resulting calculations are only approximate, though, since the equivalence principle holds only locally. Another redshift argument, which Einstein used, is that light loses energy traveling against a gravitational field. But he would need his final theory for exact, predictive calculations.)

If gravity relativistically redshifts light to a longer wavelength, it must do the same for other vibrations, too—including the ticking of a clock.

Which means that distant observers see clocks run more slowly in a stronger gravitational field (because there is a longer time between ticks). This fact is now taken into account—along with special relativity's time dilation, courtesy of the Lorentz transformations—in calibrating GPS directions. In 1911, though, Einstein's focus was on predictions that might be tested later, when the technology was up to it: in fact, he'd come up with the redshift argument to test the slowing of time idea, which he'd already intuited in 1907.

• • •

In 1911 Einstein began to explore his ideas in more depth. He needed to be able to transfer his intuitive thought experiments into testable equations, and since gravity can slow down the apparent speed of light, he wondered if this variable light-speed might represent the "gravitational potential" in his new theory. I mentioned in chapter 6 that Joseph-Louis Lagrange had formulated Newton's inverse-square law of gravity in terms of a scalar "potential," V. It comes out easily starting from Newton's laws, as you can see in this endnote,[7] or from Gauss's flux law (analogously to Maxwell's derivation of $\nabla \cdot \mathbf{E} = 4\pi\rho$ in chap. 6). Either way, Newton's law of gravity can be expressed as

$$\frac{\partial^2 V}{\partial x^2} + \frac{\partial^2 V}{\partial y^2} + \frac{\partial^2 V}{\partial z^2} \equiv \nabla^2 V = 4\pi G\rho.$$

Here ρ is the density of the matter producing the gravitational field. It's analogous to the charge density in Maxwell's equations, with the same caveats.[8] G is the gravitational constant of proportionality in Newton's inverse square law.

The original advantage of this formulation was that it treats the potential as a continuous scalar "field," to use Faraday-Maxwell language, rather than the action-at-a-distance view that seemed implicit in the inverse-square law, which takes into account the numerical distance between the gravitating particles but not the nature of the whole space between them. In fact, since the inverse square law gives the force exerted by a given mass on a unit test mass placed at any point in the surrounding space, it can be interpreted as defining a vector field; still, the calculations are often simpler using the scalar potential. So, Einstein wondered if his variable speed of light could play a similar role in his new theory as the potential V plays in Newton's—and he went on to create a "static" gravitational theory. "Static" means unchanging in time, so it describes a quiescent star, say, or the weak, essentially uniform gravitational field here on Earth. But Einstein was still feeling his way, and he assumed (wrongly) that in a static gravitational field, the spatial part of space-time would be Euclidean, as it is for everyday calculations on Earth. This meant that while the Minkowski metric describes the space-time in special relativity, the relevant metric describing space-time in Einstein's static gravitational field would be "almost Minkowskian":

$$ds^2 = dx^2 + dy^2 + dz^2 - (c(x, y, z))^2 dt^2,$$

where c, the speed of light, is now a function of the spatial coordinates. This is all a mistake, as he'll realize later. Meantime, he published this static theory in 1911.

All up, in the little more than a year that he spent in Prague, he published six papers on relativity (and five other papers). He was putting ideas out there, inviting feedback that would help refine them—although he got a little more than he bargained for from the sharp-tongued Max Abraham. Abraham was an expert on electromagnetism—he had published Germany's leading textbook on Maxwell's theory, and he'd also created a pioneering (but now obsolete) model of the newly discovered electron. But he was a firm believer in Hendrik Lorentz and Henri Poincaré's semirelativistic ether-based approach to electromagnetism, and he'd been mightily upset when Einstein's special relativity dispensed with it. Now, in 1911, he applied his sharp mind, and his unfortunately sharp pen, to public criticisms of Einstein's attempts to generalize relativity. Yet he was no mere carping critic, and surprisingly, in early 1912 he published his own relativistic theory of gravity.

While the metric in Einstein's static theory was "almost" the Minkowski metric, Abraham's theory was fully Minkowskian—and initially Einstein was captivated by it. (There had already been other attempts to fit gravity into special relativity, including by Poincaré and Minkowski.) As Einstein told his old friend Michele Besso, a fellow graduate from the Swiss Polytechnic and his longtime insightful sounding-board, "At first (for 14 days) I too was completely bluffed by the beauty and simplicity of [Abraham's] formulas." But then Einstein spotted "some serious mistakes in reasoning. . . . This is what happens when one operates formally [mathematically] without thinking physically!"[9]

For months he and Abraham traded arguments in the pages of *Annalen der Physik,* in a mini version of the vector wars in *Nature* two decades earlier. Abraham's caustic tongue had alienated him from most of the physics community, but the outspoken Einstein had sympathy for Abraham's tendency to run off at the mouth, and his criticisms did help sharpen Einstein's ideas. In July 1912, just before he left Prague for Zurich and the ETH, Einstein published a paper on his attempts to apply the equivalence principle in his search for a new equation of gravity, adding, "I would like to ask all of my colleagues to have a try at this important problem!" But Abraham responded snidely, "Einstein begs credit for the theory of relativity of tomorrow and appeals to his colleagues so that they may guarantee it." At which point Einstein gave up the debate, but not his own battle for a new relativistic theory of gravity—nor, it must be said, his respect for Abraham's ability as a physicist.[10]

• • •

This was, essentially, the state of Einstein's progress toward general relativity when he arrived in Zurich in August 1912. His intuitive deductions of the equivalence principle and its consequences for the measurement of time and light speed were utterly inspired—but brilliant ideas were one thing; proving them was quite another. Direct experimental proof of the gravitational redshift, for example, wouldn't come until 1960, although Einstein didn't live to see it: he died in 1955. But in 1912 proof was a minor problem compared with Einstein's struggle to find the equations for a fully relativistic theory of gravity. His genius was in physical thinking, not

Albert Einstein in 1912. ETH-Bibliothek Zürich, Bildarchiv/Photographer: Jan. F. Langhans/ Portr_05936. Public domain.

math, although he was actually a very good mathematician, if not a well-informed one—he'd skipped too many of Minkowski's classes. So, it wasn't long before he pleaded with his old classmate, "Grossmann, you must help me, otherwise I'll go crazy!"[11]

WHY EINSTEIN NEEDED HELP

The problem driving Einstein crazy was this: how to reconcile the principle of equivalence—where the observers in figures 12.1(b) and (c) can't tell if they're in a smoothly accelerating elevator or a closed room in a gravitational field—with the principle of relativity, where the laws of physics should have the same form for all observers.[12] His static theory of gravity was a consequence of the principle of equivalence, from which he'd deduced the variable speed of light that played the role of the gravitational potential. But this seemed to contradict the principle of relativity, because clearly it didn't gel with the special theory, where the Lorentz transformations we met in figure 9.3 rely on the constant speed of light. (Einstein's new general theory of relativity would have to fit with the special theory, to cover cases where the relative acceleration is zero.)

What's more, because the relative speed of any two observers is constant in special relativity, Newton's first law (of inertia) holds for both of them. As we saw in chapter 3, Newton defined force in terms of the *change* in a body's velocity, so the law of inertia says that in the *absence* of forces, an object at rest remains at rest, and an object moving with constant velocity remains moving with that constant velocity. Applying this in special relativity, if an object is at rest in the frame of observer A, then it is in constant motion according to observer B, and vice versa; either way, the law of inertia holds for both observers, and so just as in Newtonian physics, their frames of reference are "inertial." When the relative motion is accelerated, however, this is no longer the case.

This suggested to Einstein that there was no empirical reason to tie a general theory of relativity to the inertial reference frames of the special theory.[13] Which meant there was no longer a single, finite group of coordinate transformations that kept the laws of physics invariant, the way the Lorentz transformations do in special relativity. Rather, the laws of nature should be invariant under *any* smooth coordinate transformation.[14] Since there was no special group of coordinate transformations, there was no special metric that remained invariant. So, the Minkowski metric had to be replaced by a completely general one,

$$ds^2 = g_{\mu\nu} dx^\mu dx^\nu,$$

where the coefficients $g_{\mu\nu}$ are, in general, functions of the coordinates. (So, the Minkowski metric is just one special case of this general metric.)

This meant that the space-time described by this metric must, in general, be curved—although Einstein only realized this when Grossmann introduced him to Riemannian geometry and the condition for curvature we met in chapter 10 (namely, that the Riemann tensor, which is built from derivatives of the $g_{\mu\nu}$, is not zero). In Prague, Einstein *had* already figured out that space-time must be non-Euclidean, from an intuitive argument based on length contraction, but one of the things he'd been tearing his hair over was that in a completely general metric, the x^μ could be *any* kind of coordinates. Which meant that the dx^μ didn't necessarily relate directly to time and space measurements, as they do in the standard (Cartesian) form of the Euclidean and Minkowski metrics, where the dx^μ are the time and space differences dt, dx, dy, dz.

Einstein was "much bothered by this," as he later recalled[15]—until just before he returned to Zurich. I showed in chapter 9 that in special relativity, even though Cartesian coordinates do relate to time and space in Minkowski space-time, there is no agreement between relatively moving observers on actual time and distance measurements (as the Lorentz transformations in the box for fig. 9.3 show). Rather, we saw that it is the expression $x^2 + y^2 + z^2 - (ct)^2$—and, therefore, the Minkowski metric, $ds^2 = dx^2 + dy^2 + dz^2 - (cdt)^2$—that is invariant. In other words, the two observers only agreed on ds, the interval (or "distance") between events in space-time, as given by the metric. Einstein hadn't used the interval ds in his special theory—that had come from Minkowski and Gauss, as we saw in chapter 10, and Einstein had taken a while to warm to this *geometric* "metric" approach. (In his 1905 paper he'd focused on *algebraic* consequences of the Lorentz transformations.) But just before leaving Prague, he'd realized something fundamentally important about ds and the principle of equivalence.

In a gravitational field or accelerating reference frame, when no other forces are acting, material objects and light photons must travel on *curved geodesics*, rather than the straight lines they take in inertial frames. This was, finally, Einstein's way of rigorously expressing the principle of equivalence sketched in figure 12.1. For, as we saw in chapter 10, the shortest distance between two points lies on a geodesic, and this is found by minimizing (or, rather, extremizing)[16] the "distance" $s = \int ds$. As Einstein suddenly realized, this means that it is ds, or the *metric as a whole*, which has physical meaning in his new theory—not the separate coordinate differentials in the metric. And if it is to have physical meaning, all observers must agree on it, as they do with the Minkowski metric in special relativity. But how could Einstein express this invariance when he only had a general form of the metric? He was flummoxed.

So, loyal Grossmann went off to the library, to see if anyone had figured out a way to keep *general* metrics and equations invariant. And that's how he and Einstein came to read Ricci and Levi-Civita's landmark paper on tensor analysis.

EINSTEIN AND GROSSMANN TAKE UP TENSOR CALCULUS

Grossmann was passionate about mathematics, but he hadn't cared much for physics until he "caught fire" with Einstein's vision.[17] And what a remarkable vision it was, even at this early stage. We've already seen the extraordinary deductions he made from the principle of equivalence, but Einstein was also homing in on the radical idea that gravity and geometry were inextricably entwined. That's because the metric had to describe *both* the curved geometry of space-time *and* the gravitational field itself—for when there's no gravity, you're back to the Minkowski metric of flat space-time and to objects moving on straight lines instead of curved geodesics. So, gravity affected the geometry—the curvature—of space-time, and the curvature described the gravity. At least, it would if Einstein could ever find the right equations.

Once again, his starting point was Newton's equation of gravity, expressed in terms of a gravitational potential: $\nabla^2 V = 4\pi G\rho$. After all, any new theory of gravity had to fit with Newton's spectacularly successful theory in the weak, fairly uniform fields here on Earth and elsewhere in the solar system far away from strong sources of gravity. Einstein had abandoned his earlier idea that the speed of light was the gravitational potential because it didn't gel with special relativity, but now he made a daring assertion: the metric coefficients $g_{\mu\nu}$ *themselves* represent not only the metric properties of space-time but also the relativistic analog of the Newtonian gravitational potential V. (That's why he and Grossmann used the letter g that I've already cribbed for the metric coefficients of curved space-time.)

It was daring conceptually—because, as he realized later, it made explicit the link between geometry and gravity—and it was daring mathematically, because it suggested a complexity never seen before in mathematical physics. That's partly because in 4-D space-time, where the indices each take on the values from 1 to 4, a two-index tensor has $4 \times 4 = 16$ different components, although for the metric tensor, the indices should be symmetric, $g_{\mu\nu} = g_{\nu\mu}$ (because the scalar product is commutative, as we saw near the end of chap. 11)—which means there are only ten independent metric coefficients:

$$g_{11}, g_{12}, g_{13}, g_{14}, g_{22}, g_{23}, g_{24}, g_{33}, g_{34}, g_{44}.$$

But that's still *ten* gravitational potentials, so there'd have to be *ten* gravitational field equations instead of the single Newtonian equation $\nabla^2 V = 4\pi G\rho$. What a challenge! And where to begin?

Well, for a start, in the Newtonian equation the gravitational potential on the left-hand side of the equation relates to ρ, the density of the matter that is the source of the gravity represented on the right. For Einstein and Grossmann, who were now collaborating closely on the problem, the obvious analogy for the left-hand expression seemed to be some sort of ten-component tensor expression made from the second derivatives of the potentials $g_{\mu\nu}$, like the second derivatives in $\nabla^2 V$.

It wasn't long before Grossmann found that the "Ricci tensor," $R_{\mu\nu}$, was such a beast. Ricci had discovered this tensor in 1903—although he didn't name it; it is usually later researchers that bestow such honorific mathematical and scientific terminology. The Ricci tensor is the contraction of the four-index Riemann tensor I mentioned in chapter 10, and we saw in chapter 11 that "contraction" means that you set an upper and lower index equal and sum over them: $R_{\mu\nu} \equiv R^{\alpha}{}_{\mu\alpha\nu}$. It looks innocuous, doesn't it? In fact, since the Riemann tensor is itself the sum of derivatives of the metric coefficients, $R_{\mu\nu}$ is the sum of hundreds of terms! Many of them cancel or are zero, but nonetheless there's an awful lot of calculating in general relativity; by the late 1980s the increasing availability and sophistication of computer algebra packages were becoming a boon to researchers, as you can imagine.

By the way, we'll see soon that unlike vector equations, tensor equations are best written in terms of general components, because tensor indices are so vital for computations—such as the contraction that gives the Ricci tensor. So from now on I'll generally follow the common practice of writing tensors this way: $g_{\mu\nu}$ for the metric tensor, $R_{\mu\nu}$ for the Ricci tensor, $R^{\alpha}{}_{\mu\beta\nu}$ for the Riemann tensor, and so on—rather than their more correct representation using whole-tensor labels such as ***g***, ***Ric***, ***Riem***, say. So, when you see an equation using general indices such as $\mu, \nu, \ldots$, you know it holds for all values of the indices. You also know it is form-invariant (covariant) under coordinate transformations, whereas equations in terms of specific components—with indices designated by numbers as in R_{12}, or by specific coordinate indices as in v_x for the x-component of velocity, say, or B_y for the y-component of the magnetic field—are not.

• • •

With the Ricci tensor as a candidate for the left-hand side of his equation, Einstein and Grossmann needed an analog of the Newtonian mass density ρ for the right-hand side. The mass density is a way of characterizing the distribution of the matter producing a gravitational field, but obviously energy should be part of the mix, too, courtesy of $E = mc^2$. As to how to put this into a tensor, the two collaborators' former math professor, Minkowski, had, in fact, already pointed the way, when he wrote Maxwell's equations in tensor form.

In the usual vectorial form of Maxwell's equations, the electric and magnetic field vectors ***E*** and ***B*** are intertwined: in the curl equations you have ***E*** on one side of the equation and ***B*** on the other. What Minkowski did was to combine these intertwined 3-D fields into a single tensor in flat 4-D space-time. As I mentioned in chapter 9, tensors were not yet in the mainstream—and the name "tensor" wasn't generally used in this context until after Einstein published his 1916 overview of the foundations of general relativity. So, in 1910, when Sommerfeld followed Minkowski's lead and combined the electric and magnetic field vectors ***E*** and ***B*** into a single quantity, ***F*** (usually denoted by its general coordinate $F^{\mu\nu}$), he used the term "six-vector" rather than "tensor." That's because $F^{\mu\nu}$ has six independent components, since it is just another way of writing the information in $\boldsymbol{E}$ and $\boldsymbol{B}$, which each have three components, one for each of the axes x, y, z. Surprisingly, even in his 1916 overview Einstein, too, called $F^{\mu\nu}$ a "six-vector," although he used "tensor" elsewhere. But he did use the now common notation I'm using here—aside from the fact that he used ∂ rather than the modern commas for partial derivatives, which we saw at the end of chapter 11. I'll list these six electromagnetic tensor components so you can see the idea, but just take them as definitions:

$$F^{14} = -E^x, F^{24} = -E^y, F^{34} = -E^z, F^{12} = B^z, F^{31} = B^y, F^{23} = B^x;$$

in addition, $F^{\mu\nu} = -F^{\nu\mu}$, and $F^{\mu\mu} = 0$.

With these definitions, the four beautiful Maxwell equations of chapter 8 become even more economical, even more elegantly simple. I'll give them here just to show you, but we don't need the details—except that for comparison with the vector version, j^μ designates the current density or *source* of the electromagnetic field:

$$F^{\mu\nu}{}_{,\nu} = 4\pi j^\mu$$
$$F_{\mu\nu,\lambda} + F_{\nu\lambda,\mu} + F_{\lambda\mu,\nu} = 0.$$

Minkowski had checked that $F^{\mu\nu}$ is, indeed, what we now call a tensor.[18] What he'd done was quite brilliant: he'd turned the 3-D vectors $\boldsymbol{E}$ and $\boldsymbol{B}$, which are not frame independent—for instance, a moving electric charge creates a magnetic field, so $\boldsymbol{B}$ is zero in the charge's rest frame but not in a relatively moving observer's—into the 4-D space-time tensor $F^{\mu\nu}$, which is! Its *components* change under Lorentz transformations, but if the whole tensor is zero (or nonzero) in one frame, it is zero (or nonzero) in all of them.

If you skimmed chapter 11, however, you might be wondering about the position of the indices in this pair of tensor equations. But tensor indices can be raised and lowered using the metric tensor, so in Minkowski space-time $F^{\mu\nu}$ and $F_{\mu\nu}$ are just different ways of saying the same thing. (You need an upstairs index in the first equation, though, in order to sum over the repeated index ν.)

Just to show you that this beautiful pair of equations is not the last word on tensor elegance, in the later language of differential geometry, Maxwell's equations are even more concise:

$$\boldsymbol{d}^*\boldsymbol{F} = 4\pi^*\boldsymbol{J},\ \boldsymbol{dF} = 0.$$

Here the notation has swung back from index form to Heaviside's boldface notation for whole vectors, so that $\boldsymbol{F}$ is the tensor with components $F_{\mu\nu}$. The $\boldsymbol{d}$ is the gradient operator, analogous to the comma for partial derivatives in the index-form equations above. (The star denotes the "dual" of the tensor.)

As with vectors, though, you still need the component forms to do the calculations!

Today, the electromagnetic field tensor $F^{\mu\nu}$ is often called the Maxwell tensor or the Faraday tensor.[19] What Einstein and Grossmann really needed, though, was a way to combine the matter and energy that produce a gravitational field into a single tensor, too. But we are not yet done with electromagnetic theory. After Minkowski's premature death, his friend Arnold Sommerfeld developed his ideas as we saw in chapter 9. Sommerfeld was now friendly with both Levi-Civita and Einstein, and he also knew a lot about tensors. So, in his 1910 paper "On the Theory of Relativity I: Four-Dimensional Vector Algebra," he had followed Minkowski's lead and had combined the essential physical quantities associated with an electromagnetic field into a single tensor $T_{\mu\nu}$.

For example, it was known that, for the energy carried through 3-D space by a plane electromagnetic wave, the vector product $\boldsymbol{E} \times \boldsymbol{B}$ is proportional to the rate of energy flow per unit area—that is, to the energy *flux* (flux is the rate of flow, as in fig. 6.1). Since $\boldsymbol{E} \times \boldsymbol{B}$ is another vector, it has three components, which Sommerfeld labeled T_{41}, T_{42}, T_{43}. Next, he identified T_{44} with the electromagnetic energy density of the field (it's proportional to the square of the amplitude of the electromagnetic wave carrying the energy). He identified the remaining components with what he called “Maxwell stresses.”

Maxwell had identified the effect of these electrokinetic and electrostatic stresses with radiation pressure—a remarkable prediction that is validated in many applications today, including the optical tweezers I mentioned in chapter 2. Since pressure requires a force and force is a change in momentum, these “stress” components are identified with momentum flux. So today $T_{\mu\nu}$ is generally called the stress-energy tensor or the energy-momentum tensor. Analogously with the “Maxwell stresses” in figure 9.4, the indices tell you which component is acting across which surface of a little volume element. For example, in Cartesian coordinates T_{32} would be the z-component of energy-momentum acting across the surface whose normal is in the y direction (and vice versa, for it turns out that $T_{\mu\nu} = T_{\nu\mu}$).[20]

It’s quite amazing the way a lot of information can be combined into a single tensor like this: $\boldsymbol{E}$ and $\boldsymbol{B}$ into $F_{\mu\nu}$, energy, matter, and momentum into $T_{\mu\nu}$, and of course the gravitational field into $g_{\mu\nu}$. It’s another example of the power of tensors in representing data. But it also holds the key to Einstein’s requirement that all observers deduce the same physics, because Ricci had designed tensors to encode invariance. Unfortunately, Einstein and Grossmann still had a ways to go in figuring out the right equations to describe the gravitational field precisely—equations that kept their form even when the coordinate frame was changed.

ENTWURF: ON THE ROAD TO GENERAL RELATIVITY

The two friends had started out with the idea that they needed a tensor equation, linking the left and right sides of the relativistic analog of $\nabla^2 V = 4\pi G\rho$—and obviously a gravitational version of Sommerfeld’s electromagnetic $T_{\mu\nu}$ would figure on the right-hand side. But Einstein and Grossmann were having second thoughts about using the Ricci tensor for the left-hand side. In fact, they’d begun to have doubts about whether tensors were the right tools at all, because no matter how hard they tried, they simply couldn’t find a suitably form-invariant equation.

This was because they were juggling many different balls that all had to meld into the final equations. For a start, these equations had to reduce to the Newtonian limit in weak fields such as that here on Earth—and to special relativity in the "local" neighborhood of an event, where the spacetime is "locally flat." We saw the idea of "local flatness" in figure 10.1; how big this local region can be varies with the situation, as you can see in the next endnote. The equations also had to produce analogs of known physical laws such as conservation of energy and momentum, but this was proving particularly difficult—and even today the notion of gravitational energy is problematic.[21] But the point here is what a complicated business it was, trying to find a satisfactory relativistic theory of gravity!

Nonetheless, in 1913 Einstein and Grossmann published what is known today as the *Entwurf* theory—*Entwurf* is the German for "outline," and the English translation of their paper's title is "Outline of a Generalized Theory of Relativity and of a Theory of Gravitation." It contained the essential framework of general relativity, and it was a remarkable achievement, the culmination of months of intense work. The ideas were Einstein's, and he wrote the physics part of the paper; Grossmann had the mathematical expertise, and he wrote the mathematical section.

Unfortunately, although the *Entwurf* equations were tensorial, they were only covariant—form invariant—under certain restricted coordinate transformations. Yet the whole point of *general* relativity was that *any* reference frame should be allowed, to represent all sorts of relatively moving observers. That's why Grossmann had doubts about the value of tensors: general covariance didn't seem possible when it came to gravity. Einstein tied himself up in knots trying to explain away the fact that his theory didn't live up to his initial goal—and "with a heavy heart" he, too, convinced himself that a more general theory simply wasn't possible.[22]

Einstein and Grossmann published a second joint paper in 1914, deriving the *Entwurf* gravitational equations in a more elegant way, but still they weren't fully covariant. And still Einstein kept trying. Since 1907 he'd published almost a dozen papers on his journey to a relativistic theory of gravity, and there were more to come. When it was all over, he would joke to a friend, "Einstein has it easy, every year he retracts what he wrote the year before." By the end he'd be retracting or updating every week. In 1914, though, he wasn't yet laughing, for his colleagues reacted none too warmly to the *Entwurf* theory. He complained to his old friend Michele Besso, "The fraternity of physicists behaves rather passively. . . . Abraham seems to have the greatest understanding for it. To be sure, he fulminates against all relativity . . . but he does it with understanding." Einstein went on to say that Max Planck—cofounder of quantum theory, and an early champion of Einstein and special relativity—was one of those who were "not open" to his new theory, but he hoped Sommerfeld and Lorentz might be.[23]

Lorentz was always interested and supportive, but he never really gave up the idea of the ether. Sommerfeld, too, was a valued colleague—he cautiously accepted the *Entwurf* approach, and later he even advised Einstein to make sure he was known as the main author of the new theory. But Einstein had confidence in his longtime friend, replying, "Grossmann will never claim to be co-discoverer"; Grossmann's role, Einstein said, had been "to orient" him mathematically. After that, Einstein did indeed prove more than capable of innovatively applying tensor math.[24]

• • •

Although Einstein was working hard on his new theory of gravity, he always found time for his new postdoctoral assistant Otto Stern, whom we met in connection with the discovery of quantum spin. When he'd first turned up to work with Einstein, Stern had been surprised to find "a guy without a tie sitting behind a desk" and looking more like a "road-mender" than the famous man he'd expected. But, he said, Einstein "was terribly nice."[25]

Einstein still found time for good friends, too, such as Besso, who visited Zurich in June 1913. Which was just as well because it turned out to be a fruitful visit. Einstein had already figured out a way to test the *Entwurf* theory, by seeing if it could account for the missing 43 angular seconds in the precession of the perihelion of Mercury—and Besso joined him in working out the calculations. Unfortunately, they could only manage to find 18 seconds, not 43. As we'll see, though, Einstein would later successfully use the method he and Besso devised—which may be why, in 2022, their original calculations, in the so-called Einstein-Besso Manuscript, sold at auction for more than $15 million.

Besso could see some of the problems with the *Entwurf* theory before Einstein did—and not just the fact that it didn't give the right prediction for Mercury. But Einstein had other problems to contend with: in 1914, he took up Max Planck's offer of a professorship in Berlin—a move that ended his close collaboration with Grossmann, given the difficulties of long-distance communication in those days—and his marriage to Mileva Marić finally fell acrimoniously apart. Part of the reason was Einstein's work. "You see, with such fame, not much time remains for his wife," Marić had written to a friend a couple of years earlier. She went on to say that she realized she "puts on a haughty and superior air," but she didn't know if that was because of shyness or pride. What she did know was that "I am very starved for love." Einstein had not tried to penetrate her brave, taciturn front, as he might have done in happier times. Instead, he had become close to his cousin Elsa, who conveniently lived in Berlin; they would marry in 1919.[26]

Then, of course, there was the First World War, from 1914 into 1918. As a pacifist, Einstein steered clear of it, and although he was now in Berlin, he refused to sign the kaiser's "Declaration to the Cultural World," which disclaimed Germany's responsibility for the war. Max Planck and Felix Klein were among the many respected German scientists who added their signatures to the declaration, but David Hilbert also refused to sign. He'd been Minkowski's longtime friend, and a student of Klein, and he was now a professor at Göttingen, which he'd helped become a famously vibrant intellectual center. Hilbert was a mathematical genius who loved dancing, but who was also an outrageous flirt—rather like Einstein. But they both held firm to their antiwar principles, and Klein—who'd signed the document without reading it, so patriotic was he—came to regret acting so hastily.[27]

The war also disrupted another of Einstein's plans to test his theory. In 1911, he'd predicted the angle at which light would be bent as it passed by the sun, and he'd approached astronomers to see if this bending could be detected during a solar eclipse. Distant stars that appeared close to the sun when viewed from Earth would become visible during a total eclipse, and their positions could be photographed and compared with the same stars' positions months later, when the line of sight from Earth to those particular stars did not pass near the sun. (The stars themselves are so far away that during those few months, their actual positions relative to Earth remain effectively unchanged.) Several planned eclipse expeditions were abandoned or fatally disrupted by the war. Ironically, this was lucky for Einstein: like his first Mercury prediction, his 1911 light prediction would prove too small.

ENTER TULLIO LEVI-CIVITA AND DAVID HILBERT

In November 1914, Einstein published a paper further developing and polishing the work he'd done with Grossmann—and he immediately sent a copy to his worthy opponent Max Abraham. Abraham, in turn, promptly wrote to his friend Levi-Civita, telling him he didn't understand Einstein's reasoning, and suggesting that he and Levi-Civita get together to discuss it. The two men met in Padua, and Levi-Civita was so excited by their discussions that in early 1915, he reached out to Einstein for the first time. He wanted to clarify some errors in Einstein's use of tensors, and Einstein replied gratefully: "By examining my paper so carefully, you are doing me a great favor. You can imagine how rarely someone delves independently and critically into this subject."[28]

Over the next two months they exchanged some dozen letters each as Einstein struggled to find the right gravitational field tensor. When Italy entered the war, against Germany, communication between Padua and Berlin became difficult, and the correspondence faltered.[29] Still, in the summer of 1915 Einstein was feeling confident when he went to Göttingen, at Hilbert's invitation, to deliver a series of six lectures on gravity. By this stage, he had mastered tensors better than most mathematicians, for Ricci's work was still languishing in the shadows.

Einstein's confidence was not misplaced. At thirty-six he was delighted with the reception his ideas received from the mathematical elder statesmen Klein and Hilbert—Klein was almost seventy now, and Hilbert was fifty-three. They both were experts in invariant theory, and Hilbert had also initiated an axiomatic approach to geometry (the way Giuseppe Peano had done for algebra). It seems that Hilbert, who was especially impressed with Einstein's geometrical (metric tensor) approach to gravity, encouraged him not to give up his original search for covariant gravitational field equations.

One of the reasons that Einstein had abandoned the idea of general covariance was that it meant the form of the metric should be invariant under *any* coordinate transformation; but this kind of generality would include transformations that had nothing to do with relative motion—such as transforming from Cartesian to polar coordinates, where the form of the metric isn't invariant. It's analogous to expressing the equation of a circle in Cartesian and polar coordinates, respectively: $x^2 + y^2 = a^2$ and $r = a$. It's the same circle, but on the face of it, the equations look nothing alike. Einstein would eventually realize that even with these kinds of transformations, generally covariant equations could be found—assuming the transformations are homogeneous. (This means that tensors that are zero in one frame are zero in all frames, and Einstein realized that this gives a way to write invariant tensor equations.)[30]

Back in Berlin, Einstein kept working doggedly on his equations. As I mentioned earlier, he was trying to link the energy-momentum tensor $T_{\mu\nu}$ with a suitable tensor expression made from the $g_{\mu\nu}$ and its second derivatives—by analogy with the Newtonian equation $\nabla^2 V = 4\pi G\rho$. After wracking his brains for months, trying to find the right $g_{\mu\nu}$ expression, Einstein suddenly realized the *Entwurf* theory was on the wrong track. He had misunderstood the Newtonian limit of the theory when he threw aside the Riemann and Ricci tensors, but now he looked again at the work he and Grossmann had discarded three years earlier. For the Riemann and Ricci tensors are made from the very derivatives of $g_{\mu\nu}$ that he needed. But as he worked on his new approach, he received unwelcome news from Sommerfeld: Hilbert was poaching on his preserves. At least, that's how Einstein saw it.

Before Einstein's Göttingen visit Hilbert had been interested in Gustav Mie's 1912 electromagnetic theory of matter. Mie was a physics professor at the University of Greifswald in northeastern Germany, and his 1908 theory of the scattering of electromagnetic radiation by a spherical particle is still much cited. Not so his later attempt to find a unified theory of matter and electromagnetic radiation, but at the time, Hilbert was much taken with it. Einstein had been unimpressed with the gravitational analysis in Mie's theory, and so Hilbert, after hearing Einstein's lectures at Göttingen, decided to try to mesh Mie's approach to electromagnetism with Einstein's geometric approach to gravity. Following Mie, he believed that matter was electromagnetic in origin, and since matter produced gravity, Hilbert thought he could improve Mie's attempt to unify electromagnetism and gravity.

Einstein was exhausted after a decade of hard work on relativity, but perhaps the thought that Hilbert was taking up his gravitational ideas spurred him on. It would be heartbreaking if, having laid all the groundwork over the past decade, Hilbert rode across the finishing line just ahead of him. At any rate, he worked ferociously hard throughout November 1915, exchanging updates with Hilbert along the way—possibly because he wanted to stake his priority, but also to share ideas and to convey exciting progress. (As if this weren't taxing enough, at the same time he was also writing to his ex-wife and eleven-year-old son, Hans Albert—his younger son, Eduard, was only five, not yet old enough for letters, but Einstein hoped to build better relationships with them all.)[31]

In particular, he told Hilbert about a paper he was presenting to the Prussian Academy of Sciences on November 18, in which he applied his latest equations to Mercury's perihelion. This time he'd found all 43 of those missing angular seconds—a result that made him "beside [him]self with ecstasy for days," as he later told a friend. He'd immediately sent the news to Besso, of course, who'd helped him devise the method he used to find this magnificent result: "*Motions of the perihelion quantitatively explained*," he wrote triumphantly.[32] Indeed, throughout November Einstein wrote four papers in four weeks, each paper getting closer to the final, fully covariant, general theory of relativity that he submitted, with enormous excitement and relief, on November 25.

You may have heard what happened next, so before we see Einstein's famous equations, let's get the issue of priority out of the way. Five days before Einstein's final November paper, Hilbert submitted his own paper, "Foundations of Physics (first note)," to the Göttingen Academy of Sciences—a paper that apparently also contained the right gravitational field equations. And ever since the 1920s, there has been confusion in some quarters over whether the Einstein equations should, in fact, be called the Hilbert equations, or the Einstein-Hilbert equations. The confusion became controversy in the late 1990s, when science historian Leo Corry, of Tel Aviv University, discovered that the original page proofs of Hilbert's November 20 paper bore the printer's date stamp of December 6—*after* December 2 when Einstein's November 25 paper was published. What's more, these proofs did not appear to contain the Einstein equations at all—although they do contain a formulation that is implicitly equivalent, in the restricted case of Mie's electromagnetism. Unlike Einstein's, though, the underlying theory in Hilbert's proofs was not fully covariant. This, together with the fact that the explicit field equations do appear in the later published version of Hilbert's November 20 paper, with acknowledgments to Einstein's four November papers, has led several specialists to conclude that Hilbert only learned the explicit form of the equations from Einstein.[33]

On the other hand, there is part of a page missing from the proofs, which may have contained the specific equations. Perfect fodder for conspiracy theories—who tore the page?—and intemperate language has been traded by some on both sides of the debate. As far as I can tell, though, most of these scholars believe the context of the missing section indicates that Hilbert almost certainly *didn't* beat Einstein to the "Einstein equations." Besides, everyone agrees that general relativity is Einstein's, and that there would be no "Hilbert equations" without Einstein's solid foundations to build on.[34]

I've referenced many of these scholarly papers in the endnote,[35] but either way, it seems there was no real dispute between the players themselves about who found the final equations of general relativity—it was others, including Klein, who, several years later, drew attention to the similar timing of their discoveries.[36] Certainly Hilbert never claimed to be the coauthor of general relativity, although possibly he might have, had not Einstein published ahead of him.[37] As it is, he not only cited Einstein's November papers when his own paper was eventually published, he often praised Einstein, too. "Every boy in the streets of Göttingen understands more about four-dimensional geometry than Einstein," he once proclaimed; "yet, in spite of that, Einstein did the work and not the mathematicians." And after that initial bitter feeling that Hilbert had stolen his ideas, by the end of 1915 Einstein had reached out to him in friendship again.[38]

What this story really shows is that even the greatest geniuses need inspiration from each other—Einstein the greatest living physicist, who no doubt benefited from Hilbert's mathematical rigor, and Hilbert the best living mathematician, who certainly benefited from Einstein's physical insight. But it is Einstein alone who built the foundations of general relativity, and with Grossmann's initial help, put tensor analysis on the map. So let me show you how he finally did it.

EINSTEIN'S GENERAL THEORY OF RELATIVITY, AT LAST

We saw a little earlier that, following Minkowski, Sommerfeld had created a single tensor, $T_{\mu\nu}$, which contained all the essential information about the energy and momentum carried by an electromagnetic field. It turns out that the divergence of $T_{\mu\nu}$ has a very special property. "Divergence" in space-time is defined by analogy with the ordinary vector calculus operation that we saw in figure 7.1, where the divergence of a vector $V = Xi + Yj + Zk$ is $\frac{\partial X}{\partial x} + \frac{\partial Y}{\partial y} + \frac{\partial Z}{\partial z}$. It's a scalar, so it is invariant—true for any coordinates. So, if we use Ricci's notation $x \equiv x^1, X \equiv V^1$, and so on—as well as the comma notation for partial derivatives, Hamilton's nabla, and the Einstein summation convention (where repeated up and down indices indicate a sum)—the ordinary vector calculus divergence can be represented like this:

$$\nabla \cdot V = V^1{}_{,1} + V^2{}_{,2} + V^3{}_{,3} \equiv V^\nu{}_{,\nu}.$$

Extending this to Minkowski space-time, we can use the same representation, $V^{\nu}{}_{,\nu}$, where the indices now range from 1 to 4 (or 0 to 3 if the time coordinate is denoted with a 0).[39] This is a *definition* of divergence in Minkowski space-time, by *analogy* with the Euclidean vector definition—so now extend the analogy to tensors in this flat, Minkowskian space-time, and define the divergence of the energy-momentum tensor to be $T^{\mu\nu}{}_{,\nu}$. (As I've mentioned, with a metric you can raise or lower the indices of a tensor, so $T_{\mu\nu}$ and $T^{\mu\nu}$ are different ways of writing the same information.) Regardless of what you want to call this divergence expression, though, it turns out that for electromagnetism, and for many other physical situations, when you do the calculations, you always get this equation:

$$T^{\mu\nu}{}_{,\nu} = 0.$$

It also turns out that when you consider the meaning of each of these four equations—four because μ takes four values in space-time—you get the law of conservation of energy when $\mu = 4$ (or 0, whichever index you prefer for the time component), plus the three spatial component equations making up the usual law of conservation of momentum. In other words, the laws of conservation of energy and momentum are rolled into one beautifully simple-looking equation via the energy-momentum tensor $T_{\mu\nu}$. It's another example of the brilliantly economical way information is represented in tensors.

Einstein had been grappling with the conservation of *gravitational* energy since 1907. It wasn't so much about what to put into $T_{\mu\nu}$, for that would be defined from the features of the matter and energy distribution under consideration—just as Sommerfeld put the known features of an electromagnetic field into his $T_{\mu\nu}$. The problem was with the conservation law. So now it's time to answer the question Einstein had asked at the beginning of his collaboration with Grossmann: how to transfer the laws of physics from the special theory to the general one.

The tensorial conservation equation $T^{\mu\nu}{}_{,\nu} = 0$ is an example of a physical law that holds in special relativity, which means it keeps the same form for all observers moving with constant relative motion. In other words, it is form-invariant (covariant) under Lorentz transformations. It turns out that all you have to do to change a Lorentz-covariant tensor equation to a generally covariant one is to replace partial derivatives by covariant derivatives (which I defined near the end of chap. 11). In a marvelous example of the power of symbolism, this definition means you simply have to change all the commas in the above equations to semicolons (and similarly for the commas in Maxwell's equations that we saw earlier, written in terms of the tensor $F^{\mu\nu}$).[40] So, in general relativity the (local) law of conservation of energy-momentum equation is $T^{\mu\nu}{}_{;\nu} = 0$.

Actually, Einstein didn't know this rule, or use this semicolon notation—many of the niceties of tensor analysis came later; but notation aside, he obtained the result $T^{\mu\nu}{}_{;\nu} = 0$ anyway. (You can see the way he wrote this equation in the box.)

TRANSLATING EINSTEIN'S NOTATION FOR THE CONSERVATION EQUATION INTO THE MODERN FORM

In his 1916 overview, *The Foundation of the Theory of General Relativity*, in his equation (57a), Einstein wrote the conservation of energy-momentum equation in the equivalent form:

$$\frac{\partial T^{\alpha}_{\sigma}}{\partial x_{a}} = -\Gamma^{\beta}_{\alpha\sigma} T^{\alpha}_{\beta}.$$

He's used α instead of my ν to indicate the sums, but that's an arbitrary choice, as is his choice of σ and my choice of μ to label T: as I mentioned in chapter 11, the labels on arbitrary vector and tensor components are themselves arbitrary, just like our time-honored use of x for the unknown in algebra. As we'll see below, though, Einstein used $T_{\mu\nu}$ in his final equations, which is why I've chosen μ and ν for my labels.

The term on the right-hand side in this equation is the curvature part Ricci added to the ordinary partial derivative to define a covariant derivative. (The Γ symbol is the Christoffel symbol.) So, adding this term to both sides, and using semicolon notation to define the covariant derivative, Einstein's equation here becomes:

$$T^{\alpha}{}_{\sigma;\alpha} = 0.$$

With one final lesson from chapter 11, you can raise and lower indices here with the metric tensor—$g^{\sigma\gamma}T^{\alpha}{}_{\sigma;\alpha} = T^{\alpha\gamma}{}_{;\alpha}$—and since this tensor is symmetric in its indices, then what Einstein wrote is entirely equivalent to the modern representation:

$$T^{\alpha\gamma}{}_{;\alpha} = T^{\gamma\alpha}{}_{;\alpha} = 0 \equiv T^{\mu\nu}{}_{;\nu} = 0,$$

where the equivalence $\equiv$ comes from relabeling the indices.

Yet for various complicated reasons, this conservation law didn't seem to gel with his field equations, which—as we saw earlier by analogy with the Newtonian law of gravity—related an expression using second derivatives of $g_{\mu\nu}$ on the left-hand side to $T_{\mu\nu}$ on the right-hand side. In November 1915, when Einstein finally returned to the Ricci tensor for the left-hand side, he found it wasn't enough simply to write $R_{\mu\nu} = kT_{\mu\nu}$ (where k is the proportionality constant). That's because when he took the divergence of both sides, he ended up with an equation that placed physically unrealistic restrictions on the gravitational field. We'll see more on this in the next chapter.

Finally, by November 25, 1915, Einstein had discovered the correct gravitational field equations, which he wrote in the form

$$R_{\mu\nu} = k(T_{\mu\nu} - \frac{1}{2} g_{\mu\nu}T),$$

where $T = T^{\mu}_{\mu}$, which is a scalar—because the up and down indices are the same, indicating the components are summed, giving a scalar as we saw in chapter 11—and units are usually chosen so that $k = 8\pi$.[41] Today Einstein's equations are often written in the equivalent[42] form,

$$R_{\mu\nu} - \frac{1}{2} g_{\mu\nu} R = kT_{\mu\nu}.$$

This is the form Hilbert used in the published version of his November 20 note.[43]

• • •

I've given a lot of beautiful equations in this book, but the equations of general relativity are standouts. I say "equations," because this single "equation" represents ten equations (because $R_{\mu\nu}$ and $T_{\mu\nu}$ are symmetric, so there are ten independent components, just as we saw for $g_{\mu\nu}$). Yet this elegant handful of symbols is the key that has unlocked so many mysteries of the universe—from the precession of Mercury's perihelion to the Big Bang singularity. From the bending of light—which was eventually confirmed during the eclipse expeditions of 1919 and 1922, and by that spectacular first direct image of the shadow of a black hole in 2019—to the gravitational waves that were first detected in 2015. From the effect of gravity on clocks and GPS to more esoteric things such as "gravitomagnetism," "frame-dragging," black holes, and gravitational lenses—and so far, in every test the equations have stood firm.[44] This might change before I finish writing this book—observational technology is improving all the time! And it is well known that we don't yet have a complete theory of quantum gravity, so a new theory will eventually evolve. But it will surely have to mesh with general relativity at the current level of accuracy, just as Einstein's theory does with Newton's.

Einstein did not live to see most of these spectacular confirmations of his theory, but he knew he'd found something special—his successful Mercury perihelion and light bending calculations were proof of that. Looking back years later, he poignantly recalled the monumental effort it had taken:

> In the light of knowledge attained, the happy achievement seems almost a matter of course, and any intelligent student can grasp it without too much trouble. But the years searching in the dark, with their intense longing, their alternations of confidence and exhaustion and the final emergence into the light—only those who have experienced it can understand that.[45]

• • •

On the first page of his 1916 overview of his new theory, Einstein paid tribute to the mathematicians whose work had made general relativity possible: Minkowski for formulating the idea of space-time; Ricci and Levi-Civita for tensor calculus; Gauss, Riemann, and Christoffel for their work in non-Euclidean geometry, which Ricci had generalized; and, of course, faithful Grossmann, who not only tracked down and deciphered the relevant papers but who "also helped me in my search for the field equations of gravitation."[46]

General relativity made Einstein a superstar, after his light-bending prediction was confirmed in the 1919 eclipse expedition—for when the expedition's results were released, headlines trumpeted them around the world. "REVOLUTION IN SCIENCE, New Theory of the Universe: Newtonian Ideas Overthrown," proclaimed the *London Times* on November 7, 1919. And on November 10, in a poetic allusion to the fact that the bending light path from the stars shifts their apparent positions in the sky, the *New York Times* ran with "LIGHTS ALL ASKEW IN THE HEAVENS, Men of Science More or Less Agog over Results of Eclipse Observations. EINSTEIN'S THEORY TRIUMPHS."

Einstein's success also quietly immortalized Ricci—in the mathematical physics community, at least. Critics such as Beltrami had complained that tensor analysis had no practical benefit compared with existing methods, but Einstein and Grossmann had changed all that. As Ricci's dear friend Levi-Civita put it, the essential role of tensors in general relativity "also became for Ricci 'the just dispenser of glories,'" and his "great contribution" was officially recognized at last.[47]

(13)

WHAT HAPPENED NEXT

As we saw with Einstein's thought experiment about falling people and accelerating elevators, a surprising amount of intuition goes into finding a law about how nature works. Logic alone is simply not enough. Yet even when a theory's equations work spectacularly well, it is often necessary for others to come along and tighten up the mathematical foundations, or to shine a light more closely on the theory's assumptions. Which is why David Hilbert and Felix Klein needed Emmy Noether.

Noether arrived in Göttingen in 1915. She'd earned her doctorate in 1907 at Erlangen, where Klein had taught earlier, and she was an expert on invariant theory—so Klein and Hilbert had invited her to Göttingen, hoping she could help sort out their questions about general relativity and energy. And help them she did, for in 1917 Klein told Hilbert, "You know that Miss Noether advises me continually regarding my work, and it is only thanks to her that I have understood these questions."[1]

Despite Einstein's legendary intuition, he'd struggled when he tried to interpret the relationship between mathematical covariance—the idea that if you write a tensor equation, it will keep the same form when you transform to another set of coordinates—and the physical principles of relativity and equivalence that had guided him to his theory. His struggle helped later mathematicians sharpen the distinction between transformations of coordinates, frames of reference, and points or events, and between relativity as a theory of invariance and symmetry groups as opposed to one of covariance—but if you're confused by such subtleties, don't worry: these distinctions are still being debated today.[2] All that matters, really, is that Einstein's tensor equations give a marvelously accurate description of gravity, from which so many extraordinary consequences have been correctly predicted. So, what I want to focus on here are the issues that Noether helped resolve, during the debates that took place immediately after Einstein published his theory at the end of 1915, and his longer 1916 overview, *The Foundation of the General Theory of Relativity*.

By the way, this 1916 paper is a master class in Ricci's tensor calculus. Because this language was still new to most physicists and mathematicians, Einstein took great care in setting out, economically but clearly, the tensor rules we saw in chapter 11.

In November 1915 Einstein had finally ended up with the conservation of energy-momentum equation $T^{\mu\nu}{}_{;\nu} = 0$, as we saw in the previous chapter. He'd come at it via a circuitous route that was different from Hilbert's, and in May 1916 he asked if Hilbert thought there might be some deep principle underlying their two separate approaches. Hilbert replied that he thought there probably was, and that he'd already asked "Miss Noether" to investigate the issue. Hilbert and Klein, and colleagues such as Einstein, had the highest respect for Noether, yet her presence was so singular they couldn't help calling her "Miss" rather than "Dr." Perhaps they thought it was more polite than just using her surname, as they often did among themselves.

The questions Klein and Hilbert were trying to resolve concerned the physical and mathematical meaning of the energy-momentum conservation law in general relativity. The traditional route to conservation equations in ordinary mechanics had been the calculus of variations. This involves minimizing the "action"—that is, the integral of a "Lagrangian" L, which is a function of position and velocity. Leonhard Euler and Joseph-Louis Lagrange had pioneered this method, and then our vector pioneer

William Rowan Hamilton introduced a useful alternative, in which L is expressed in terms of momentum rather than velocity (so it is a new function, now called the Hamiltonian, denoted by H). Sometimes it is easier to work in terms of momentum rather than velocity, but either way, when L and H are expressed in terms of kinetic and potential energy, the equation of motion arising when the integral is minimized has a solution that gives the usual conservation of energy equation. The details of this method aren't important here: for an individual particle the equation of motion found in this way is equivalent to Newton's second law of motion, and we'll see shortly how this leads to conservation equations.

Meantime, to find their conservation of energy-momentum equations, Hilbert had used an elegant Lagrangian approach while Einstein had used a mix of tensors and a Hamiltonian. They were breaking new ground, because first they had to figure out how to define *gravitational* energy—and how to choose the right L or H to express this new kind of energy. When that was done, there was still a nagging question: How did the resulting equation, $T^{\mu\nu}{}_{;\nu} = 0$, gel with the traditional idea of energy-momentum conservation?

EMMY NOETHER AND THE CONSERVATION OF ENERGY-MOMENTUM

Through 1916 and 1917, Einstein, Hilbert, Klein, and Noether exchanged ideas on this brand-new problem of gravitational energy. Each added to the debate, and several others joined in—notably Klein and Hilbert's former student Hermann Weyl. But it was Noether who tied it all together, in the famous "Noether theorems" proved in her 1918 paper "Invariante Variationsprobleme."[3] These two theorems relate conservation laws to "symmetries." And symmetries, as we've seen, relate to "invariance"—the idea that certain things are unchanged when you change your frame of reference. We also saw this kind of invariance in figures 9.1 and 9.3, and of course it is the defining feature of tensor equations. Since invariance has to do with things remaining the same, it also has to do with conservation—for if a physical quantity doesn't change, it is "conserved."

In ordinary mechanics—the study of the way objects move under forces—this connection had been known since Lagrange. For example, we've seen that gravitational force $\boldsymbol{F}$ can be expressed in terms of a potential V: $\boldsymbol{F} = \nabla V$. But since Newton's second law says that the force acting on an object equals the rate of change of the object's momentum vector, $\boldsymbol{p}$, we have,

$$\boldsymbol{F} = \frac{d\boldsymbol{p}}{dt} = \nabla V \equiv \frac{\partial V}{\partial x}\boldsymbol{i} + \frac{\partial V}{\partial y}\boldsymbol{j} + \frac{\partial V}{\partial z}\boldsymbol{k}.$$

In terms of components, this equation gives $\frac{dp_x}{dt} = \frac{\partial V}{\partial x}$, and similarly for the y and z components. Now for the invariance, and let's consider the group of coordinate transformations that are translations by various amounts in the x-direction. If the potential is invariant under these transformations, then it doesn't change—it has the same value at the point (x, y, z) as it does at $(x + a, y, z)$:

$$V(x + a, y, z) = V(x, y, z), \text{ for every value of } a.$$

Since it doesn't matter which value of x is used, V must be independent of x. Which means $\frac{\partial V}{\partial x} = 0$, and this, in turn, means that

$$\frac{dp_x}{dt} = \frac{\partial V}{\partial x} = 0 \Rightarrow p_x = C, \text{ where C is the constant of integration.}$$

And *this* means that the x-component of the momentum is constant—it is conserved.

The same equations of motion, and the same "conservation" laws, are found using the Lagrangian or Hamiltonian methods, as I mentioned. This would be like using a sledgehammer to crack a walnut for this simple case, but these methods are often easier for complicated situations with many particles.

At the simplest level of her theorems, Noether showed that this is a general result: when a function is independent of a particular variable, it indicates that something is conserved. In general relativity, for example, we saw in figure 12.1 that free-falling objects move along geodesics rather than the straight lines of Euclidean geometry, so the equation of motion is an equation about geodesics—and we saw in chapter 12 that geodesics depend on the metric that describes the curved space-time. We also saw that Einstein chose the components of the metric to play the role of the potential, analogous to V. So, like the Newtonian example we just saw, if there is a frame in which the metric tensor $g_{\mu\nu}$ is independent of a particular space-time coordinate, then that component of momentum is conserved along the particle's path. In space-time, though, "momentum" is a 4-D vector: its spatial components are the ordinary spatial momentum components, but the time component is defined to be the energy.[4]

I've been talking about the energy and momentum of an individual moving object here, not the energy-momentum tensor used to describe the energy and momentum carried by a gravitational field—but labeling the time component of the particle's momentum as "energy" is analogous to what we saw for the time components T_{41}, T_{42}, T_{43}, T_{44} in chapter 12. So, there are two conservation issues to address: conservation of a moving object's energy and momentum in a gravitational field—as we've just seen, this is related to the symmetry (or invariance or coordinate-independence) of the potentials, V and $g_{\mu\nu}$—and conservation of the energy-momentum of the gravitational field itself, which we've seen is $T^{\mu\nu}{}_{;\nu} = 0$. Noether was the first to show mathematically that we're actually talking about different types of conservation law.

I've sketched the essence of how Noether did it in the endnote.[5] The upshot is that in classical mechanics *and* special relativity, the resulting conservation laws are clearly physical—they arise from physical conditions such as those described by Newton's laws. They also involve divergences, as we saw for electromagnetism in chapter 12.[6] The equation $T^{\mu\nu}{}_{;\nu} = 0$ seems to involve a divergence, but as Noether proved, it is not a physical divergence but a mathematical analogy.

This is not to say that $T^{\mu\nu}$ is unphysical in general relativity. True, there are problems in defining local energy density, and unless there are certain

Emmy Noether, circa 1900. Photographer unknown. Wikimedia Commons, public domain.

symmetries in the space-time, there's no global conservation law. Still, an observer can define energy density at a point, and the total gravitational energy of a system can be defined, and so can the energy flux carried away by gravitational waves.[7] So $T^{\mu\nu}$ plays a vital role in the study of cosmology and gravitational radiation, for example. And, whatever you want to call it, the equation $T^{\mu\nu}{}_{;\nu} = 0$ is needed to find the relationship between energy-matter sources and the curvature of space-time.[8]

IT'S A LOT SIMPLER WITH TENSORS

We saw in chapter 12 that the gravitational field equations are:

$$R_{\mu\nu} - \frac{1}{2} g_{\mu\nu} R = kT_{\mu\nu}.$$

And we met, in chapter 11, covariant derivatives, and the fact that you can raise and lower tensor indices without changing their essential content[9]—so

I'm going to write the gravitational field equations with upstairs indices, and then take the covariant derivative of both sides:

$$(R^{\mu\nu} - \frac{1}{2} g^{\mu\nu} R)_{;\nu} = kT^{\mu\nu}{}_{;\nu}.$$

Now, both Einstein and Hilbert found that conservation of energy-momentum means the right-hand side of the equation is zero. Which means the left-hand side must be zero too. What neither Einstein nor Hilbert knew, nor Klein nor Noether, was that the left-hand side is zero quite independently of the conservation equation. That's because of what are known as "the contracted Bianchi identities":

$$(R^{\mu\nu} - \frac{1}{2} g^{\mu\nu} R)_{;\nu} = 0.$$

(The plural is because this "equation" is really four equations, one for each component μ. The repeated index ν indicates a sum.) A mathematical "identity" is an equation that is always true—because it must be true, by definition. The full Bianchi identities follow directly from the definition of the Riemann tensor, as Ricci had known in the 1880s, although it was his nemesis Luigi Bianchi who first published the result, when he rediscovered the identities in 1902. (Bianchi won the Italian Royal Mathematics Prize that Ricci had entered in the 1890s, and Bianchi was a none-too-sympathetic judge when Ricci entered again in 1901.) As we saw in chapter 12, the Ricci tensor is formed from the contracted Riemann tensor, so that's how the "contracted" Bianchi identities give rise to this equation.[10]

Noether had spelled out the relationship between conservation and symmetries, and had shown in detail why $T^{\mu\nu}{}_{;\nu} = 0$ is just a mathematical analogy to traditional conservation laws—but Hilbert and Klein had assumed that this conservation equation forced $(R^{\mu\nu} - \frac{1}{2} g^{\mu\nu} R)_{;\nu} = 0$ to be true. So, they regarded this latter equation as a consequence of the variational (Lagrangian and Hamiltonian) methods that had led Einstein and Hilbert to the conservation equation. In fact, as Tullio Levi-Civita noted in 1917, it's much simpler to argue the other way, for the conservation equations follow directly from the field equations, via the tensorial Bianchi identities![11]

By the way, I mentioned in chapter 12 that Einstein had first tried $R^{\mu\nu} = kT^{\mu\nu}$ for his field equation. If you differentiate both sides of this equation, you get $R^{\mu\nu}{}_{;\nu} = kT^{\mu\nu}{}_{;\nu}$, and Einstein rightly rejected this as unphysical.[12] But the Bianchi identities show immediately that this is the wrong equation if conservation of energy-momentum is to hold, so that $T^{\mu\nu}{}_{;\nu} = 0$. Had Einstein known these identities, he'd have saved himself from quite a few headaches!

• • •

Noether's theorems are far more complicated than I've indicated here, of course. They were overlooked for many decades, because aside from their mathematical complexity, their significance lies in their generalizing a host of already-known conservation results. These known results had been built up slowly and separately over the centuries, and it took physicists some time to appreciate the importance of Noether's discovery that a single fundamental principle united mathematical symmetries and physical conservation laws. But the generality of this principle also means that her theorems are more widely applicable than general relativity: in recent times they have found applications from quantum mechanics to elasticity to fluid mechanics, for example, and in pure mathematics and numerical analysis. Back in 1918, though, even the link with general relativity was not fully understood.

In fact, it wasn't until 1924 that Jan Schouten and Dirk Struik highlighted the connection between Noether's Lagrangian identities and the beautiful tensor identities $(R^{\mu\nu} - \frac{1}{2} g^{\mu\nu} R)_{;\nu} = 0$. Noether used the symmetries of the Lagrangian, while the Bianchi identities depend on the symmetries of the Riemann tensor (which remains invariant when various combinations of its indices are interchanged, just as $g_{\mu\nu}$ and $T_{\mu\nu}$ are invariant when you swap the order of their indices: $g_{\mu\nu} = g_{\nu\mu}$ and $T_{\mu\nu} = T_{\nu\mu}$).[13]

Struik's interest in relativity had arisen early, for while he was a student in Leiden, Einstein had given a guest lecture there—Struik's professor was Einstein's close friend Paul Ehrenfest, who'd worked with Klein at Göttingen. Half a century later Struik still remembered the excitement of Ehrenfest's lectures, where science felt "alive"—an exciting, contemporary activity emerging "from conflict and debate" between famous living scientists and mathematicians, including, Struik recalled, Klein, Abraham, and Einstein.[14]

If Struik's phrase "conflict and debate" brings to mind Marxist dialectics, you'd be right: Struik later became a historian of science, and in 1936 he cofounded the Marxist journal *Science and Society,* which is still going today. He was motivated by the idea that science both shapes and is shaped by society, and that scientists have a social as well as a scientific responsibility for their work—an idea that has finally found its way into science classes and ethics committees.

NOETHER'S STRUGGLE FOR ACCEPTANCE

In 1924–25, Struik and his new wife, Dr. Ruth Ramler, who was also a mathematician, spent a year at Göttingen. "You had to have a thick skin to survive," he recalled, for "the Göttingen mathematicians were known for their sarcastic humor." Einstein would have agreed: "The people of Göttingen sometimes strike me," he'd declared, "not as if they want to help one formulate something clearly, but as if they want only to show us physicists how much brighter they are than we." Struik noted that Noether, "who was shy and rather clumsy, was often the butt of some joke." But it wasn't just sexism, for Struik added that "the good-natured Erich Bessel-Hagen" was also treated to the same "humor."[15]

Sexism was, however, the reason Noether found it hard to forge an academic career. In 1915, with the support of Klein and Hilbert, she had applied for habilitation as a private tutor or *Privatdozent.* I mentioned earlier that when Einstein presented his special relativity paper in his own first step on the academic ladder, it was rejected as "incomprehensible." But Noether was rejected because of misogyny. At Göttingen, mathematics was part of the philosophy faculty, so most of the members had no idea of Noether's mathematical brilliance—they just saw that she was a woman, and that it was unthinkable that a woman be allowed to teach. "What will our soldiers think when they return to the University and find that they are expected to learn at the feet of a woman?" Hypatia could have told them a thing or two about that! As it was, Hilbert responded that he didn't see that "the sex of the candidate" was an issue: the university is "not a bathhouse," he retorted.[16]

In May 1918, after Einstein had studied a paper of Noether's on invariants, he wrote to Hilbert saying how impressed he was with the generality of her approach, adding, "It would have done the Old Guard at Göttingen no harm to be sent back to school under Miss Noether. She really seems to know her trade!" Later that year he studied her conservation theorems hot off the press, and he was so impressed he wrote to Klein, "I once again feel that refusing her the right to teach is a great injustice." He offered to approach the relevant ministry himself if Klein was too busy. Fortunately, the war finally ended, and a new, more democratic government was established in Germany—and in June 1919, Noether was finally allowed to become a *Privatdozent*.[17]

Her habilitation thesis had been her 1918 conservation theorems, but in her lifetime, she was much more famous for her later work on abstract algebra—she was so cutting-edge she's been dubbed "the mother of modern algebra."[18] But she was also the first woman to play a role in relativity theory, and she's been an inspiration to many mathematically inclined young women who have come after her. I remember my first international conference on general relativity, where there were about four hundred men and ten women—including University of Paris professor Yvonne Choquet-Bruhat, who was a living legend for proving, from the 1950s on, some landmark theorems in general relativity. We are still a minority, but things are improving all the time, and women are increasingly visible. For example, in 2022, the Australian National University's Professor Susan Scott won the European Academy of Sciences' Blaise Pascal Medal for her work on gravitation, including her role in the 2015 detection of gravitational waves. And on a related topic, Katie Bouman was a Harvard postdoctoral fellow when she famously played a key role in devising the algorithms that led to the first direct image of a black hole in 2019.

Black holes and gravitational waves are predictions of general relativity, although Einstein himself was ambivalent about the possibility of their physical existence. But that doesn't matter: it's all there in those extraordinary little tensor equations,

$$R_{\mu\nu} - \frac{1}{2} g_{\mu\nu} R = k T_{\mu\nu}.$$

At least, it's there for those who know how to solve these equations, and today this often requires the use of numerical algorithms and computer power. Tensors such as the Ricci and metric tensors are there at the heart of it, although not necessarily in the numerical methods themselves. But for those of us working on exact solutions, those tensor indices and their symmetry properties help lighten the load of our computational tasks.

THE MEANING OF "PARALLEL"

There's another interesting tensor problem that was solved in the wake of general relativity. In the parallelogram rule for vector addition, which we saw back in figure 3.1, vector ***A*** is added to ***B*** by translating or "transporting" it to align with the end of ***B***, all the while keeping ***A*** parallel to itself. Alternatively, you can move ***B***, keeping *it* parallel to itself. On a curved surface such as that of a globe, you cannot move vectors around like this: as I've mentioned, meridians of longitude that are parallel at the equator are no longer so at the poles. Evidently, then, the usual idea of parallelism makes sense only locally, when a curved line is approximately straight.

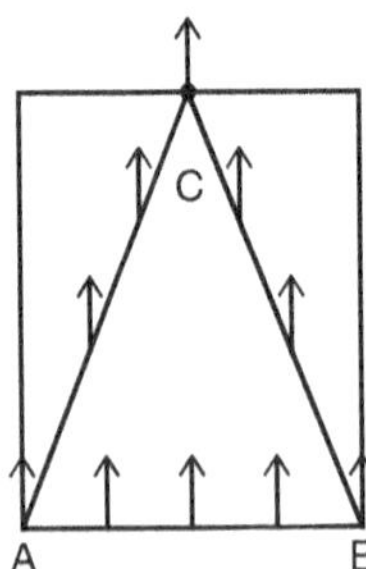

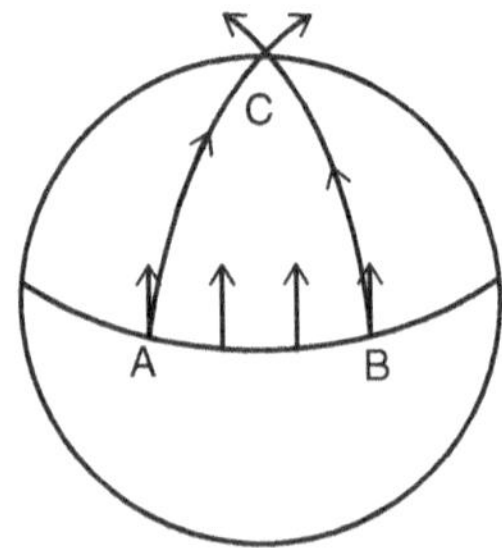

FIGURE 13.1. Parallel transporting a vector that starts out being vertical at point A is unproblematic in flat space—but on a curved surface, the notion must be carefully defined. I've explained this a little more in the related box.

It was Levi-Civita who figured out, in 1917, how to "parallel transport" vectors along curved surfaces—and he did it with tensors. His "parallel transport" equation (in the box) helps show how much the curvature causes two initially parallel geodesics to deviate with respect to each other—like the converging meridians on a globe. In a hugely curved space-time such as that around a black hole, the geodesics converge so quickly and drastically that you would be crushed if you came too close. (The inverse square law shows that you'd also be torn apart, because the gravity changes so dramatically over the distance between your head and your feet!)

PARALLEL TRANSPORT AS A WAY OF CHARACTERIZING CURVATURE

To see how Levi-Civita's idea of parallel transport relates to curvature, it's easiest to start with the idea in flat space. Imagine a flat piece of paper with a triangle drawn on it (as in fig. 13.1). Hold a pencil—standing in for a vector—vertically at the point *A*, and then, keeping it parallel to itself, move it along the flat page to the point *C*. Now do the same thing, but travel to *C* in the opposite direction, so you're transporting the vertical pencil along the line from *A* to *B* and then up to *C*. No surprises there—the pencil remains vertical all the time. The same thing happens if you take an intrinsic, ant's-eye view, keeping your vector in the 2-D plane of the page: begin with, say, a vector tangent to the side *AC*. It will remain parallel to itself when you transport it from *A* to *C* in either direction.

Now imagine holding the pencil vertically at a point *A* on the "equator" of a ball or orange. In this vertical position the pencil is tangent to the ball at *A* and pointing upward. Move it up toward the pole at *C*, keeping it tangent to the sphere but holding it straight (make sure you don't twist it). Now go the other way, parallel transporting the pencil along the curve from *A* to *B* and then up to *C*. This time the pencil—or vector—ends up at an angle of 90 degrees from the same vector transported directly from *A* to *C*. This is noncommutativity in action!

Hamilton would be amazed that his "shocking" notion that mathematical operations are not always commutative has found so many applications—this time, as a measure of curvature. It is the kind of intrinsic curvature our 2-D ant-alien could discover, simply by crawling along the surface with a pencil and noting how it changed direction at *C*.

CALCULUS, TOO

When you differentiate vectors in ordinary vector analysis, intuitively speaking you compare the vector at two nearby points and divide by the distance between them—and you implicitly assume the vector remains parallel when you calculate its value at one point and then the other. So, vectors must be able to be moved or "transported" along a curve in some sort of parallel way, if we are to understand how to define derivatives in curved spaces (that is, what Ricci called "covariant" derivatives). It's not surprising, then, that Levi-Civita's idea of parallel transport connects covariant derivatives and curvature.

It turns out that the definition of the parallel transport of a vector $\boldsymbol{V}$ along a curve with tangent vector $\boldsymbol{U}$ is $U^{\mu} V^{\nu}{}_{;\mu} = 0$. (The semicolon denotes the covariant derivative.)

As so often happens in science and mathematics, someone else had independently discovered the same ideas as Levi-Civita, at around the same time—although he hadn't yet published them. This someone was Jan Schouten. His colleague Dirk Struik recalled the day Schouten had come bursting into his office waving Levi-Civita's paper. "He also has my geodesically moving systems," Schouten told Struik, "only he calls them parallel." Struik recalled that Levi-Civita's approach was much simpler than Schouten's; still, he mused, "Few people realize that Schouten barely missed getting credit for the most important discovery in tensor calculus since its invention by Ricci."[19]

Struik worked with Levi-Civita in 1923 and described him as vivacious, gentle, and charming. The influential Scottish algebraic geometer William Hodge wrote that Levi-Civita was "one of the personally best known and best liked mathematicians of his time."[20] And like Struik, Levi-Civita had married a math graduate, Liberia Trevisani—his former student. She had hoped to teach mathematics, but ended up traveling the world with her husband, who was in great demand on the lecture circuit. Tragically, all that would change in the 1930s, when the Fascists and Nazis began their persecution, and Levi-Civita, Einstein, Noether, and so many other Jewish academics were stripped of their academic positions, if not of their lives.

The likes of Arnold Sommerfeld and David Hilbert were appalled at the direction their country had taken in both world wars. Soon after the first war, Arthur Eddington, a leader of the 1919 eclipse expedition, had seen hope in the symbolism of a British team confirming a German theory—Einstein's light-bending prediction. But hostility between the former enemy nations still simmered, and when Italian mathematicians organized the first postwar mathematics congress in 1928, they took pains to invite their German colleagues. Many German mathematicians refused to go, but when Hilbert triumphantly led a delegation of his countrymen into the opening session, they received a standing ovation. Hilbert addressed the gathering in terms that would certainly have pleased Eddington. There are no limits in mathematics, he said, including national ones. "It is a complete misunderstanding of our science to construct differences according to people and races. . . . For mathematics, the whole world is a single country."[21] It's a noble sentiment—and it is true in the sense that international collaboration has always been important in the advancement of mathematics, which is now an international language. But war changes everything. In 2022, for example, in the wake of Vladimir Putin's disastrous invasion of Ukraine, there was a similar debate among mathematicians, after a journal refused to publish papers from Russian institutions, and the International Mathematics Union stripped St. Petersburg of its right to host the 2022 International Congress of Mathematicians. Some felt as Hilbert did, that there should be no discrimination, and others felt that sanctions sent a message about institutional *and* individual responsibility for war (such as Hilbert and Einstein had shown when they refused to sign the kaiser's waiver of responsibility during World War I).[22]

• • •

Levi-Civita's invariant tensor definition of parallelism was important for the math of tensors—especially for understanding the idea of covariant differentiation. But in 1929 Einstein began corresponding with the French mathematician Élie Cartan on the possibility of spaces with a broader definition of parallelism. Cartan is the one who did the most to put all the ideas of his predecessors—including the long-neglected Hermann Grassmann, whom Cartan had studied well—into the modern "differential form" version of Ricci's tensor calculus. It was Cartan who, for example, clarified the idea of one-forms as duals of vectors, which I mentioned briefly in chapter 11.

The content of the letters between Cartan and Einstein has to do with Cartan's idea of "absolute" parallelism, a notion that allowed torsion or twisting of "parallel" vectors. As I mentioned with the pencil illustration in figure 13.1, this isn't allowed in Levi-Civita's definition of parallel transport—and therefore it isn't used in general relativity (because non-twisting parallel transport relates to the covariant derivatives in Einstein's equations). Einstein had hoped adding torsion might be a way to unify electromagnetism and gravity—but like Hilbert and Gustav Mie, he didn't succeed. Nonetheless, space-times with torsion, such as those in the Einstein-Cartan theory, are still being explored today—as a possible way of avoiding the Big Bang singularity, for example, or to account for the intrinsic spin of matter.

Although the details of the Einstein-Cartan correspondence are beyond my scope here, a cursory glance shows that tensors were the mathematical language they were speaking. It also shows the respectful way the two men interacted. Einstein, who was fifty in 1929, wrote such things as, "I am very fortunate that I have acquired you as a coworker. For you have exactly that which I lack: an enviable facility in mathematics. . . . I am both touched and delighted that you have taken so many pains over the problem." And sixty-year-old Cartan wrote, "I'm very proud my letters may be of some interest to you. . . . I consider it to be a privilege," he added, "that you are willing to spare me some of your time, which is so precious for science."[23]

Einstein's letter of June 13, 1931, is especially poignant. He informs Cartan that his old friend Marcel Grossmann has published a paper "rudely" criticizing the idea of absolute parallelism, but he wants Cartan to know that Grossmann is seriously ill with advanced multiple sclerosis. "I tell you all this to urge you not to answer him publicly," Einstein said loyally, adding that Grossmann was too ill to be accountable for the unpleasant tone and misguided content of his critique. When Grossmann died five years later, after his long and debilitating illness, Einstein wrote movingly to his widow, telling her what a treasured friend Marcel had been.[24]

Grossmann may never have realized what a crucial role he had played in putting tensors on the map. But since 1975 his legacy in general relativity has been honored in the Marcel Grossmann Meetings, which, every three or four years, bring together researchers from all over the world to discuss the latest developments. And in honoring Grossmann, these meetings also honor—implicitly, at least—the mathematical brilliance of Ricci and Levi-Civita, and the genius of Einstein.

EPILOGUE

The success of general relativity—especially after the 1919 eclipse expedition—brought tensors into the mainstream. For example, Paul Dirac used them, along with vectors and matrices, in his equations of quantum electrodynamics (QED). In this theory, which dates from 1927, Dirac united Maxwell's theory of electromagnetism, the quantum mechanics of the electron (described by "Dirac's equation"), and special relativity. It was a spectacular achievement—and Dirac was only twenty-five years old.

Dirac's QED helped launch quantum field theory (because QED deals with interactions between charged particles and electromagnetic fields). This is the kind of quantum theory that is needed to analyze the particles in particle colliders, for example, because these particles are accelerated to such high speeds that relativistic effects are significant. Particle accelerators such as CERN's famous Large Hadron Collider aim to find new particles predicted by various theories—and the detection of the Higgs boson in 2012 was big news at the time. Nowadays, some physicists wonder if the cost has been worth it.[1] Back at the end of 1929, however, Dirac used his theory to predict the existence of the bizarre phenomenon of antimatter—in particular, the existence of "positrons," positively charged particles that are "mirror images" of electrons, which carry a negative charge. Three years later, positrons were discovered experimentally, using a cloud chamber, and today they are used routinely—for example, in medical imaging processes such as PET (positron emission tomography).[2]

Dirac's equation described the behavior of a single electron, but in 1936, the remarkable mathematical physicist Bertha Swirles, a Girton graduate, extended Dirac's relativistic quantum analysis to two electrons, including their electron spin interactions.[3] Like Noether and Chisholm Young, Swirles spent time at Göttingen, where she worked with Max Born and Werner Heisenberg, two of the pioneers of quantum theory. In 1940, she married Cambridge geophysicist Harold Jeffreys, with whom she wrote a textbook on mathematical physics. She also continued her own research in quantum theory, publishing many papers, and took up a mathematics lectureship at her former college, Girton.

Today relativistic quantum mechanics (including QED) is important in many other areas, including quantum chemistry, materials science, and other nanotech applications. And we've already seen that electron spin has many applications, from MRIs to the qubits in quantum computers. As for tensors and quantum mechanics, we saw in chapter 11 that vectors and tensor products are important in figuring out the quantum state of a collection of particles or qubits. Tensor products are also used to prove that it is impossible to copy a qubit (the no-cloning theorem), and in analyzing such tantalizing things as quantum teleportation and entanglement.

Then there is relativistic cosmology, which Einstein pioneered soon after he'd finished his theory of general relativity. Tensors are key, of course—especially the metric, and the other tensors in Einstein's equations. We saw earlier that Einstein had created a static space-time, such as might describe the space-time around an object such as Earth or a quiescent star, whose gravitational field doesn't change in time. In 1917 he set out to do the same for the cosmos itself.

The idea in cosmology is to make some assumptions about the spatial distribution of matter in the universe—assumptions that will give essential features that must be reflected in the space-time metric. Einstein assumed his cosmological metric was independent of time, for like nearly everyone else at the time, he believed the universe was static. After all, the sun kept on burning and the stars kept to their usual positions and paths in the sky—as far as astronomers could tell, anyway. Einstein also assumed the universe was spherically symmetric, because, overall, the stars and galaxies appear to be symmetrically distributed. But the problem was that gravity causes matter to attract other matter—which you'd think would mean that all the stars and galaxies would be continually drawn together, ultimately coalescing. Since they don't do this, and since the cosmos appeared to be static, with everything staying happily in its right place, Einstein added a new term to the field equations. This extra term is the infamous "cosmological constant," which Einstein added to counteract gravity's natural attractive tendency.

Einstein's static model was the very first relativistic model of the cosmos. Much to his annoyance, though, in the 1920s mathematicians such as Alexander Friedmann found that general relativity predicted the universe should be expanding (or contracting). It was only in 1929, when Edwin Hubble found definitive evidence that the universe is indeed expanding, that Einstein abandoned his belief in a static cosmos—abandoning the cosmological constant in the process. He famously called this constant his biggest blunder, but today he has been redeemed: the cosmological constant is back—at least tentatively. Since it was designed to "push back" on gravity, it seems to be an ideal tool in the search for the mysterious dark energy causing the universe not just to expand, but to do so at an accelerating rate.

• • •

I could go on, but I must finally stop! It's enough to say that in mathematical physics today, vectors and tensors are vital. And they are still evolving, as we saw with the modern interpretation of tensors in chapter 11, and with Cartan's development of modern differential geometry. But it's not just in mathematics and physics that vectors and tensors are useful. They're important in engineering and chemistry, too—from modeling airflow over the blades in turbines and aircraft engines to crystallography, for example. They're important in digital technologies such as machine learning, search engines, computer vision, and natural language processing, as we saw in chapter 11, and in the neural networks we saw in chapter 9.

Tensors are important in all these areas for two main reasons: for solving problems involving invariance—in general relativity, quantum mechanics, and neural networks, for example—and for representing and handling data. When ancient Mesopotamian mathematicians etched tables of data into clay tablets, they surely never imagined the huge amount of data that tensors need to handle in data science today. And when ancient Chinese mathematicians used matrix-like arrays to solve systems of linear equations, they would have had no idea how complex and diverse these systems are today—covering everything from optimizing business costs and game strategies to Schrödinger's wave equation to the Google page-rank algorithm we saw in chapter 4.

In complex problems such as these—as well as in physics and math when equations are too difficult to solve exactly, or where you want to simulate something such as the merging of two black holes, or a future climate scenario—it is often necessary to use numerical methods. At the simplest level, these methods involve guessing a solution, and then using an algorithm that tweaks this solution, giving a closer and closer approximation. Newton was one of the pioneers of this approach, but today there's a whole field devoted to sophisticated computational math, including numerical linear algebra (NLA). Tensors, or algorithms based on tensorial ideas, are used in many of these computational methods in a variety of ways to solve a dizzying array of problems.[4]

In many of these modern applications, tensor ideas are not only used in the traditional sense of Ricci's calculus but are often adapted in clever new ways. For instance, in chapter 11 we saw the idea of "ragged tensors" in data science. A different example of this kind of adaptation is Dirac's remarkable realization that his tensor equations could only make physical sense if the coefficients were *matrices*, rather than ordinary functions like the metric tensor's components $g_{\mu\nu}$. We've seen this kind of evolution of ideas over and over in this story: in the way mathematicians went from numbers to vectors, and the way ancient tables morphed into matrices, and then vectors and matrices morphed into tensors; the way algebra developed into the calculus of functions, and then into the calculus of vectors and tensors; and the way mathematicians went from algebraic transformation equations to tensor operators.

To take just one more example of the adaptation of tensor ideas, tensor networks (TN) use the modern idea of tensors as multilinear maps—an idea I implied in chapter 11, when I spoke of the modern definition of a tensor as something that operates on vectors and one-forms to give a scalar. TNs are used to map and label arrays of pixels for image classification, say, or to classify entanglement properties, to take just two examples—and we've seen how useful tensor indices are for labeling different features. To handle huge amounts of data, TNs also use tensor adaptations of matrix methods such as the decomposition I mentioned in chapter 4.

A SENSE OF WONDER

In fact, all the mathematical tools we've followed in this story are indispensable today: algebra, calculus, imaginary numbers, vectors, quaternions, matrices, and tensors. The technology they've enabled helps make our lives more comfortable and interesting, and I wouldn't want to be without it. Of course, like many of us these days I do want it to be better regulated, for we now know that technology has its dark side when it's misapplied or used without taking care of people and the planet.

We need more than high-tech comfort, though, if we want a meaningful life. We need a sense of wonder, too. Our ancestors gazed in awe at the night sky for a hundred thousand years. They sought solace in a beautiful landscape, and they learned how to live with and from the earth. They were curious, too, trying to understand nature's secrets—and how to use this secret knowledge to make life easier. All this is what makes us human. But as we now know all too starkly, somewhere along the way that very scientific knowledge—and that ever-increasing desire for "comfort," which in the developed world now also means owning lots of "stuff" and making huge profits—helped disrupt the delicate balance of nature. We've cut down too many forests, poisoned too many rivers, and burned too many fossil fuels, so that we are losing all that was beautiful and wondrous about our planet. We are losing an essential part of our humanity.

Yet nature can be terrible as well as beautiful, and it was science that helped us to understand and mitigate some of that terror. It explained eclipses that once seemed malign portents from the gods, for example, and it makes weather forecasts to help us prepare for floods and fires, and electricity to protect us from the dark and cold. And while the effects of producing that electricity have now made nature even more terrifying, science and technology have become key weapons against cataclysmic climate change, and there is an exciting, greener future ahead. This future cannot be all about technology, of course. We need our ancient link with nature, too. Still, these new technologies are full of marvelous examples of human ingenuity that also connect us with our past.

Indeed, there is wonder aplenty in the magnificent intellectual creation that is mathematics and mathematical science. I've written this book to share some of these ideas with you—and to show how the mathematical concepts in this story have evolved over their extraordinary five-thousand-year multicultural journey through the human imagination. Such an odyssey connects us with our forerunners, just as marveling at a starry sky does.

When Einstein finally found his field equations, he told a friend he thought they were "beautiful beyond comparison."[5] In this story we've seen some beautiful equations come into being. Einstein's were the crowning achievement of Ricci's tensor analysis, and we also saw how Maxwell's equations evolved into their beautiful vector and tensor forms. We saw that Hamilton's quaternion rules scratched on Broome Bridge are delightfully economical, Euler's equation is elegant simplicity itself, Newton's laws are amazingly concise and potent, and much more. I think it is important to be able to appreciate this kind of intellectual beauty, just as it is important to appreciate beauty in music and the arts.

We've met some fascinating and dedicated people, too. Some of them were motivated primarily by the desire to solve practical problems—and, despite our recent technological overreach, one of the things I've wanted to bring out in this story is the way mathematical ideas have evolved hand in hand with these practical needs. On the other hand, many of these pioneers were driven simply by the desire to know and understand, or by the thrill of following through an intriguing pattern or proof. This story has shown that we need both kinds of thinker. It is truly awesome that, thanks to all of them, we humans not only have the possibility of safe, comfortable, and interesting lives; we can also comprehend so much about our vast and mind-bending universe. Such is the power of mathematics!

TIMELINE

ca. 3000 BCE, Mesopotamia and Egypt: The first known cuneiform and hieroglyphic tables of data are made.

ca. 2000 BCE, Mesopotamia: Cuneiform multiplication tables and documents show calculations for the area of agricultural fields; geometric methods are used to solve quadratic equations by completing the square.

ca. 1700 BCE, Mesopotamia: Plimpton 322 demonstrates Pythagoras's theorem is known before Pythagoras.

ca. 1650 BCE, Egypt: Ahmes papyrus includes approximate methods for finding the circumference and area of a circle.

ca. 300 BCE, Egypt/Greece: Euclid's *Elements* gives the rules of geometry governing flat ("Euclidean") space.

ca. 250 BCE, China: *Jiuzhang Suanshu* (*Nine Chapters on the Mathematical Art*) includes the algorithmic method known today as Gaussian elimination.

ca. 200 BCE, Sicily/Greece: Archimedes comes close to the idea of calculus.

ca. 150 CE, Egypt/Greece: Ptolemy, drawing on forerunners whose works are no longer extant—notably Eratosthenes and Aristarchus—writes his *Almagest* and *Geography*, which together make a landmark compilation of mathematical astronomy, trigonometry, and the use of coordinates (celestial and terrestrial latitude and longitude). Trigonometry and coordinates will be fundamental in the creation of vectors.

ca. 400 CE, Egypt/Greece: Hypatia writes learned commentaries on *Almagest* and other works; 415 CE, Hypatia is brutally murdered by zealots.

Seventh century CE, India: Mathematics flourishes; Brahmagupta considers both positive and negative numbers.

Ninth century CE: Arabic mathematics flourishes, including Mohammed ibn-Mūsā al-Khwārizmī at Caliph al-Ma'mūn's House of Wisdom in Baghdad; ca. 830 CE, al-Khwārizmī writes the textbook *Al-Jabr wa'l muqābalah* (from which the name "algebra" is later derived). But algebra is still done using words, for algebraic symbolism will not exist until the seventeenth century.

ca. 1200, Middle East (modern Iran): Sharaf al-Dīn al-Ṭūsī does pioneering work on solving cubic equations.

1540, Italy: Girolamo Cardano's *Ars Magna* (*The Great Art*) contains the first algorithms for solving cubic equations. He is flummoxed by the "imaginary" numbers showing up in some of these solutions.

1572, Italy: Rafael Bombelli's *Algebra* pioneers the math of complex numbers; these numbers will play a key role in the discovery of quaternions and vectors.

1619, England: Thomas Harriot's work on the mechanics of collisions prefigures the idea of vectors and uses the parallelogram rule correctly.

1631, England: Thomas Harriot's *Artis Analyticae Praxis* (*Practice of the Analytical Art*) is posthumously published—the first algebra textbook where the equations are fully symbolic and use essentially modern symbolism.

1637, France: René Descartes's *Discourse on Method* introduces the algebraic symbols x and y and pioneers what would become "Cartesian" coordinates.

1670s, England/Germany: Isaac Newton and Gottfried Leibniz independently create general, algorithmic differential and integral calculus.

1687: England: Newton publishes his *Philosophiae Naturalis Principia*

Mathematica (*Mathematical Principles of Natural Philosophy*); it includes his general calculus algorithms and his theory of gravity and introduces to physics the idea that forces and velocities are vectorial. Most of his proofs, however, are done geometrically because he (rightly) feels calculus is not yet rigorous enough. (It needs the definitions of limits and continuity that will come in the nineteenth century.)

Early 1700s, Switzerland: Johann (aka Jean) Bernoulli defends Leibniz in the calculus priority dispute with Newton; later he begins to translate Newton's proofs into algebraic (but Leibnizian) calculus.

1759, France: Émilie du Châtelet's French translation of *Principia* is published—the first translation outside Britain and still the definitive French version today. It includes a technical appendix, in which she helps pioneer the process of writing Newton's geometrical proofs in terms of modern (Leibnizian) calculus.

1785, France: Charles Coulomb shows that the force from a static electric charge obeys an inverse square law like Newton's law of gravity.

1780s, France: Joseph-Louis Lagrange reformulates Newton's law of gravity in terms of a gravitational "potential." Einstein will use this idea in his own theory of gravity (general relativity).

1788, France: Lagrange publishes *Mécanique Analytique* (*Analytical Mechanics*), which goes a long way toward updating Newtonian mechanics and expressing it in modern calculus form. Half a century later, Hermann Grassmann will be inspired by this book, when he sets out on his idiosyncratic journey to codiscovering vector analysis.

1799, France: Pierre-Simon Laplace publishes the first volume of his monumental *Traité de Mécanique Céleste* (*Treatise on Celestial Mechanics*). This also inspires Grassmann and is influential on William Rowan Hamilton's contemporaneous development of vector analysis. It will also inspire the self-taught Mary Somerville, whose explicated, expanded English version of Laplace's first two volumes, *Mechanism of the Heavens,* will be published in 1831.

1790s, France: Lagrange and Laplace are members of the Committee on Weights and Measures that introduces the metric system.

1800, Italy: Alessandro Volta invents the first electric battery, enabling electric currents to be produced for the very first time.

1801, England: Thomas Young's double-slit experiment shows the wave nature of light.

Early 1800s, Europe: Argand (France), Gauss (Germany), Wessel (Norway), and Warren (England) independently represent complex numbers on a plane. Multiplication by *i* is soon interpreted as a 90° rotation around this plane.

Early 1800s, France: Sophie Germain corresponds with Carl Gauss; in 1816 she wins a prestigious mathematical prize from the French Academy of Sciences, for a paper pioneering the mathematics of vibrating surfaces.

1820, Denmark: Hans Øersted discovers that a changing electric current affects a magnet.

1821, France: André-Marie Ampère applies Øersted's discovery by inventing the first telegraphic system. He goes on to be an important contributor to electromagnetic theory and experiment.

1821, England: Michael Faraday applies Øersted's discovery by inventing the first prototype of an electric motor.

1822, France: Joseph Fourier derives the heat equation; he applies it in an 1827 paper that initiates climate science.

1828, Germany: Carl Friedrich Gauss develops non-Euclidean geometry. (Janos Bolyai in Hungary and Nicolai Lobachevsky in Russia are also working on this.) Gauss defines a curved surface's intrinsic geometry, defining the distance measure (line element or metric) on a curved surface, and the intrinsic curvature of the surface.

1831, England: Michael Faraday discovers that a moving magnet can generate an electric force; as a result, he invents the prototype of the electric generator. Faraday's and Øersted's discoveries amount to the discovery of electromagnetism. Subsequently, Faraday conceives the idea of electric and magnetic fields.

1831, Scotland/England: Mary Somerville's astronomy textbook *Mechanism of the Heavens* is rapturously received by leading scientific men. She uses Leibnizian calculus.

1832, Ireland: William Rowan Hamilton uses calculus to predict conical refraction, one of the very first mathematical predictions of a previously unknown physical phenomenon.

1833, 1840, England: William Whewell coins the terms "scientist" and "physicist."

1837, Ireland: Hamilton is made president of the Royal Irish Academy; he wants to include literature as well as science and asks the venerable writer Maria Edgeworth for advice. He takes most of it but doesn't heed her suggestion to make meetings available to women.

1843, Ireland: Hamilton invents quaternions and vectors; October 16, 1843, he scratches their basic rules on Broome Bridge. A month later he formally announces his discovery in a paper read to the Royal Irish Academy.

1843, England: William Wordsworth is made poet laureate. He is friendly with Hamilton, who loves and composes poetry.

1844, Germany: Hermann Grassmann publishes his book on what we now call vectors and vector spaces, *Ausdehnungslehre*.

Early 1840s, England: Arthur Cayley corresponds with George Boole on invariants; Cayley laments the lack of transport between them (the first railways are still being laid).

1845, United States: Writer Henry David Thoreau seeks a simpler life in the woods by Walden Pond. His book *Walden* will help pioneer the environmental movement and show how individuals can use fewer resources.

1850s, United States/Ireland: Eunice Newton Foote and John Tyndall independently pioneer the science of carbon dioxide's heating (greenhouse) effect.

1855, Scotland/England: James Clerk Maxwell publishes his first paper on Faraday's idea of electric and magnetic fields. He and William Thomson (on whose paper Maxwell drew) are the only ones to make significant attempts at putting Faraday's idea into mathematical (vectorial) language.

1857, Scotland/Ireland: Peter Guthrie Tait (Maxwell's childhood friend) begins to apply Hamilton's vector calculus operator, ∇, and (with the help of his assistant) names it "nabla." He and Hamilton soon begin corresponding with each other.

1858, England: Arthur Cayley formalizes the algebra of matrices; he has been inspired by Hamilton's discovery of noncommutative quaternion algebra.

Late 1850s, Scotland/England: William Thomson (the future Lord Kelvin, and a friend of Maxwell and Tait) helps lay the first submarine telegraphic cable under the Atlantic Ocean—he is a director and scientific advisor of the North Atlantic Telegraph Company.

1854, 1861, Germany: Bernhard Riemann applies his former professor Gauss's analysis of curvature to spaces of arbitrary dimension, beyond the usual three we're used to. He finds the condition for identifying curvature is expressed in what will become known as the Riemann tensor. (If the surface is flat, the Riemann tensor is zero, and vice versa.)

January 1865, England: Maxwell's theory of the electromagnetic field, first presented at the end of 1864, is published. It is vectorial, but the equations are in component form, not yet in full, whole-vector calculus form. Maxwell deduces from his equations that light is an electromagnetic wave and suggests the possible existence of other electromagnetic radiation.

1867, Scotland/England: Thomson and Tait publish their successful physics textbook *A Treatise on Natural Philosophy* (known colloquially as *T* and *T′*); it has no quaternions in it, because Thomson thinks quaternions and whole vectors are useless since it is the components one needs to calculate with. In the same year, Tait publishes his *Elementary Treatise on Quaternions*, which further develops Hamilton's vector calculus.

1870, Scotland: Maxwell pioneers the vector calculus terms "divergence," "grad," and "curl."

1873, England: Maxwell's *Treatise on Electricity and Magnetism* is published; volume 2 includes the whole-vector form of his equations of electromagnetism (in quaternionic notation). It also defines what would become known as the stress tensor; Cauchy and Thomson had done something similar, but Maxwell gave it a two-index symbol, which later would become standard.

1870s, England: William Kingdon Clifford creates a "geometric algebra" based on a synthesis of Hamilton's and Grassmann's creation of vectors. He is friendly with Tait and Maxwell and with the famous novelist George Eliot. (Her masterwork *Middlemarch* has just been published, in 1871–72.)

1880s, England/United States: Oliver Heaviside and Josiah Gibbs independently create modern vector analysis. Their inspiration is, first, Maxwell's use of vectors to describe electromagnetism in his *Treatise*, and then Hamilton and Tait's works on quaternions (which they found from the references Maxwell gave).

1880s, Italy: Gregorio Ricci develops his "absolute differential calculus," which is known today as tensor calculus. Gibbs, extending Grassmann's work, also has the idea of tensor products. As with whole vectors, mainstream mathematicians see the elegance but not the practical point of tensors.

1888, Germany/England: Heinrich Hertz and Oliver Lodge announce the first-ever deliberate creation of radio waves, spectacularly confirming Maxwell's mathematical prediction of their existence.

1888, Italy: Giuseppe Peano devises the modern axiomatic definition of a vector space.

Late 1880s–early 1890s, United Kingdom, United States, Australia: The Vector "Wars": Heaviside and Gibbs for vector analysis vs. Tait, Alexander McAulay, and others for quaternions; Cayley and Thomson vs. all of them, favoring components over whole vectors and quaternions. Vector analysis—the math of real, whole vectors—wins (until quaternions show up again in the late twentieth century).

Late 1890s, Switzerland: Albert Einstein and Mileva Marić, who are in the same physics class, become lovers. Marić and fellow student Marcel Grossmann are the first to recognize and support Einstein's genius. Grossmann goes on to a successful career as a mathematics professor, but Marić, the only girl in their class, does not qualify to graduate (sexism?); she tries again, but ultimately loses her hard-fought dream of becoming a scientist.

1895, Germany: Grace Chisholm, who received an unofficial degree from Cambridge and then worked on her doctorate in mathematics at Göttingen with Felix Klein, is the first woman to receive an official doctorate in Germany, and one of the first anywhere in the world. Women will not be able to take official degrees from Oxford until 1920 or from Cambridge until 1948.

1900, Italy/Germany: Gregorio Ricci and his former student Tullio Levi-Civita write an overview of Ricci's "absolute differential calculus"—today called tensor calculus—at the request of Felix Klein, Ricci's former mentor and an editor of *Mathematische Annalen,* where the Ricci–Levi-Civita paper is published. Twelve years later, it will inspire Einstein and Grossmann in their search for the mathematical foundations of the general theory of relativity.

1905, Switzerland: Einstein, working as a patent officer, publishes five groundbreaking papers, including *On the Electrodynamics of Moving Bodies,* better known today as the paper that created the special theory of relativity. Hendrik Lorentz (1904) and Henri Poincaré (1905) also have similar theories, but they are ether-based and are not fully relativistic. Lorentz uses whole-vector notation, but Poincaré and Einstein use components.

1907, Germany: Hermann Minkowski, Einstein's former math lecturer, uses the special theory of relativity to create the concept of space-time.

1908–10, Germany: Minkowski begins to write Maxwell's equations in tensor form, a task then taken up after Minkowski's sudden death by his friend Arnold Sommerfeld. They create the concept of the energy-momentum tensor, which will be a key to the general theory of relativity.

1912, Switzerland: Einstein and his old friend Grossmann are now colleagues at their old school, the Swiss Polytechnic or ETH. They discover Ricci's tensor calculus and collaborate on the mathematical foundations of general relativity.

1914–18, World War I: In addition to the horrors on the front, food is scarce, and scientific communication is interrupted (including that between Einstein and Levi-Civita on tensor theory).

1914, Germany: Einstein takes up a professorship in Berlin. The Einstein-Marić marriage breaks down. Einstein has been working incredibly hard developing the general theory of relativity. He is also courting his cousin Elsa Einstein.

1915, Germany: David Hilbert and Einstein collaborate (and ultimately compete) on the final steps toward the general theory of relativity. It is completed in November. After a little friction, the two men remain friendly colleagues. Einstein had created the concept of the curvature of space-time as a representation of the effects of gravity, and with Grossmann's help, he did it with tensors. Today, general relativity remains one of the intellectual wonders of the world.

1916–18, Germany: Emmy Noether works with Klein and Hilbert—who are in touch with Einstein—on tightening the interpretation of energy conservation in general relativity theory. She finds a key in her two 1918 theorems (now known as "Noether's theorems"), which show the interconnection between mathematical symmetries and physical conservation laws.

1917, Italy: Levi-Civita shows the tensor connection between general relativity and conservation of energy-momentum, via the contracted Bianchi identities. He also uses tensors to define the notion of "parallel" when taking derivatives on curved surfaces.

1919, England: Arthur Eddington and colleagues announce their findings on the bending of light, which had required careful measurements during a total solar eclipse. It is the first successful new test of general relativity, and Einstein becomes a superstar. His success puts tensors into the scientific mainstream.

1920s, Germany, Netherlands, England; 1975, Australia: In 1922, Otto Stern and Walther Gerlach find the angular momentum of electrons magnetically deflected is quantized; soon afterward, George Uhlenbeck and Samuel Goudsmit perform an experiment that suggests Stern and Gerlach had measured "spin" angular momentum, not the orbital angular momentum of electrons orbiting nuclei. In the mid-1920s Paul Dirac provides theoretical support for spin in his relativistic theory of quantum mechanical electron behavior—he uses Pauli spin matrices to describe electron rotations, and Wolfgang Pauli had shown that the math of these has exactly the same structure as Hamilton's quaternion rotations. In 1975, Tony Klein and Geoff Opat show that spin is physical, not just a mathematical analogy.

1924, Netherlands: Jan Schouten and Dirk Struik highlight the relationship between Noether's theorems and the tensorial Bianchi identities that are important in general relativity.

1920s, France: Élie Cartan helps develop modern differential geometry. This work, by Cartan and others over the twentieth century, puts tensors (and vectors) into modern form, as operators and linear mappings, rather than Ricci's rules showing how components transform under those mappings and coordinate transformations.

1960, United States: The gravitational redshift predicted by Einstein is experimentally detected by Harvard University's Robert Pound and Glen Rebka. (Less certain evidence, from astronomical spectra, had been found by Daniel Popper in 1954.)

1960s–70s, United States: Following the lead of Gerry Salton, vectors and matrices are used in programming search engines.

1981, United States: NASA starts routinely using quaternions to help guide its spacecraft.

Late 1990s, United States: Larry Page and Sergey Brin use vectors and matrices in their Google PageRank algorithm.

Early 2000s, United States: Quaternions are used in robotics, CGI, molecular dynamics, the rotations of our mobile phone screens, spacecraft control, and much more.

2011, United States: NASA announces that its Gravity Probe B satellite results confirm Einstein's predictions of the warping of space-time (the geodetic effect) and frame-dragging (the amount by which a spinning object pulls space and time with it).

2015, international LIGO collaboration: Gravitational waves, as predicted by general relativity, are detected for the first time. New detections are revealed over subsequent years.

2017: The International Astronomical Union acknowledges ancient cultures by including eighty-six indigenous star names in a new star-naming system.

2018, France: An improved test confirms the existence of gravitational redshift (equivalence principle).

2019, international collaboration (Event Horizon Telescope): The first direct image of (the shadow of) a black hole is created.

2020, England/Sweden: Roger Penrose shares the Nobel Prize for Physics, for showing why general relativity predicts the existence of black holes.

Ongoing: Increasingly stringent tests confirm the (tensor) theory of general relativity. For example, in 2019, a team using the LARES and LAGEOS satellites confirms to higher accuracy the general relativistic prediction of frame-dragging; in 2021, a team using NASA's Nuclear Spectroscopic Telescope Array and the European Space Agency's XMM-Newton Telescope observes X-rays from the back of a black hole (another test of Einstein's light-bending prediction); in 2022, the MICROSCOPE collaboration announces new accuracy for the equivalence principle; in 2022, the Event Horizon Telescope reveals the first image of the supermassive black hole at the center of our galaxy; in 2023, the Atacama Cosmology Telescope (ACT) collaboration uses gravitational lensing to map dark matter, and the Parkes radio telescope Murriyang finds new evidence of gravitational waves; and much more.

Today, worldwide: Quaternions, vectors, and tensors are fundamental in a huge variety of applications in physics, engineering, IT (including AI, CGI, and search engines)—in fact, in just about everything that requires pinpointing objects in space or representing and processing information.

ACKNOWLEDGMENTS

I want to begin by acknowledging the Traditional Custodians of the land on which I live and work—the Bunurong people—and paying my respects to their Elders past and present.

This book owes its inception to the insight and encouragement of the incomparable Joe Calamia, at the University of Chicago Press. I'm immensely grateful to Joe, not just for his bold and creative suggestion to write a story of vectors and tensors but for his ongoing editorial skill and unflagging support. It has been an absolute pleasure to work with him.

It's been fun (and superbly professional) working with the rest of the team at the Press, too, including Matt Lang, Nicholas Lilly, and Caterina MacLean. Special thanks go to Susan Olin, whose meticulous eye for detail has smoothed out my grammar and clarified aspects of the story and whose patience, encouragement, and good humor have made the difficult copyediting stage of this project a happier experience for me than I was expecting. I'm also extremely grateful to the anonymous referees for their kind and invaluable suggestions for improving both the historical approach to the storyline and the technical details. And thank you Tobiah Waldron, for loving math and taking on the task of preparing the index.

Many thanks are also due to the people and organizations that have generously permitted me to use the images in this book. The "kindness of strangers" in this process really lifted my spirits, so huge thanks to Steven Archer, of the Trinity College Library at Cambridge University, for young Maxwell; Lord Max Egremont of Petworth House and archivist Abigail Hartley of the West Sussex Records Office, for Harriot's manuscript; Michelle Tobin of the Dublin Institute of Advanced Studies and David Malone of the Hamilton Institute at Maynooth University, for helping me track down Hamilton and his son, and to Meabdh Murphy of the Royal Irish Academy for providing the image; to Joe Calamia for the beautiful statue of Maxwell; and to Catherine Booth of the James Clerk Maxwell Foundation (JCMF), who, with help from the webmaster and the curator at JCMF, carried on Maxwell's reputation for kindness by stepping in at the eleventh hour to provide the image of Tait. (I can highly recommend a visit to the JCMF Museum at Maxwell's birthplace in India Street, Edinburgh.) I also thank Justine Kent of Peterhouse College, Cambridge, for her generosity, and Heike Hartmann of ETH-Bibliothek, Zurich, for directing me to the images of Einstein, Marić, Grossmann, and Minkowski. I'm also most grateful for my affiliation with the School of Mathematics at Monash University, and to Monash's marvelous library.

I also offer my gratitude to all the readers who have enjoyed my books, especially the kind strangers who have written to me over the years. It is humbling and wonderful for a solitary writer to feel part of such a far-flung community. And to those closer to home, I especially thank my friends Gina Ward and Ika Willis, whose support and perceptive feedback on early chapters of this book came at just the right time—and Gina, our long friendship and ongoing literary interests have sustained me from the beginning. A big thank you, too, to Ursula and Werner Theinert, for your special friendship and valuable support of my work.

Many thanks, too, to Joe Mazur, Carolyn Landon, Cheryl Hingley, Elizabeth Finkel, Erica Jolly, Margaret Harris, Vera Ray (RIP), Penny and Molly Anggo, Bet Sibley, Catherine Watson, Phil Henshall, Anne and Phil Dempster, Peter and Anne-Marie Biram, John and Mary Mutsaers, Liane Arno, Matt Stone, John and Helen Laing, Annie Bain, Karin Murphy-Ellis, John Di Stefano, Elizabeth and Ian Fraser, Ingmar Quist, Tricia Szirom,

Sandra Shotlander, Susan Hawthorne, Harry Freeman, Gill Heal, Michael and Gail Box, Margaret Pitt, and John and Peter Snare. Over the years you have each been supportive of my work in some special way—from collegial support to taking a genuine interest in what I'm writing just when I needed it, to reading (even buying) my books and offering generous feedback, or to understanding that when I'm engrossed in a book or rushing to meet deadlines I can be out of social action for a while.

Finally, huge, heartfelt thanks to my darling Morgan, whose endless support and encouragement have been beyond compare. I dedicate this book to you.

NOTES

PROLOGUE

1. Letter from William Rowan Hamilton to P. G. Tait, published in the preface to the third edition of P. G. Tait, *An Elementary Treatise on Quaternions* (Cambridge: Cambridge University Press, 1890).
2. Maxwell didn't use the term "vector field," but that's what he'd created (as he made clear a few years later, in his 1873 *Treatise on Electricity and Magnetism*); I'll explore what this means—and how it influenced later physicists—in subsequent chapters.
3. *Ancient tabular mathematics*: see, e.g., Eleanor Robson, "Mathematical Cuneiform Tablets in the Ashmolean Museum, Oxford," *SCIAMVS* 5 (2004): 2–65; Duncan J. Melville, "Computation in Early Mesopotamia," in *Computations and Computing Devices in Mathematics Education before the Advent of Electronic Calculators*, ed. A. Volkov and V. Freiman (Switzerland: Springer Nature, 2018), 25–47. Melville highlights the difficulties in interpreting such ancient documents and their intended uses; and Robert Middeke-Conlin points out that the debate over the utility of some of the Mesopotamian mathematical tablets is ongoing, in "The Mathematics of Canal Construction in the Kingdoms of Larsa and Babylon," *Water History* 12 (2020): 105–28. Daniel Mansfield's 2021 study, however, adds significant clarity to some of these interpretations; see his "Plimpton 322: A Study of Rectangles," *Foundations of Science* 26 (2021): 977–1005.

4. I'm indebted for the information on Mesopotamian land use and surveying, and on Plimpton's role, to Mansfield, "Plimpton 322." Mansfield discovered the Pythagorean nature of Si427, and he also gives new insights on Plimpton 322 as well as details about Mesopotamian multiplication tables. On his claims for Mesopotamian trigonometry, however, see, e.g., Evelyn Lamb, "Don't Fall for Babylonian Trigonometry Hype," *Scientific American* (blog), August 29, 2017.
5. *Australian indigenous astronomy*: Duane Hamacher, "Stories from the Sky: Astronomy in Indigenous Knowledge," *The Conversation*, December 1, 2014; Ray P. Norris, Cilla Norris, Duane W. Hamacher, and Reg Abrahams, "Wurdi Youang: An Australian Aboriginal Stone Arrangement with Possible Solar Indications," *Rock Art Research* 30, no. 1 (2013): 55–65.
6. Some of Ptolemy's brilliant but now mostly lost sources are Eudoxus of Cnidus, perhaps the first to begin the process of geometric modeling in astronomy, and who pioneered the protocalculus method of exhaustion that we'll meet in chap. 2; Eratosthenes of Cyrene, who apparently used a basic kind of latitude and longitude, and who deduced the size of the earth extraordinarily accurately, given he used a shadow stick and simple measuring rods; Hipparchus of Nicaea, who seems to have been the first to systematically use a 360° circle to make precise geometrical representations of planetary motion, and whose math and astronomy were so good that he discovered the precession of the equinoxes; and Apollonius of Perga, who used a kind of post-hoc coordinate system in his mathematical analysis of conics, and whose epicycle and eccentric models of planetary motion Ptolemy built directly upon.
7. For example, if you were traveling, say, *northeast* at 35 mph, your vector would point in the direction at 45° to the axes, but its components would be such that when you add them using Pythagoras's theorem, the magnitude is still 35. So, it would be the vector $\left(\frac{35}{\sqrt{2}}, \frac{35}{\sqrt{2}}\right)$.

   The vector $\left(\frac{35}{\sqrt{2}}, \frac{35}{\sqrt{2}}\right)$ is found from fig. 0.2, and the fact that when working out the components you need $\sin 45° = \cos 45° = \frac{1}{\sqrt{2}}$; if you're rusty on trigonometry, you can take a look ahead at fig. 3.4 to see why we need sin 45° and cos 45° for the vertical and horizontal components. Then, using Pythagoras's theorem, you have $\left(\frac{35}{\sqrt{2}}\right)^2 + \left(\frac{35}{\sqrt{2}}\right)^2 = 35^2$, so the vector is $\left(\frac{35}{\sqrt{2}}, \frac{35}{\sqrt{2}}\right)$ and the magnitude is 35.

8. The components of four-velocity are defined by analogy with ordinary velocity, as the derivatives of the (space-time) coordinates with respect to (proper) time. The proper time is the directly measurable "clock time," that is, from a clock at rest with respect to the observer.

9. *The Mad Hatter parody conjecture* refers to a consequence of one of Hamilton's new types of multiplication (noncommutativity, which we'll see in chaps. 1 and 4); it was suggested by Victorian literature expert Melanie Bayley, "Alice's Adventures in Algebra: Wonderland Solved," *New Scientist*, December 16, 2009. Michael Brooks followed up the story with Bayley in his *The Art of More* (Melbourne: Scribe, 2021), 194–96.

Brooks and Bayley paint Dodgson as a conservative, mediocre mathematician (see also, e.g., Michael Deakin, "Lewis Carroll—Mathematician?," *Function* 18, no. 1 [February 1994]: 10–18)—one who couldn't have understood Hamilton's work. This may well be true. Dodgson/Carroll's mathematical fame today, however, rests not on work published in his lifetime but on posthumously discovered later manuscripts on methods of voting and on symbolic logic (in which he used absurd propositions and symbolic algebra to teach the rules of logic); cf. Francine Abeles, "Logic and Lewis Carroll," *Nature* 527 (November 19, 2015): 302–4; Amirouche Moktefi, "Why Make Things Simple When You Can Make Them Complicated? An Appreciation of Lewis Carroll's Symbolic Logic," *Logica Universalis* 15 (2021): 359–79—and, for a taste of Carroll's logical puzzles, see, e.g., the University of Hawaii website http://math.hawaii.edu/~hile/math100/logice.htm.

Note, however, that in Carroll's *Symbolic Logic* (New York: Dover, 1958; originally published 1897, three decades after *Alice*), he sets up a *commutative* algebra of symbolic logic (e.g., 35, 70): Carroll's equation is about logic, not algebra, but it is interesting that he chose to emphasize "commutativity"—perhaps he did have a problem with Hamilton's noncommutative multiplication. Or perhaps he simply wanted to distinguish his symbolic logical propositions from vector products.

CHAPTER 1

1. In the preface to his *Lectures on Quaternions*, Hamilton said he was led to quaternions because he wanted to "connect *calculation* with *geometry*," and to move these calculations "from the plane to space." Later he shows how to do 3-D rotations; see *Lectures on Quaternions* (Dublin: Hodges and Smith, London: Whittaker, and Cambridge: Macmillan, 1853), 269 (art. 282).

2. Melanie Bayley ("Alice's Adventures in Algebra") pointed out this and other examples that suggest, to her, that Carroll was parodying Hamilton. But it's possible that Carroll might simply have been distinguishing the rules of logic from those of Hamilton's algebra.
3. *Hamilton's electrical analogy* is from a letter to his son Archibald (quoted in Michael J. Crowe, *A History of Vector Analysis* [Notre Dame, IN: University of Notre Dame Press, 1967], 29–30). He gave a similar description to P. G. Tait in an 1858 letter: https://www.tcd.ie/library/manuscripts/blog/tag/moon-landing/. The commemorative art installation is by Emma Ray; see the website of the Royal Irish Academy, which cocommissioned the work, and especially the YouTube video of the preparation and installation, at https://www.google.com/search?client=safari&rls=en&q=Commemorative+art+installation+Hamilton+Broombridge+Luas&ie=UTF-8&oe=UTF-8#fpstate=ive&vld=cid:a9bfbdde,vid:1nQct3p3184.
4. While visiting the Old Library at Trinity College, Dublin, Armstrong paused beside a marble bust of Hamilton and explained to his guide how quaternions help spacecraft navigation: Estelle Gittins, July 19, 2019, https://www.tcd.ie/library/manuscripts/blog/tag/moon-landing/.
5. I have adapted a diagram—possibly attributed to Pythagoras—given in T. L. Heath, *Translation of Euclid's Elements* (Cambridge: Cambridge University Press, 1925) reproduced in John Stillwell, *Mathematics and Its History* (New York: Springer-Verlag, 1989), 7. For Euclid's more sophisticated proof, see Carl Boyer, *A History of Mathematics*, rev. Uta Merzbach (New York: John Wiley and Sons, 1991), 108.
6. There are different versions, but see, e.g., Library of Congress, https://www.loc.gov/item/2021666184/.
7. Modern mathematicians tend to prefer defining i as the (principal) solution to $x^2 + 1 = 0$, rather than specifying it as $\sqrt{-1}$; in other words, i is usually defined in terms of its square, $i^2 = -1$, rather than as a square root. That's because the latter can lead to conundrums such as this:

$$-1 = i \times i = \sqrt{-1} \times \sqrt{-1} = \sqrt{(-1)(-1)} = \sqrt{1} = \pm 1,$$

and if you take the positive root, you have $-1 = 1$, which is clearly wrong!
8. *Descartes on imaginary numbers*: Brian E. Blank, "Book Review: An Imaginary Tale by Paul Nahin," *Notices of the AMS* (November 1999): 1233.
9. Al-Khwārizmī quoted in Boyer, *History of Mathematics*, 229; his geometrical way of completing the square, 231. Translation of *Al-jabr . . .*, and geometric

example: Raymond Flood and Robin Wilson, *The Great Mathematicians* (London: Arcturus, 2011), 46–47.

10. For more about Harriot's extraordinary life and work, see my *Thomas Harriot: A Life in Science* (New York: Oxford University Press, 2019), and the references therein. Note that his posthumous book, *Praxis*, was put together by his friends, but evidently they were not such good mathematicians as he: his papers offer more sophisticated work than that which they published, including the use of imaginary numbers.

11. For a fascinating history of the evolution of algebraic symbolism, see Joseph Mazur, *Enlightening Symbols: A Short History of Mathematical Notation and Its Hidden Powers* (Princeton, NJ: Princeton University Press, 2014).

12. In addition to his special relativity and $E = mc^2$ papers, in 1905 Einstein also published two important papers on Brownian motion and the size of molecules as well as a pioneering paper on the quantum theory of light. For an introductory overview, see my *Young Einstein and the Story of $E = mc^2$* (Sydney: Ligature, 2014).

13. This problem is from tablet CBS 43, as translated in Eleanor Robson, "Mathematical Cuneiform Tablets in Philadelphia, Part I: Problems and Calculations," *SCIAMVS* 1 (2000): 11–48; for my illustrative purpose, I have changed the right-hand side of the problem to 21—the tablet (shown on 39) has 41, but as Robson says (42), the symbols are not entirely clear on the tablet, and as deciphered do not yield the kind of simple integer or two-place (in the sexagesimal system) solution used at the time. My diagram of the Old Babylonian geometrical method of completing the square is adapted from p. 42.

14. *Canals, etc.*: Robert Middeke-Conlin, "The Mathematics of Canal Construction in the Kingdoms of Larsa and Babyon," *Water History* 12 (2020): 105–28.

15. Note that earlier mathematicians—including the legendary early twelfth-century Persian poet Omar Khayyam, who was also a mathematician—had found a purely geometrical way of solving some cubic equations with positive roots via the intersection of two curves: Boyer, *History of Mathematics*, 241; Flood and Wilson, *The Great Mathematicians*, 49. On al-Ṭūsī, see J. J. O'Connor and E. F. Robertson's MacTutor entry for him, at https://mathshistory.st-andrews.ac.uk/Biographies/Al-Tusi_Sharaf/.

16. *Cardano's underlying algorithm* (based on Tartaglia's) for solving an equation of the form $x^3 = cx + d$ is this: choose new variables u, v and set $x = u + v$, $uv = c/3$. Put these into the original equation, and you'll get $u^3 + v^3 = d$; eliminate

v and this becomes a quadratic equation in u^3, which can be solved using the quadratic formula. Put this solution for u^3 into $u^3 + v^3 = d$ and solve for v^3. Take the cube roots of u^3 and v^3 to find u, v, and hence $x = u + v$. It's ingenious, and all created without the modern symbolism that makes it easier to keep track of your thought processes. The example I gave, $x^3 = 6x + 40$, and Cardano's algorithm for solving it—together with his geometric completion of the cube—is in chap. 12 of his *Ars Magna*, reprinted in R. Laubenbacher and D. Pengelley, "Algebra: The Search for an Elusive Formula," in *Mathematical Expeditions*, Undergraduate Texts in Mathematics (New York: Springer, 1999), 230; https://doi.org/10.1007/978-1-4612-0523-4_5.

17. For instance, Schrödinger's equation describes the dynamics of fundamental particles such as photons, electrons, and other subatomic particles—and it contains i. Electromagnetic waves, too, are easier to handle mathematically using the complex form, so i is behind all sorts of modern technology.

18. *Wallis on Harriot*: quoted in Jacqueline Stedall, "Rob'd of Glories: The Posthumous Misfortunes of Thomas Harriot and His Algebra," *Archive for History of Exact Sciences* 54, no. 6 (June 2000): 490. *Harriot first to algebraically (symbolically) solve cubics*: The great mathematician Lagrange first made this observation; see Seltman, "Harriot's Algebra: Reputation and Reality," in *Thomas Harriot*, vol. 1, *An Elizabethan Man of Science*, ed. Robert Fox (Aldershot: Ashgate, 2000), 185.

19. Similarly, a quadratic equation has two solutions, a quartic has four solutions, and so on. The German mathematician Peter Roth suggested this link between degree and number of solutions at around the same time (in his 1608 *Arithmetica Philosphica*), but he did not write his equations symbolically or explore complex roots. A rigorous proof of the "fundamental theorem of algebra" suggested by Harriot's "factor" construction came two hundred years later—Harriot himself didn't claim as much. An example of his use of factors and symbols to get complex solutions is found in, e.g., British Library Manuscript 6783, fols. 157, 156.

20. Following Euler (or figs. 3.4 and 3.6 and related discussion in chap. 3), you can write a complex number $a + ib$ as $r(\cos\theta + i\sin\theta) = re^{i\theta}$, where $r = \sqrt{(a^2 + b^2)}$ and θ is found from the inverse cosine and sine accordingly. From De Moivre's theorem (or simply from the index laws), the cube root of this number is $\sqrt[3]{(re^{i\theta})} = r^{1/3}e\dfrac{i(\theta+2k\pi)}{3}$, where $k = 0, 1, 2$ gives the three different roots. Applying this to $\sqrt[3]{(2+11i)} + \sqrt[3]{(2-11i)} = r^{1/3}e\dfrac{i(\theta+2k\pi)}{3}$ $+r^{1/3}e\dfrac{i(-\theta-2k\pi)}{3} = 2r^{1/3}\cos\dfrac{(\theta+2k\pi)}{3}$, you get the three solutions of Carda-

no's equation, $x = 4, -2 + \sqrt{3}, -2 - \sqrt{3}$. It is a bit fiddly, but all the steps use only senior high school or freshman university math.

21. *Harriot's quotation*: British Library Additional Manuscript 6783 fol. 186. See also Jacqueline Stedall, "Notes Made by Thomas Harriot on the Treatises of François Viète," *Archive for Exact Sciences* 62, no. 2 (March 2008): 179–200.

22. *Seltman's quotation* is in her "Harriot's Algebra," in *Thomas Harriot*, vol. 1, *An Elizabethan Man of Science*, ed. Robert Fox (Aldershot: Ashgate, 2000), 184, my emphasis. Regarding Harriot's use of complex and negative solutions, Seltman gives a fine analysis of the superiority of Harriot's manuscripts to the posthumously published *Artis Analyticae Praxis*, which was put together by his less capable friends, on the basis of what they understood from his papers. This includes their rejection of Harriot's use of imaginary and negative numbers.

CHAPTER 2

1. Sound waves are changes of air pressure, and the pebbles' shock waves travel through the water in the pond—but light travels from the sun through *empty space*. So what could possibly be rippling in a light wave? The mysterious, undetectable "ether" had long been postulated, but Maxwell would eventually provide the answer (chap. 6). Mary Somerville's recollection is in her memoir, *Personal Recollections from Early Life to Old Age of Mary Somerville*, edited by her daughter Martha Charters Somerville (London: John Murray, 1873), 132.

2. *Optical tweezers* use the laser-beams' radiation pressure to move the tiny particles, and this is another example of a mathematical prediction: it was Maxwell who mathematically predicted the existence of radiation pressure, which was experimentally confirmed three decades later, in 1901. *Maxwell on radiation pressure: Treatise on Electricity and Magnetism* (Oxford: Clarendon Press, 1873, 3rd edition (1891) reprinted in 1954 by Dover), 2:440–41 (arts. 792–93).

3. Ahmes wrote what is now known as the Rhind papyrus, after the collector who bought it in Egypt in the 1850s; it's now in the British Museum.

4. *On Newton*: Richard S. Westfall, *Never at Rest: A Biography of Isaac Newton* (Cambridge: Cambridge University Press, 1980); Westfall also wrote the entry on Newton in the *Encylopaedia Britannica*.

5. The description of Leibniz is from the first page of the introduction to Philip P. Wiener, ed., *Leibniz Selections* (New York: Charles Scribner's Sons, 1951).

6. *Attempts at defining infinitesimals and limits*: Leibniz: "A differential is less than any given quantity," and "If one preferred to reject infinitesimally small quantities, it was possible instead to assume them to be as small as one judges necessary in order that they should be incomparable and the error produced should be of no consequence, or less than any given magnitude." Newton: "Quantities, and the ratios of quantities, which in any finite time converge continually to equality, and before the end of that time approach nearer to each other than by any given difference, become ultimately equal."

   Modern definition: Defining $f(x)$ in a suitable domain, $\lim_{x \to \alpha} f(x) = L$ if, given any number $\varepsilon > 0$, we can find a number $\delta > 0$ such that $f(x)$ satisfies $L - \varepsilon < f(x) < L + \varepsilon$ whenever $a - \delta < x < a + \delta$. For limits where x approaches infinity, we have $\lim_{x \to \infty} f(x) = L$ if for any number $\varepsilon > 0$ we can find a number M such that $L - \varepsilon < f(x) < L + \varepsilon$ when $n > M$. This definition arises from work by the likes of Augustin Louis Cauchy and Karl Weierstrass two centuries after Newton's attempt at defining a limit.

7. *Wallis acknowledged Harriot* (and also Oughtred and Descartes) in his *Arithmetica Infinitorum*—quoted in Boyer, *The History of Calculus and Its Conceptual Development* (New York: Dover, 1959), 170; see also 168–69. For more details on Wallis's debt to and exceptionally informed admiration of Harriot, see Stedall, "Rob'd of Glories," 481–90. *Newton's first published account of his calculus* was in *Principia*, bk. 2. Although he gave most of his proofs geometrically, he sometimes gave the algorithms of calculus in terms of algebraic symbolism, e.g., in bk. 2, sec. 2, lemma 2. When he gave geometrical diagrams, he sometimes used symbolic algebra to explain the algorithm or construction, as in bk. 2, prop. 10, problem 3. Even in bk. 1, calculus concepts are often clearly evident in the relevant geometrical constructions (such as his proof of prop. 39). In his first manuscripts on calculus, however, he'd used algebra rather than geometry.

8. *Wallis vs. Fermat et al.*: Jacqueline Stedall, "John Wallis and the French: His Quarrels with Fermat, Pascal, Dulaurens, and Descartes," *Historia Mathematica* 39 (2012): 265–79. *Descartes and Harriot*: Wallis's claims were exaggerated, but they were not original to him for there is some uncertainty about whether or not Descartes had seen Harriot's *Praxis* before he wrote his famous *La géometrie*; of course independent codiscovery is not uncommon, and Descartes went much further than Harriot, but Descartes was notoriously vague about his sources, and even his compatriot, Viète's editor Jean Beaugrand, noted similarities between Descartes's work and Harriot's. See

Stedall, "Rob'd of Glories," 488–89, and Jacqueline Stedall, "Reconstructing Thomas Harriot's Treatise on Equations," in *Thomas Harriot*, vol. 2, *Mathematics, Exploration, and Natural Philosophy in Early Modern England*, ed. Robert Fox (Farnham, Surrey: Ashgate, 2012), 62, and also Carl Boyer, *The Rainbow: From Myth to Mathematics* (Princeton, NJ: Princeton University Press, 1987), 203, 211.

9. *Wallis*: excerpt from his biography in John Stillwell, *Mathematics and Its History* (New York: Springer-Verlag, 1989), 110–12.

10. *Wallis's political fortunes*: See the Bodleian Library's description of J. Wallis, *A Collection of Letters and Other Papers*, MS e Mus. 203, at https://archives.bodleian.ox.ac.uk/repositories/2/resources/5805; and J. J. O'Connor and E. F. Robertson, https://mathshistory.st-andrews.ac.uk/Biographies/Wallis/.

11. *Einstein's assessment of Newton* is from his *Ideas and Opinions* (1954; New York: Three Rivers Press, 1982), 254–55.

12. *Hooke's work on planetary motion*: e.g., Michael Nauenberg, "Robert Hooke's Seminal Contributions to Orbital Dynamics," *Physics in Perspective* 7 (2005): 1–31. Nauenberg has made a detailed study of Hooke's work and how Newton made use of it. It is important to give Hooke the credit he deserves, although I tend to think Nauenberg overplays Hooke's mathematical ability, given that his argument is based on a single construction, made in 1685 after he had most likely seen Newton's preliminary paper *De Motu*. Either way, Hooke constructed the orbit resulting from a force that varies directly with distance, and while he did it in a novel way, it is *one* calculation compared with the *hundreds* in *Principia*, for all kinds of forces and motions.

13. *Newton "dry calculators"*: letter to Halley quoted in Nauenberg, "Robert Hooke," 7. Nauenberg calls this a "diatribe," which suggests to me, in light of the breadth of *Principia* compared with Hooke's contribution (cf. previous note), that he is making his case for Hooke a little too vehemently. *Modern mathematics, computation versus creativity*: Patrick Bangert's whimsical report—published in the Australian Mathematical Society's *Gazette* 32, no. 3 (July 2005)—suggests that many mathematicians think their subject has more to do with patterns, language, art, or logic than applications. Similarly, in July 2021, Ole Warnaar reported in the *Gazette* (vol. 48, no. 3) on feedback from the society's members regarding proposed revisions to the Australian mathematics curriculum: criticisms included its excessively utilitarian approach. It's certainly exciting, and socially vital, to apply mathematics usefully, and Warnaar applauds teaching such skills; but he also laments that

"not enough effort has been made to try to convey the intrinsic beauty of mathematics and the enjoyment one can derive from learning and understanding new mathematical concepts."

14. *During evaporation*, heat raises the kinetic energy of the water molecules so they can escape the electrical bonds that bound the molecules together as a liquid. The violet cloth reflects cooler violet light to our eyes, absorbing the other warmer colors. So it dries fastest because it absorbs heat more quickly than the other colors on the bedsheet. Du Châtelet's result, in an expanded version of her 1738 *Essay on Fire* (*Dissertation sur la nature et la propagation du feu*), was published in 1744 (Paris: Chez Prault Fils).

15. *Newton's geometrical version of calculus* in *Principia*: For example, try translating into modern symbols Newton's proof of proposition 39 in book 1. He wants to find the velocity of a falling body under a centripetal force, and he defines force as proportional to the increment of velocity (I) divided by the increment of time, a geometric/differential version of dv/dt; but he also finds the derivative of v^2 in an almost modern way, writing in effect $[(v+I)^2 - v^2]/\Delta y$ to get $2vdv/dy$. (He uses v^2 because he's effectively proving a theorem about kinetic energy as the work done by the force.)

16. For more on du Châtelet's Newtonian work, see my *Seduced by Logic: Émilie du Châtelet, Mary Somerville and the Newtonian Revolution* (New York: Oxford University Press, 2012).

CHAPTER 3

1. *SI unit*: the Système international d'unités—the International System of Units—is abbreviated internationally as SI.

2. On *Questiones Mechanicae* and its influence: David Marshall Miller, "The Parallelogram Rule from Pseudo-Aristotle to Newton," *Archive for History of Exact Sciences* 71 (2017): 157–91, esp. 161–66. By modern standards, the *Questiones* contains only a proto-parallelogram rule, but even then, few grasped its importance.

3. *Tartaglia*: his 45° calculation assumes you're firing from the ground rather than from a height. *"Vituperative" work*: quoted in Michael Brooks, *The Art of More* (Melbourne: Scribe, 2021), 94. Here and in the next paragraph, I'm indebted to Brooks, and also to J. J. O'Connor and E. F. Robertson's University of St Andrews MacTutor article on Tartaglia, https://mathshistory.st-andrews.ac.uk/Biographies/Tartaglia/#:~:text=Quick%20Info&text=Tartaglia%20was%20an%20Italian%20mathematician,published%20in%20Cardan%27s%20Ars%20Magna.

4. Miller, "Parallelogram Rule," 164.

5. Joseph Jarrett—in "Algebra and the Art of War: Marlowe's Military Mathematics in Marlowe's 'Tamburlaine I and II,'" *Cahiers Élizabéthain* 95, no. 1 (2018): 19–39—suggests that Marlowe was able to represent huge battle scenes on a small stage by applying a kind of compact representation similar to the algebraic symbolism pioneered by Harriot. Jarrett also alerted me to Altdorfer's battle paintings.
6. Galileo and Harriot got falling motion and horizontal projection right, but they treated the oblique motion of a projectile as a single decelerated motion rather than in terms of independent components: Matthias Schemmel, "Thomas Harriot as an English Galileo: The Force of Shared Knowledge in Early Modern Mechanics," in *Thomas Harriot*, vol. 2, *Mathematics, Exploration and Natural Philosophy in Early Modern England*, ed. Robert Fox (Farnham, Surrey: Ashgate, 2012), 89–111, esp. 95, 97.
7. *Galileo, Stevin, and Descartes and the parallelogram rule*: Miller, "Parallelogram Rule," 166, 167, 170, 183, 186.
8. *Harriot's calculations (fig. 3.3)*: In each diagram the solid circles *a* and *A* on the left show the starting points of the two balls. They collide in the middle and rebound to the positions shown with dotted circles. The calculations below the diagram take account of the velocities and masses of the balls: Harriot is looking at what happens in a given time interval *x*, so the velocities are expressed in terms of the directions and lengths of the lines between the balls, analogously to what we would do with vectors today.

    Harriot explains his use of the parallelogram rule carefully. For instance, he says that if there had been no collision, then in the second time interval *x* the ball *a* would have continued moving with the same speed in the same line *ab*—so he has intuited the first law of motion. On collision, though, this motion is "translated" to the parallel line *fc*—not physically, he stresses, but for the purposes of "compos[ing] the apparent motion." Similarly for the second ball, the motion *AB* is translated to *FC*.

    The points *f* and *F* are found from the composition of the motion of each ball before and after the collision. For example, for the first ball, *bf* equals the sum of (what we would call) two vectors: (minus) the vertical component of the ball's initial motion, *bd*—that is, the motion as if it had simply rebounded from a stationary ball of equal mass—plus the vertical component of the extra motion imparted by the larger ball's momentum (as we would put it), *df*(=*gb*).

    It can be done more easily using conservation of momentum and kinetic energy, but these concepts weren't available to Harriot. The symmetry of his diagram, however, shows that he is assuming the conservation of "impetus," and the calculations below the diagram—where his $\bar{b}$, $\bar{B}$ represent the velocities of the two balls and b, B their masses—are essentially the same as in our conservation of momentum calculations. Note that

$$\begin{array}{c} \bar{b}+\bar{B}\,| \\ b+B\,| \\ \hline B \end{array}$$

is his notation for $(\bar{b} + \bar{B})(b + B)/B$. (He's made a slight error in this term.)

Harriot's minor error (plus analysis of his paper): Johannes Lohne, "Essays on Thomas Harriot," *Archive for History of Exact Sciences* 20, nos. 3/4 (1979): 189–312, and Jon V. Pepper, "Harriot's Manuscript on the Theory of Impacts," *Annals of Science* 33, no. 2 (1976): 131–51, DOI: 10.1080/00033797600200191.

9. "*As if*" is Harriot's wording, just as in modern accounts such as Miller, "Parallelogram Rule," 158. Harriot's paper was first published in the 1970s: see Lohne, "Essays on Thomas Harriot," for an English translation from the original Latin.

10. *Wallis and Harriot*: Jacqueline Stedall, "Rob'd of Glories: The Posthumous Misfortunes of Thomas Harriot and His Algebra," *Archive for History of Exact Sciences* 54, no. 6 (June 2000): 483. *Wallis and Fermat*: Miller, "Parallelogram Rule," 171–72, 186–87.

11. Wallis was trying to find a way to think about complex solutions to quadratic equations of the form $x^2 + 2bx + c^2 = 0$—and if you remember the quadratic formula that you can deduce by completing the square, you'll know that $x = -b \pm \sqrt{(b^2 - c^2)}$. When $b \geq c$, Wallis found a way to represent the two solutions as points on the real number line—so he tried something similar for the complex solutions you get when $b < c$. He avoided having to deal explicitly with i by considering $x = -b \pm \sqrt{(c^2 - b^2)}$, and he'd used cumbersome constructions with triangles to represent his solutions rather than representing them as points on a complex plane as we would do today. John Stillwell shows Wallis's attempt and its flaws in fig. 13.3, *Mathematics and Its History* (New York: Springer-Verlag, 1989).

12. The modern version of Euler's definition of e is $\lim_{x\to\infty}(1+\frac{1}{n})^n$. The idea of it first arose in a study of compound interest by Jakob (or Jacques) Bernoulli (Johann/Jean's older brother), and even earlier, although implicitly, in John Napier's logarithms and Thomas Harriot's unpublished calculations for continuously compounding interest and meridional parts. But it was Euler who recognized the power of this number and brought it to proper attention. He also wrote it as a Taylor series, which is the way decimal approximations of e are worked out, such as the one on your calculator (mine gives 2.718281828).

13. *On Euler and his identity*: Ed Sandifer, "How Euler Did It," *MAA Online*,

August 2007. See also Carl Boyer, *History of Mathematics*, rev. Uta Merzbach (New York: John Wiley and Sons, 1991), 443–44 (and 441–42 on Euler and college math). *De Moivre and Newton*: Orlando Merlino, "A Short History of Complex Numbers," University of Rhode Island, January 2006.

14. *Euler and Fermat's last theorem*: Later mathematicians found that Fermat hadn't proven one of the steps in his proof, but the step itself was correct; this, and Euler's use of complex numbers in the proof, is explained well in Harold M. Edwards, "Fermat's Last Theorem," *Scientific American* 239, no. 4 (October 1978): 104–23.

15. *Euler/d'Alembert*: Stillwell, *Mathematics and Its History*, 202. *Sophie Germain* didn't publish her work on Fermat's theorem, but Legendre credited her with a result he used in his proof when $n = 5$.

16. *Hamilton's apples and oranges*: Karen Hunger Parshall, "The Development of Abstract Algebra," in *The Princeton Companion to Mathematics*, ed. Timothy Gowers et al. (Princeton, NJ: Princeton University Press, 2010), 95–106, esp. 105.

17. Gauss quoted by Christian Gérini, "Argand's Geometric Representation of Imaginary Numbers," University of Toulon, January 2009, English trans. Helen Tomlinson, April 2017, online at http://www.bibnum.education.fr/sites/default/files/21-argand-analysis.pdf. See this also for Argand's "directed lines." *De Morgan quoted* in Raymond Flood and Robin Wilson, *The Great Mathematicians* (London: Arcturus, 2011), 143. *For more context on De Morgan*: Morris Kline, *Mathematics: The Loss of Certainty* (New York: Oxford University Press 1982), 155–56.

18. Babbage quoted in Dirk Struik, *A Concise History of Mathematics* (New York: Dover, 1967), 168.

19. Hamilton quoted in Janet Folina, "Newton and Hamilton: In Defense of Truth in Algebra," *Southern Journal of Philosophy* 50, no. 3 (2012): 515.

20. *Hamilton's process*, from negative numbers/science of time to complex couples to quaternions, is explained in detail by Teun Koetsier, "Explanation in the Historiography of Mathematics: The Case of Hamilton's Quaternions," *Studies in History and Philosophy of Science Part A* 26, no. 4 (1995): 593–616. *Hamilton's "steps" as vectors*: see his *Lectures on Quaternions* (Dublin: Hodges and Smith; London: Whittaker; and Cambridge: Macmillan, 1853), 3ff.

    Hamilton's importance re: complex numbers: Parshall, "Development of Abstract Algebra," 105; Boyer, *History of Mathematics*, 583. *Hamilton citing Newton* on algebra of time: Folina, "Newton and Hamilton," 513.

21. *De Morgan on Hamilton's couples*: quoted in Diana Willment, "Complex Numbers from 1600 to 1840" (master's thesis, Middlesex University, 1985), 102. *Hamilton, symbolism, De Morgan*: Koetsier, "Explanation," 610.

22. *Literary friends*, including Wordsworth, admiring Hamilton: Michael J. Crowe, *A History of Vector Analysis* (Notre Dame, IN: University of Notre Dame Press, 1967), 22; and Daniel Brown, *The Poetry of Victorian Scientists: Style, Science and Nonsense* (Cambridge: Cambridge University Press, 2013), 1. *Maria Edgeworth on Hamilton*: "Miss Edgeworth Advises," Royal Irish Academy (RIA) (blog), June 24, 2018, https://www.ria.ie/news/library-library-blog/miss-edgeworth-advises.
23. *Maria Edgeworth, Peacock, and Somerville*: see my *Seduced by Logic: Émilie du Châtelet, Mary Somerville and the Newtonian Revolution* (New York: Oxford University Press, 2012), 195, 197–98.
24. *Maria Edgeworth and Royal Irish Academy*: RIA (blog), "Miss Edgeworth Advises," and Clare O'Halloran, "'Better without the Ladies': The Royal Irish Academy and the Admission of Women Members," *18th–19th Century Social Perspectives (History Ireland)* 19, no. 6 (November/December 2011): 42–46.
25. *Hamilton on vectors as directed lines*: e.g., his *Lectures on Quaternions*, 35.

CHAPTER 4

1. *Papa, can you multiply triplets?* Hamilton's letter to Archibald, 1865, in Robert P. Graves, *Life of Sir William Rowan Hamilton*, 3 vols. (Dublin: Hodges, Figgis, 1882, 1885, 1889), 2:434–35, widely quoted, e.g., in Michael J. Crowe, *A History of Vector Analysis* (Notre Dame, IN: University of Notre Dame Press, 1967), 29. *Explaining to "my boys"*: letter to De Morgan, 1852, Graves, *Life of Hamilton* (1889), 3: #59, 307–8.
2. *Laws of algebra/arithmetic*: For example, $(3 + 2) + 5 = 5 + 5 = 10$; but in this case, it doesn't matter where the brackets go, because $3 + (2 + 5) = 3 + 7$ which also equals 10. This is called the associative law for addition, and there's a similar one for multiplication. Similarly, $2 \times 3 = 3 \times 2$, and $2 + 3 = 3 + 2$; this is the famous commutative law for multiplication and addition. Peacock also introduced the distributive law, $a(b + c) = ab + bc$.
3. *Law of moduli*: With an ordinary complex number $z = x + iy$, Hamilton had shown that you can define the "modulus" (the magnitude or "absolute value") of this number by taking the square root of

$$(x + iy)(x - iy) = x^2 + y^2,$$

where the second factor on the left-hand side is the "conjugate" of the first. (This may have been known as early as Euler.) It also turns out that *the modulus of the product of two complex numbers equals the product of the two moduli*—what Hamilton referred to as "the law of moduli," written symbolically in today's textbooks as $|zw| = |z||w|$.

For example, let $z = x + iy$, $w = a + ib$. Then

$$|zw| = |(x+iy)(a+ib)| = |(xa-yb) + i(xb+ya)| = \sqrt{(xa-yb)^2 + (xb+ya)^2}$$

and

$$|z||w| = \sqrt{(x^2+y^2)(a^2+b^2)} = \sqrt{(xa-yb)^2 + (xb+ya)^2}.$$

So $|zw| = |z||w|$, and the law of moduli holds in two dimensions.

If you try this with $x + iy + jz$ and $a + ib + jc$, however, for the law of moduli to hold you have to make simplifying assumptions about the relationships between x, y, a, b—which Hamilton tried, but which destroys the generality of the law of moduli—or about the relationships between i, j, ij, and ji. See the narrative for what Hamilton did next. And see Teun Koetsier, "Explanation in the Historiography of Mathematics: The Case of Hamilton's Quaternions," *Studies in the History and Philosophy of Science Part A* 26, no. 4 (1995): 593–616, for Hamilton's process, including letters to Graves.

4. Hamilton explained this process in the preface to his *Lectures on Quaternions*.
5. Augustus De Morgan, *Essays on the Life and Work of Newton*, edited, with notes and appendices, by Philip Jourdain (Chicago: Open Court, 1914). *Biographical notes on De Morgan*: Leslie Stephen, *Dictionary of National Biography* 14 (1885–1900), s.v. De Morgan; incidentally, Stephen was the father of the famous novelist Virginia Woolf. Also see, e.g., Carl Boyer, *History of Mathematics*, rev. Uta Merzbach (New York: John Wiley and Sons, 1991), 581.
6. *"Deeply reverential"*: Alexander MacFarlane, *Lectures on Ten British Mathematicians* (London: Chapman and Hall, 1916), chap. 3 (from a lecture delivered in 1901).
7. Recent scholarship has traced the evolution of the gossip about Hamilton, and put it in the context of women's roles and social constraints and painting a more positive picture of his and Helen's life: Anne van Weerden and Stephen Wepster, "A Most Gossiped about Genius: Sir William Rowan Hamilton," *BSHM Bulletin* 33, no. 1 (2018): 2–20. They also give a more balanced account than earlier records suggested, e.g., putting an episode where Hamilton was supposed to be violently drunk into the context of the temperance movement.

8. *De Morgan and Lovelace*: two papers by Christopher Hollings, Ursula Martin, and Adrian Rice: "The Early Mathematical Education of Ada Lovelace," *BHSM Bulletin* 32, no. 3 (2017): 221–34, and "Lovelace-De Morgan Correspondence: A Critical Re-appraisal," *Historia Mathematica* 44 (2017): 202–31. Note that Babbage's "difference engine" was ahead of its time and never went into production.

9. De Morgan quoted in Janet Folina, "Newton and Hamilton: In Defense of Truth in Algebra," *Southern Journal of Philosophy* 50, no. 3 (2012): 511. Note that Hamilton and De Morgan had different approaches to algebraic foundations (511–12).

10. *Hamilton to De Morgan, 1841*, quoted, e.g., Michael J. Crowe, *A History of Vector Analysis* (Notre Dame, IN: University of Notre Dame Press, 1967), 27.

11. *Invoking* $k = ij$: For triples $a + ib + jc$ and $x + iy + jz$, the law of moduli says that

$$|(a + ib + jc)(x + iy + jz)| = |a + ib + jc||x + iy + jz|.$$

The right-hand side (RHS) is just

$$(a^2 + b^2 + c^2)(x^2 + y^2 + z^2).$$

Now for the left-hand side (LHS): assume $ij = -ji$, and consider the LHS when you've expanded the brackets:

$$|ax - by - cz + i(ay + bx) + j(az + cx) + ij(bz - cy)| =$$
$$(ax - by - cz)^2 + (ay + bx)^2 + (az + cx)^2 + (bz - cy)^2.$$

But you only get this last term—which you need to balance with the RHS—by incorporating the conjugate of $ij(bz - cy)$ into the modulus definition, *as if ij were a complex number like i and j*. This is what led Hamilton to suppose that to solve his problems of triplet multiplication, he had to invoke a *third* imaginary vector $k = ij$.

Hamilton's "electric circuit": I've slightly modified the tense on "closed"; cf. Hamilton's 1865 letter to Archibald, quoted in Crowe, *History of Vectors*, 29. *Hamilton to Graves*: quoted in B. L. van der Waerden, "Hamilton's Discovery of Quaternions," *Mathematics Magazine* 49, no. 5 (November 1976): 227–34, esp. 230.

12. De Morgan quoted in Folina, "Newton and Hamilton," 505. *Wordsworth* on Hamilton's mediocre poetry: Daniel Brown, *The Poetry of Victorian Scientists: Style, Science and Nonsense* (Cambridge: Cambridge University Press, 2013), 1–2. *Schrödinger*: Crowe, *History of Vector Analysis*, 17. Schrödinger

was referring to what is now called Hamiltonian dynamics, an alternative, coordinate-free form of the laws of motion, equivalent to Newton's approach but more flexible.

13. *Playing with Hamilton's graffiti* only works if you assume that you can cancel terms only at the beginning or end of the string, and if products of pairs are anticommutative. So, to find j from $j^2 = ijk$, rewrite as $j^2 = -jik$, and then, canceling j from both sides, you have $j = -ik = ki$.

14. *Scalar products* just multiply the components, so that $\boldsymbol{p} \cdot \boldsymbol{q} = p_1q_1 + p_2q_2 + p_3q_3$ is a number, or scalar, not a vector. (In Hamilton's full quaternion product, there's a minus sign in front of $\boldsymbol{p} \cdot \boldsymbol{q}$, which will prove controversial, as we'll see in chap. 7.)

    Vector products give vectors; the component form of $\boldsymbol{p} \times \boldsymbol{q}$ is easiest to remember and calculate from the determinant $\begin{vmatrix} \boldsymbol{i} & \boldsymbol{j} & \boldsymbol{k} \\ p_1 & p_2 & p_3 \\ q_1 & q_2 & q_3 \end{vmatrix}$.

15. *Cayley and quaternions*: Crowe, *History of Vector Analysis*, 35. *Cayley's life*: Tony Crilly, "Arthur Cayley: The Road Not Taken," *Mathematical Intelligencer* 20, no. 4 (1998): 49–53; Crilly also wrote the *Britannica* entry on Cayley.

16. A computer algorithm for Gaussian elimination is given, e.g., in Erwin Kreyszig, *Advanced Engineering Mathematics* (New York: Wiley, 1993), 976.

17. *Sylvester "invariants"*: Crilly, "Arthur Cayley: The Road Not Taken," 51.

18. *Boole and Cayley*: Tony Crilly, "The Rise of Cayley's Invariant Theory," *Historia Mathematica* 13 (1986): 241–54.

19. Eunice Foote, "Circumstances Affecting the Heat of the Sun's Rays," *American Journal of Science and Arts* (1856): 382–83. John Tyndall, apparently independently, put the physics into Foote's empirical discovery just a few years later: see Roland Jackson, "John Tyndall: The Forgotten Cofounder of Climate Science," *The Conversation*, July 31, 2020. Jean-Baptiste Fourier was the first to look at the heating effect of the atmosphere, in 1820, but he did it in the context of calculating Earth's temperature.

20. *Search engines*: The form of the scalar product needed to deduce the angle between two vectors a and b is $\boldsymbol{a} \cdot \boldsymbol{b} = |\boldsymbol{a}||\boldsymbol{b}|\cos\theta$. For my description I've drawn on Amy Langville's excellent introduction, "The Linear Algebra behind Search Engines: Focus on the Vector Space Model," *Convergence*, Mathematical Association of America (December 2006); https://www.maa.org/press/periodicals/loci/joma/the-linear-algebra-behind-search-engines-focus-on-the-vector-space-model.

21. *Google PageRank algorithm*: I've drawn on Cornell University's informative lecture, http://pi.math.cornell.edu/~mec/Winter2009/RalucaRemus/Lecture3/lecture3.html.

22. *Critiques of AI, social media, and search algorithms*: See, e.g., Cathy O'Neill, *Weapons of Math Destruction: How Big Data Increases Inequality and Threatens Democracy* (New York: Crown Publishing, 2017); Safiyah Umoja Noble, *Algorithms of Oppression* (New York: NYU Press, 2018); Shoshana Zuboff, *The Age of Surveillance Capitalism: The Fight for a Human Future at the New Frontier of Power* (London: Profile Books, 2019); and many others.

23. Sarah Flannery's algorithm failed the safety protocol (cf. her book *In Code* [London: Profile Books, 2001]), but researchers still believe noncommutative multiplication will prove a valuable cryptographic tool.

24. *Quaternion rotations*: Hamilton briefly outlined the following approach in his *Lectures on Quaternions*, 269 (art. 282):

To take a simple example, to rotate a vector $\boldsymbol{p}$ about the i-axis, you could choose the unit quaternion $U = \cos\theta + i\sin\theta$, by analogy with the complex numbers in my fig. 4.3. (This is the quaternion $U = (\cos\theta, \sin\theta, 0, 0)$, because the j and k (or y and z) components of the axis of rotation in this case are zero.) Then, using Euler's theorem, you have $U = e^{i\theta}$. This form makes multiplications easier—the index laws turn multiplications into additions—and shows clearly how they are related to rotations, as we saw also in figure 3.8.

Now form a new vector,

$$\boldsymbol{a} = U\boldsymbol{p}U^{-1} = e^{i\theta}\boldsymbol{p}\,e^{-i\theta} = e^{i\theta}(ix + jy + kz)e^{-i\theta}.$$

It must have taken Hamilton a bit of experimenting to come up with this combination, for you don't need to also multiply by U^{-1} when you're in the Argand plane (cf. figs. 4.3, 3.8). (By the way, for unit quaternions, the inverse is the complex conjugate, so I could have written $\boldsymbol{a} = U\boldsymbol{p}U^{*}$ rather than $\boldsymbol{a} = U\boldsymbol{p}U^{-1}$.) Geometrically, this U^{-1} factor is needed to counteract an extraneous rotation that happens because it's taking place in a 4-D hyperspace, but that is beyond my scope here. Algebraically we're talking about the quaternion analog of a matrix *similarity transformation*. But you don't need these technicalities to carry out the simple algebra that results from this "machinery":

If you replace k by Hamilton's definition ij, you get

$$\begin{aligned}\boldsymbol{a} &= e^{i\theta}(ix + jy + kz)e^{-i\theta}\\ &= e^{i\theta}(ix + (y + iz)j)e^{-i\theta}\\ &= e^{i\theta}(ix)e^{-i\theta} + e^{i\theta}(y + iz)je^{-i\theta}.\end{aligned}$$

Now comes the nifty part: remembering that $e^{-i\theta} = \cos\theta - i\sin\theta$, the $je^{-i\theta}$ in the last term becomes

$$j(\cos\theta - i\sin\theta) = j\cos\theta - ji\sin\theta.$$

But Hamilton defined $ij = k = -ji$, so we have

$$j\cos\theta - ji\sin\theta = j\cos\theta + ij\sin\theta = (\cos\theta)j + (i\sin\theta)j = (\cos\theta + i\sin\theta)j = e^{i\theta}j.$$

(I wrote $j\cos\theta = (\cos\theta)j$ because $\cos\theta$ is just a real number or scalar; it is only when both numbers are complex that multiplications are not necessarily commutative—as when $ij = k = -ji$. Similarly for $j\sin\theta = (\sin\theta)j$.)

Finally, using index laws and Hamilton's rules for products of the i, j, k, we have the rotated version of the vector $\boldsymbol{p}$:

$$\boldsymbol{a} = xi + e^{i2\theta}(y + zi)j = xi + e^{i2\theta}(yj + zk).$$

This looks right, because $\boldsymbol{p}$ has been rotated about the i-axis, so its i-component doesn't change, for it moves only in the j-k plane. And the $e^{i2\theta}$ factor shows that $\boldsymbol{p}$'s j and k components have, indeed, been rotated in the j-k plane, by an angle of 2θ. (So if you want to rotate by θ, choose the unit quaternion to be $U = e^{i\theta/2}$.)

To rotate about an arbitrary axis in the direction of a unit vector $\boldsymbol{u} = ai + bj + ck$, rather than just i as in the above calculation, put $U = \cos\theta + \boldsymbol{u}\sin\theta = e^{u\theta}$, by analogy with Euler's formula. (You can prove Euler's formula by comparing the series for the expressions on each side of the equation.) My account in the above is an expanded version of Hamilton's outline, and of the example in the lecture notes, "Introducing the Quaternions," by John Huerta, Fullerton College.

Alternatively, you can do the above calculations by expanding $\boldsymbol{a} = U\boldsymbol{p}U^{-1}$ using the *scalar and vector products* that come from quaternion multiplication, as I showed in the narrative:

$$PQ = wa - \boldsymbol{p}\cdot\boldsymbol{q} + w\boldsymbol{q} + a\boldsymbol{p} + \boldsymbol{p}\times\boldsymbol{q}.$$

25. For an example of matrices giving gimbal lock, see Justin Wyss-Gallifent's MATH431 lecture "Gimbal Lock," November 3, 2021: http://www.math.umd.edu/~immortal/MATH431/book/ch_gimballock.pdf.

26. *Spectral lines*: If an atom absorbs energy, its electrons jump to a higher energy state; when they return to their original, more stable state, the atom emits a photon, which shows up as a colored spectral line. The color corresponds to

the emitted light's wavelength, which in turn is related to the size of the energy jump and the makeup of the atom. Which is why the pioneering female astronomer Annie Jump Cannon had been working at Harvard College Observatory since 1896, painstakingly classifying the spectra of stars to determine their chemical composition. She continued doing this till 1941, the year she died.

27. Actually, the connection with Stern-Gerlach was made a little later. Anyway, Uhlenbeck and Goudsmit's values were $\pm h/4\pi$; the sign depends on whether the spin axis is aligned with or against the magnetic field, and the magnitude is *half* the "normalized" Planck constant $h/2\pi$ (which is represented as $\hbar$, pronounced "h-bar"). That's why the electron is now said to have a spin of ½.

28. *Ehrenfest, Lorentz*: Goudsmit gave an account of the discovery of electron spin in a delightful paper he read for the golden jubilee of the Dutch Physical Society in April 1971; https://www.lorentz.leidenuniv.nl/history/spin/goudsmit.html.

29. *Pauli matrices and quaternions*: What the relationship between Pauli matrices and quaternions also shows is that you can have a "vector space" of vectors that are actually matrices. It is the rules of vectors that matter: if something behaves like a vector, it can be treated as one. *Pauli on quaternions*: W. Pauli, *General Principles of Quantum Mechanics*, Springer-Verlag, Berlin/Heidelberg, 115; he noted that i times the matrix obeys the rules of unit quaternion multiplication. *Dirac*: P. A. M. Dirac, "The Quantum Theory of the Electron," *Proceedings of the Royal Society of London*, series A, 117, 778 (February 1, 1928): 610–24; and Paul A. M. Dirac, "Theory of Electrons and Positrons," 1933 Nobel Prize speech, https://www.nobelprize.org/uploads/2018/06/dirac-lecture.pdf.

30. *Quaternion and spin rotations*: Dirac's new theory showed that all the building blocks of matter—electrons, protons, and neutrons—have spins that contain a ½; that is, they are odd-number multiples of ½, and they are called fermions. (Photons and other so-called bosons have integer spins.)

It's the ½ in a particle's spin that causes the strange behavior in which you need *two* 360° rotations to bring a fermion back to its original state. This discovery came out of the math you need to rotate a quantum particle's spin axis—similar math to quaternion rotations, as it happens. Which isn't really surprising, given that each gives the same counterintuitive type of rotation. I've given a brief summary here:

As I implied in fig. 4.4 and the endnote on quaternion rotations, to rotate a vector through an angle θ around the x-axis, the unit quaternion is $U = e^{i\frac{\theta}{2}}$.

Similarly, to rotate a spin half angular momentum vector through θ about the x-axis, you need a unitary operator $U = e^{-i\frac{\theta}{2}\sigma_x}$, where units are chosen so that $2\pi/h = 1$, and σ_x is a Pauli spin matrix. *Note the $\theta/2$ in both cases*—it's why you need two full 2π rotations to get back to the original unrotated ($\theta = 0 + 2n\pi$) state.

By contrast, to rotate the orbital rather than the spin angular momentum through an angle θ around the x-axis, the unitary operator is $U = e^{-i\theta J_x}$. Rotate through 2π and you *are* back where you started. Note, too, that in both quaternion and quantum rotations, you also need to postmultiply by the inverse or conjugate; this is called a "similarity transformation."

Another way of describing the connection between quaternions and spin is that both quaternion and spin half rotation matrices are elements of the group $SU(2)$. *Group theory* deals with underlying structures, so by studying group properties of various mathematical or physical structures, you can sometimes spot similarities between two apparently very different things.

31. Klein and Opat used neutrons rather than electrons, because the charge on electrons would interfere too strongly with the external magnetic field in the experiment. They used a ferromagnetic crystal to diffract a beam of neutrons (traveling as a matter wave) into two parts, one of which interacted with the external magnetic field. (In quantum mechanics the "matter wave" or "wave function" describes the probability of a particle being detected at a certain time and place.) What they found was that when one beam—one part of the wave—has been rotated by an odd-integer multiple of 360° or 2π radians—which, of course, includes 2π itself—it interfered *destructively* with the other, nonrotated half, producing a distinctive interference pattern. For even multiples of 2π, the destructive interference disappeared. So in the odd-integer 2π case, you need to rotate the spin again, so that all up it goes through an even multiple such as 4π to get the interference pattern back to normal. A. G. Klein and G. I. Opat, "Observation of 2π Rotations by Fresnel Diffraction of Neutrons," *Physical Review Letters* 37, no. 5 (August 2, 1976): 238–40.
32. The other teams were headed by Helmut Rauch and Sam Werner, but instead of being rivals the three teams subsequently worked together, as Klein describes in "Neutron Interferometry: A Tale of Three Continents," available at http://www.europhysicsnews.org or http://dx.doi.org/10.1051/epn/2009802.
33. Hamilton's *Ode* is analyzed by Brown, *Poetry of Victorian Scientists*, 7–9.
34. *Octonions today*: For a technical overview of current research, see Peter Rowlands and Sydney Rowlands, "Are Octonions Necessary to the Standard

Model?," *Journal of Physics: Conference Series* 1251, 012044 (2019), DOI 10.1088/1742-6596/1251/1/012044. One of these researchers is a young Canadian woman, Cohl Furey; another is University of California mathematician John Baez. In 2021, Baez gave an update on the situation at https://math.ucr.edu/home/baez/standard/.

CHAPTER 5

1. Robert P. Graves, *The Life of Sir William Rowan Hamilton*, 3 vols. (Dublin: Hodges, Figgis, 1882, 1885, 1889), 2:585–86.
2. Graves, *Life of Hamilton*, 2:586.
3. For Grassmann's early life and work here and in much of the following, I'm indebted primarily to Michael J. Crowe, *History of Vector Analysis* (Notre Dame, IN: University of Notre Dame Press, 1967), chap. 3 (for more detail see Hans-Joachim Petsche, *Hermann Grassmann* [Basel: Birkhäuser, 2009]); and for an overview of the discovery of vectors, including Hamilton and Grassmann, Jean-Luc Dorier, "A General Outline of the Genesis of Vector Space Theory," *Historia Mathematica* 22 (1995): 227–61.
4. *Grassmann "astounded"*: 1847 letter to Saint-Venant, quoted in Crowe, *History of Vector Analysis*, 56.
5. Crowe, *History of Vector Analysis*, 70–72.
6. The fanatical new authorities had hesitated over Italian-born Lagrange—foreigners were usually stripped of their posts and possessions—but in the end a special decree enabled him to stay on, and as head of the committee, no less. This was largely thanks to the efforts on behalf of foreign scientists by pioneering chemist Antoine Lavoisier—he and his wife Marie have been dubbed the "father and mother" of modern chemistry. Lavoisier famously and tragically ended up at the guillotine for his business interests in a tax-collecting company, and Lagrange famously declared that while it took the mob but a moment to cut off his head, it might take a century to see its like again. Eighteen months later the government declared Lavoisier innocent of any wrongdoing by the tax company.
7. Some sources say two children survived: either way, how tragic—especially for his poor mother! No wonder Lagrange himself had no children.
8. If you multiply the vectors representing the two directed lines defining a parallelogram via the *vector product*, you get another vector. (This makes the vector product "closed.") The product vector is perpendicular to the plane of the two original vectors. But if you multiply the two sides of the

parallelogram via *Grassmann's outer product*, you get not another "line"—Grassmann didn't use the term "vector"—but a directed *area*. So the outer product is conceptually different from the vector product. (It is related to a tensor product, though, as we'll see.) However, they are equivalent in that the vector you get from the vector product of the two sides is in the same direction as Grassmann's directed area and has the same magnitude.

9. *Hamilton's letter* to Mortimer O'Sullivan was published in Graves, *Life of Hamilton*, 2:683.
10. *Herschel to Hamilton*, quoted in Graves, *Life of Hamilton*, 3:121.
11. Möbius, Apelt, Baltzer, quoted in Crowe, *History of Vector Analysis*, 78–80.
12. Möbius, Apelt, Baltzer, quoted in Crowe, *History of Vector Analysis*, 78–80.
13. Möbius, Apelt, Baltzer, quoted in Crowe, *History of Vector Analysis*, 78–80.
14. *Ampère vs. Grassmann*: For a contemporary and well-judged assessment, see Maxwell's *Treatise on Electricity and Magnetism*, 1891 (3rd edition of the 1873 original, Clarendon Press or Dover reprint), arts. 482, 509–10 (511–25 for Maxwell's own analysis), 526 (Maxwell favors Ampère because Grassmann's formula violated Newton's third law), but 687 for Maxwell's conclusion that it was impossible to decide *experimentally* between the two formulae.

Recent attempts to experimentally and conceptually decide: Christine Blondel and Bertrand Wolff, trans. Andrew Butricia, "Ampère's Force Law: An Obsolete Formula?" *Histoire de l'Électricité et du Magnetisme* (May 2009; trans. 2013, rev. 2021), http://www.ampere.cnrs.fr/histoire/parcours-historique/lois-courants/force-obsolete/eng.

This paper gives a brief nonpartisan overview of recent research. For a detailed, ultimately pro-Ampère account: A. K. T. Assis and J. P. M. C. Chaib, *Ampère's Electrodynamics* (Apeiron, 2015), chaps. 14, 16.4, and conclusion, 491; still, in 1996 Assis had shown (with Marcelo A. Bueno) that Grassmann's and Ampère's formulae were equivalent in the experimental set-up espoused by Ampère: "Equivalence between Ampère and Grassmann's Forces," *IEEE Transactions on Magnetics* 32, no. 2 (March 1996): 431–36.

Note that some modern authors are using the Ampère vs. Grassmann debate to question the field theory approach (into which Grassmann's result was integrated post-Maxwell); cf. Ampère's action-at-a-distance. Maxwell himself said it was always a good idea to have more than one way of seeing things!

Grassmann's original paper: Hermann Grassmann, "Neue Theorie der Elektrodynamik," *Annalen der Physik und Chemie* 1 (1845): 1–18.

15. *Leibniz and Grassmann*: Joseph Kouneiher, "Broken Symmetry, Pointless Space and Leibniz's Legacy: The Origin of Physics," *Advanced Studies in Theoretical Physics* (September 2015), accessed from ResearchGate, https://www.researchgate.net/publication/281526332_Broken_symmetry_Pointless_Space_and_Leibniz%27s_Legacy_the_origin_of_physics.
16. Graves, *Life of Hamilton*, 3:424.
17. Graves, *Life of Hamilton*, 3:441–42.
18. William Rowan Hamilton, *Lectures on Quaternions* (London: Whittaker, and Cambridge: Macmillan, 1853), e.g., 59, for multiplication by *j* and changing the orientation of a telescope.
19. Crowe has made a convincing case for the historiography of vector analysis proceeding from Hamilton and his successors rather than Grassmann (*History of Vector Analysis*, 77, chap. 4). Much later, as we'll see, Grassmann will influence Cartan, Clifford, and *their* followers.

CHAPTER 6

1. For an accessible account of Maxwell's life and work, see my *Einstein's Heroes: Imagining the World through the Language of Mathematics* (St. Lucia: University of Queensland Press, 2003; New York: Oxford University Press, 2005).
2. *Tait's excited letter*: Cargill Gilston Knott, *The Life and Scientific Work of P. G. Tait* (London: Cambridge University Press, 1911), 9. *On the Tripos*: the figure of sixteen exams over eight days was for 1854 (Maxwell's year): D. O. Forfar, "What Became of the Senior Wranglers?" *Mathematical Spectrum* 29, no. 1 (1996); available at www.clerkmaxwellfoundation.org.
3. *Maxwell at Cambridge*: Lewis Campbell and William Garnett, *The Life of James Clerk Maxwell* (London: Macmillan, 1882), 94–95. (There is a 1997 digital edition by Sonnet Software.)
4. The poem—whose full title is "A Vision of a Wrangler, of a University, of Pedantry, and of Philosophy"—is published in Campbell and Garnett, *Life of Maxwell*, 307.
5. *Maxwell as examiner*: Campbell and Garnett, *Life of Maxwell*, 175.
6. *Tait on Maxwell*: Obituary, *Proceedings of the Royal Society of Edinburgh* 10 (1878–80): 331–39. *His father's letter to Maxwell*: Campbell and Garnett, *Life of Maxwell*, 109. *His old teacher*: David O. Forfar and Chris Pritchard, "The Remarkable Story of Maxwell and Tait," *James Clerk Maxwell Commemorative Booklet* (Edinburgh, 1999), 3.

7. *On Stokes's theorem*: Maxwell credits proof to Thomson and Tait (and also notes its first appearance in the Smith's Prize exam) in his *A Treatise on Electricity and Magnetism* (Oxford: Clarendon Press, 1873), 1:27. Thomson is the theorem's likely originator, for he had included it in a letter to Stokes back in 1850. Victor J. Katz ("The History of Stokes' Theorem," *Mathematics Magazine* [MAA] 52, no. 3 [May 1979]: 146–56) credits Hermann Hankel with the first published proof (in 1861), but it wasn't as general as Thomson's (1867).

8. *Area of a circle as a surface integral*: imagining a tiny element dS of the surface bounded by the circle, the idea is to integrate dS over the whole surface to find the area. Here the x and y axes represent the two dimensions that define the surface, so you can imagine little line segments in each direction, dx and dy, which border a rectangular elemental area of the surface—so in the surface integral you are integrating with respect to $dS = dxdy$. (Because the surface is a plane, this is actually just a double integral rather than a surface integral, where dS is more complicated than $dxdy$ because it requires the use of vectors to find the normal to the surface.) Transforming this to polar coordinates (as in fig. 2.2a), and not forgetting the Jacobian factor for changing the coordinates, you get $dS = dx\,dy = rdr\,d\theta$. Integrate this around the circle of radius R:

$$\text{Area} = \int_0^{2\pi}\int_0^R r\,dr\,d\theta = \int_0^{2\pi}\frac{1}{2}R^2\,d\theta = \pi R^2.$$

If the Jacobian is not familiar, for polar coordinates you can think of a tiny sector of a circle with angle $d\theta$ and radius r: the arc length s of the sector is $rd\theta$ (since $\frac{s}{2\pi r} = \frac{d\theta}{2\pi}$, by definition of a radian), and a radial element has length dr, so the element of area is $rdr\,d\theta$.

9. The survival of the grounds, along with the restoration of Glenlair House and outbuildings, is thanks largely to the efforts of the estate's current owner, Captain Duncan Ferguson. I've had the pleasure of meeting Duncan at Glenlair, and you can see more about his and the Glenlair Trust's work on behalf of Maxwell and Glenlair at http://www.glenlair.org.uk.

10. *Maxwell at British Association meeting*, recollected by William Swan, in Campbell and Garnett, *Life of Maxwell*, 236.

11. *Stationary gravity?* The sun and planets are moving, of course, but at any given point in the orbit they are stationary with respect to each other and are at a given distance apart. Newton's law treats the force arising just from this distance and the two masses. *Coulomb's law*: Maxwell described more accurate experiments establishing the inverse square law in his *Treatise on Electricity and Magnetism*, 1:34, 75.

12. *The (simplified) math of Lagrange's potential*: Work = force times distance, so if the distance moved is in the vertical direction (y, say), then $W = f \times y$. This formula is fine if the force remains constant, but for forces that change with distance, such as gravity, you need integral calculus to "add up" the force times the incremental distance *at each point* as the force moves the object.

Newton had given a geometrical calculus definition of this in terms of the area under the force curve, but in Leibnizian notation we'd have $W = \int_a^b f\,dy$, which gives $F(b) - F(a)$, where F is the antiderivative of f. (Actually, this is true only for "conservative" forces, including gravity, which depend only on the endpoints of the integral; for other forces, such as friction, you need a *line integral*, to take account of the characteristics of the entire path between the starting point a and finishing point b.)

What Lagrange showed, in effect, was that $f = \frac{dF}{dy}$. This is just the fundamental theorem of calculus taught in introductory calculus classes, but Lagrange extended it to three dimensions, to allow for the force and distance moved to be in any direction, not just up and down—so he found that the force has components $\frac{\partial F}{\partial x}, \frac{\partial F}{\partial y}, \frac{\partial F}{\partial z}$. (The "curly d" notation wasn't standard then, but I'm using it so as not to confuse modern mathematical readers.) Following Newton, he had the idea of force as a vectorial quantity, but like everyone before Hamilton and Grassmann, he dealt only with components, not with whole vectors.

Drawing on the relationship between work and potential energy, F is called the "potential" associated with the force f. Generally force is written with an uppercase F, so a common symbol for potential is V. (George Green was the first to use the term "potential," in 1828.)

13. *Partial derivatives*: in the $\frac{\partial F}{\partial x}$ term, F is differentiated only with respect to x, so this term tells how F changes in the x-direction while y and z remain fixed—and similarly for the other two terms.

14. *"Newton of electricity"*: Maxwell, *Treatise on Electricity and Magnetism*, 2:175.

15. Such forces, which depend only on the endpoints and not on the nature of the path between them, are called "conservative," because they lead to conservation of energy. For Newtonian gravity, for example, the motion is radial, so the inverse square law can be written as $m\ddot{r} = -\frac{GmM}{r^2}$; writing $\ddot{r} = \dot{r}d\dot{r}/dr$, integrate to find the work done by the force in moving an object from point 1 to point 2:

$$m\int_{\dot{r}_1}^{\dot{r}_2} \dot{r}\,d\dot{r} = -GMm\int_{r_1}^{r_2} \frac{1}{r^2}\,dr \Rightarrow \frac{1}{2}m\dot{r}^2 - \frac{GMm}{r} = \text{constant},$$

which means the sum of the kinetic and potential energies is conserved; the constant is found from the endpoints in the definite integrals.

16. Maxwell, *Lecture on Faraday's Lines of Force*, a talk he presented in 1873, in his collected works, *The Scientific Letters and Papers of James Clerk Maxwell*, ed. P. M. Harman, 2 vols. (Cambridge: Cambridge University Press, 1990, 1995), 803.

17. *Thomson and Faraday on fields*: Ernan McMullin, "The Origins of the Field Concept in Physics," *Physics in Perspective* 4 (2002): 13–39 (esp. 14). This paper gives a detailed overview of the field concept and its evolution up till Maxwell gave the first full-blown field theory.

18. *The Marischal professorship*: Forfar and Pritchard ("Remarkable Story," 3–4) noted that College records were not extant, but that it was believed Tait was a candidate for the job that Maxwell got—and later John S. Reid, of the University of Aberdeen, stated that Tait was a candidate, in "James Clerk Maxwell's Scottish Chair," *Philosophical Transactions of the Royal Society A* (2008), 366, 1661–84, DOI:10.1098/rsta.2007.2177. If so, as Forfar and Pritchard note, Maxwell and Tait's correspondence shows there were no hard feelings between them—in 1856 or in 1860 (when Tait beat Maxwell to a job at Edinburgh). *Cayley applying*: Crilly, "Arthur Cayley: The Road Not Taken," 52.

19. *Maxwell explaining his choice of language*: He alludes to it at the beginning of his 1865 paper (J. Clerk Maxwell, "A Dynamical Theory of the Electromagnetic Field," *Philosophical Transactions of the Royal Society London* 155 [1865]: 459–512), and explains it fully in his *Treatise on Electricity and Magnetism*, 1:98–99 (art. 95), and vol. 2 (3rd ed.), 176–77 (art. 529). He says that ordinary integrals, and line and surface integrals over finite spaces, suit the action-at-a-distance approach, while partial differential equations, and (volume) integrals throughout all of space, are the natural language for fields.

20. *Maxwell's definitions of current and their relation to flux*: He included two types of current: the conventional one in conductors such as a loop of wire—where the current is the flux of the current density—and the effective current in a capacitor, which he called the "displacement current," and which is proportional to the changing flux of the electric force between the capacitor plates.

21. *Converting integrals to derivatives in Maxwell's field equations*: The "first fundamental theorem of integral calculus" links integrals and derivatives:

$$\int_a^b f(x)dx = F(b) - F(a), \text{where } F(x) \text{is the antiderivative of } f(x).$$

In other words, $f(x) = \dfrac{dF(x)}{dx}$, assuming the relevant functions are integrable/differentiable! What this means is that you can go from $f(x)$ to $F(x)$

via integration, or from $F(x)$ to $f(x)$ via differentiation. *Stokes's theorem* is an extension of this idea, where you can go from (single) line integrals to (double) surface integrals, and vice versa. Similarly, you can go from surface integrals to volume (triple) integrals and back via what is now, post-vectors, known as the "divergence theorem." For example, this is how Maxwell deduced the differential form of Gauss's laws for static electricity and magnetism (in his *Treatise on Electricity and Magnetism*, 1:68, 79, 98–99):

It was known experimentally that the amount of electric charge e contained in a given volume could be written as the volume integral of the charge density ρ:

$$e = \iiint \rho \, dx \, dy \, dz \ldots \text{ I'll call this equation (1).}$$

It was also known (from Coulomb's law) that the force R exerted on a charge e by a unit test charge is $R = e/r^2$, and that the electric flux through a closed surface was

$$\iint R \cos \varepsilon \, dS = 4\pi e, \ldots (2)$$

where ε is the angle of the direction of the force. Maxwell, adapting Faraday, called $R\cos\varepsilon dS$ the "induction" through the surface. Maxwell labeled the components of R as X, Y, Z, which he linked to the following theorem (now known as the divergence theorem, but it didn't have a name then, and it was only known in component form as shown):

$$\iint R \cos \varepsilon \, dS = \iiint \left(\frac{dX}{dx} + \frac{dY}{dy} + \frac{dZ}{dz} \right) dx \, dy \, dz \ldots (3)$$

So then Maxwell multiplied (1) by 4π and equated the result with (2), to get

$$\iint R \cos \varepsilon \, dS = 4\pi \iiint \rho \, dx \, dy \, dz. \ldots (4)$$

Finally, equate (3) and (4), and take the closed surface from (3) as an element of the volume in (4), to get:

$$\frac{dX}{dx} + \frac{dY}{dy} + \frac{dZ}{dz} = 4\pi\rho \ldots (5)$$

If you are familiar with vector calculus already, you'll recognize the left-hand side is the *divergence* of R, but we'll come to this in the narrative when Tait and Maxwell (and Heaviside) put it into vector form.

Meantime, as Maxwell then explained, if you can write the electric force in terms of a potential V, then (5) becomes Poisson's extension of Laplace's equation. The vector form of (5) is the way Coulomb's law appears in Maxwell's equations today. The result for static magnetism follows in a similar way.

Maxwell's working that I've shown here makes the links between flux/surface integrals and divergence clear—and it is analogous to the way Maxwell used Stokes's theorem to express Ampère's and Faraday's laws as differential equations (see *Treatise on Electricity and Magnetism*, 2:29, 45, 147–48, 233, 251, 255). It involves another vector calculus operation, not divergence but *curl*, and in the next chapter we'll meet both these vector operations.

22. *Einstein's quote* is from Albert Einstein, *Ideas and Opinions* (1954; New York: Three Rivers Press, 1982), 327. *Maxwell's great guns* is from a letter to Charles Cay, reprinted in Campbell and Garnett, *Life of Maxwell*, 169.
23. This is because the general wave equation is differential; but to get out the electromagnetic wave equation you also need Maxwell's theoretical change to Ampère's law, i.e., the addition of the "displacement current."
24. Quotes are from Anne van Weerden, *A Victorian Marriage: Sir William Rowan Hamilton* (Stedum, Netherlands: J. Fransje van Weerden, 2017), 10, 56, 326.
25. Van Weerden, *A Victorian Marriage*, 326.
26. *On the Edinburgh posting (and Barrie)*: Forfar and Pritchard, "Remarkable Story." *Quote from Barrie*: Raymond Flood, "Thomson and Tait: The Treatise on Natural Philosophy," in Raymond Flood, Mark McCartney, and Andrew Whitaker, *Kelvin: Life, Labours and Legacy*, Oxford Scholarship Online (May 2008): DOI: 10.1093/acprof:oso/9780199231256.001.0001. *Gill on Maxwell's teaching*: Reid, "Maxwell's Scottish Chair," 1673.

CHAPTER 7

1. *Tait to Thomson*: in R. Flood, "Thomson and Tait: The Treatise on Natural Philosophy," in Raymond Flood, Mark McCartney, and Andrew Whitaker, *Kelvin: Life, Labours and Legacy* (Oxford: Oxford University Press, 2008) and Scholarship Online (2021), 176. DOI: 10.1093/acprof:oso/9780199231256.003.0011.
2. *Tait's study and list*: Cargill Gilston Knott, *The Life and Scientific Work of P. G. Tait* (London: Cambridge University Press, 1911), 33, 43.
3. *On the origins of Maxwell's nickname*: The equation appears in section 162 of Tait's *Sketch of Thermodynamics* (Edinburgh: Edmonston and Douglas, 1868). See also M. J. Klein, "Maxwell, His Demon, and the Second Law of Thermodynamics," in *Maxwell's Demon: Entropy, Information, Computing*,

ed. Harvey Leff and Andrew Rex (Princeton, NJ: Princeton University Press, 1990), 85–86. Klein's article is from 1970; Maxwell's "demon" was a thought experiment that helped clarify the nature of thermodynamics.

4. *Maxwell's review*: *Scientific Papers of James Clerk Maxwell*, ed. W. D. Niven (Cambridge: Cambridge University Press, 1890), 326–27. *Maxwell to Tait*, December 21, 1871, Knott, *Life of Tait*, 150.

5. *Maxwell to Tait* (with my emphasis), November 14, 1870, in Michael J. Crowe, *A History of Vector Analysis* (Notre Dame, IN: University of Notre Dame Press, 1967), 132; Maxwell to Campbell, October 19, 1872, in Lewis Campbell and William Garnett, *The Life of James Clerk Maxwell* (London: Macmillan, 1882), 186. Maxwell's "Classification" paper was published in *Proceedings of the London Mathematical Society* (March 9, 1871): 224–33.

6. *Maxwell to Tait* about vector calculus names: November 7, 1870, quoted in Knott, *Life of Tait*, 167.

7. Maxwell defined "convergence," the negative of divergence, in his *Treatise on Electricity and Magnetism* (Oxford: Clarendon Press, 1873), 1:28 (art. 25). The whole-vector equations in the narrative are given in Maxwell's *Treatise*, vol. 2 (3rd ed.), 252, 259. Minus sign aside, you can see that his version is the same as ours when you look at his component versions, in articles 77 (or my fig. 7.1) and 612. Maxwell has an additional proportionality constant K in his definition of $\mathfrak{D}$, but he notes that for air $K = 1$. So, to keep the vector ideas foremost, I'll generally write his equations assuming units are chosen to make various electrical and magnetic constants equal to 1. (Some modern texts have also scaled out the 4π in the div$\boldsymbol{E}$ equation.) $\mathfrak{D}$ is the electric displacement; Maxwell's definition given in the narrative is for isotropic substances. Note that in his whole vector equation in vol. 2 he uses e instead of ρ, which he'd used in article 77 of vol. 1, and which is used today—so I've used ρ in my narrative.

8. *Maxwell to Tait* (with my emphasis), November 2, 1871, in Crowe, *History of Vectors*, 133.

9. The analogous equations in electromagnetism and general relativity include the Bianchi identities, which we'll meet briefly in chapter 13. In electromagnetism, these identities include the div$\boldsymbol{B}$ equation reflecting that there are no magnetic monopoles. Tony and I have not yet written up our partial results, but an early paper is R. Arianrhod, A. W.-C. Lun, C. B. G. McIntosh, and Z. Perjés, "Magnetic Curvatures," *Classical and Quantum Gravity* 11 (1994): 2331–34.

10. Maxwell writes (what we would call) the div$\boldsymbol{B}$ equation in component form in his *Treatise*, 2:248 (art. 604), noting it follows from equation (A) art. 591

(233). Bruce Hunt (*The Maxwellians* [Ithaca, NY: Cornell University Press, 1991], 245) mentions the component form (which Hunt labels A') but says that Maxwell wrote this as $S.\,\nabla\mathfrak{B} = 0$. This is certainly the equivalent in Hamiltonian notation of the component equation on 248, although I can't find it in my third edition copy of the *Treatise*.

11. *Vector potential*: Maxwell, *Treatise*, vol. 2, arts. 405, 422–23, 592: he defines the vector potential $\boldsymbol{A}$ so that the line integral of $\boldsymbol{A}$ equals (via Stokes's theorem) the surface integral of the magnetic field $\boldsymbol{B}$ (which is the curl of the vector potential $\boldsymbol{A}$). He also gives it a physical interpretation, in terms of magnetic moments (art. 405) and electromagnetic momentum, arts. 590, 592, 618—although this is via mathematical analogy rather than by a direct physical match; for example, he chooses the term "electromagnetic momentum" because mathematically it is the time integral of a force (art. 590)—that is, its time-derivative is a force, just like ordinary Newtonian momentum.

12. *The whole-vector equation*: *Treatise on Electricity and Magnetism*, vol. 2 (3rd ed.), 258 (component form, 233, 248). *Potential*: modern notation varies, but A is used fairly widely, e.g., Luciano Maiani and Omar Benhar, *Relativistic Quantum Mechanics* (Boca Raton, FL: CRC Press, 2016), 56; Ray D'Inverno, *Introducing Einstein's Relativity* (Oxford: Clarendon Press, 1992), 160; Walter Strauss, *Partial Differential Equations* (New York: Wiley, 1992), 342; Bernard Schutz, *A First Course in General Relativity* (Cambridge: Cambridge University Press, 1985), 211.

13. *Maxwell to Campbell*: Campbell and Garnett, *Life of Maxwell*, 186.

14. *Maxwell's Watt Lecture*: *The Scientific Letters and Papers of James Clerk Maxwell*, ed. P. M. Harman, 2 vols. (Cambridge: Cambridge University Press, 1990, 1995), 791.

15. Tait's book, *Introduction to Quaternions*, was coauthored with his former teacher Philip Kelland. *Maxwell's review* (with my emphasis): *Nature* 9 (1873): 137–38; Crowe, *History of Vectors*, 133. *Newton*: letter to Halley, in, e.g., Nicolae Sfetcu, "Isaac Newton vs Robert Hooke on the Law of Universal Gravitation," SetThings (January 14, 2019), MultiMedia Publishing, DOI:10.13140/RG.2.2.19370.26567, Creative Commons.

16. *Maxwell, Tait, and Balfour*: Knott, *Life of Tait*, 149–50. *Kovalevsky*: Sophie Kowalevski, "Sur le problème de la rotation d'un corps solide autour d'un point fixe," *Acta Mathematica* 12 (January 1889): 177–232. It won the Prix Bordin in 1888.

17. *Maxwell's "physical reasoning"*: *Treatise on Electricity and Magnetism* 1:9 (art. 11).

18. *Thomson to R. B. Hayward*, 1892, in Crowe, *History of Vectors*, 120.
19. *Cayley's portrait/Maxwell's poem*: Alexander MacFarlane, *Lectures on Ten British Mathematicians* (1916), chap. 5. *Clifford's gymnastics*: Carl Boyer, *A History of Mathematics*, rev. Uta Merzbach (New York: John Wiley and Sons, 1991), 592; see also Monty Chisholm, "Science and Literature Linked: The Story of William and Lucy Clifford," *Advances in Applied Clifford Algebras* 19 (2009): 657–71.
20. *Clifford's atheism*: Sally Shuttleworth, "Science and Periodicals: Animal Instinct and Whispering Machines," in Juliet John, ed., *The Oxford Handbook of Victorian Literary Culture* (2016), DOI: 10.1093./oxfordhb/9780199593736.013.31.
21. *Tait's review*: reprinted in Knott, *Life of Tait*, 270–72.
22. *Inverse of a quaternion* q is $q^{-1} = q^* / qq^*$, where q^* is the complex conjugate of q. If you multiply out qq^{-1} you'll find that it does indeed give 1.
23. The American mathematician David Hestenes was the first modern mathematician to recognize, in the 1960s, the importance of Clifford and Grassmann for geometric algebra, which Hestenes and others have since developed further. Wedge products are important in modern tensor analysis (they are defined in terms of the tensor products we'll see in chap. 11).
24. *Tait to Cayley*: Knott, *Life of Tait*, 155.
25. *Liberal utilitarian*: David Weinstein, "Herbert Spencer," in Edward Zalta, ed., *Stanford Encyclopedia of Philosophy* (Fall 2019), plato.stanford.edu. *Maxwell and Tait*: e.g., Knott, *Life of Tait*, 284–88.
26. *Lewes's mind-body analysis today*: Elfed Huw Price, "George Henry Lewes (1817–1878): Embodied Cognition, Vitalism, and the Evolution of Symbolic Perception," in *Brain, Mind and Consciousness in the History of Neuroscience*, ed. Chris Smith and Harry Whitaker (New York: Springer, 2014), 105–23. *Tait, Lewes, Blackwood*: Gordon Haight, ed., *The George Eliot Letters* (New Haven, CT: Yale University Press, 1955), 5:401, 417, and 9n181.
27. *Maxwell's poem*: "British Association, 1874" was published in the December 1874 issue of *Blackwood's*: Campbell and Garnett: *Life of Maxwell*, 8 (poem reprinted, 326).
28. *Tait and Blackwood at Golf (and Maxwell "better ½")*: Martin Goldman, *The Demon in the Aether* (Edinburgh: Paul Harris Publishing, 1983), 105.
29. For more on Maxwell's poem and related debates, see Raymond Flood, Mark McCartney, and Andrew Whitaker, *James Clerk Maxwell: Perspectives on His Life and Work* (Oxford: Oxford University Press, 2014).

30. *Maxwell to Campbell*: Campbell and Garnett, *Life of Maxwell*, 202.
31. Knott, *Life of Tait*, 261.
32. *Shaw to Lucy*: quoted in Chisholm, "Science and Literature Linked," 668.

CHAPTER 8

1. Bruce Hunt coined the term "Maxwellians," in his *The Maxwellians* (Ithaca, NY: Cornell University Press, 1991). *Lodge scooped*: James Rautio, "Twenty-three Years: Acceptance of Maxwell's Theory," *Applied Computational Electromagnetics Society Journal* 25, no. 12 (December 2010), 998–1006.
2. *On Heaviside*: Here and in the following paragraphs I've drawn on Bruce Hunt, "Oliver Heaviside: A First-Rate Oddity," *Physics Today* 65, no. 11 (2012): 48–54, DOI: 10.1063/PT.3.1788; Jed Buchwald, "Oliver Heaviside, Maxwell's Apostle and Maxwellian Apostate," *Centaurus* 28 (1985): 288–330; I. Yavetz, *From Obscurity to Enigma: The Work of Oliver Heaviside, 1872–1889* (Basel: Springer, 2011); and Heaviside's papers, which I'll generally cite as I go.
3. *Maxwell's Reference to Heaviside* was added to his list of errata (p. 2) for vol. 1, with reference to p. 404. *Heaviside on Maxwell's* Treatise: Rautio, "Twenty-three Years."
4. *Heaven-sent Maxwell*: Oliver Heaviside, *Electromagnetic Theory* (London, 1893; New York: Chelsea Publishing, 1971), 1:14.
5. *Worshipping quaternions*: Heaviside, *Electromagnetic Theory*, 1:136.
6. Heaviside, *Electromagnetic Theory*, 1:137, 139.
7. *Heaviside eliminating imaginary numbers* from vectors: *Electromagnetic Theory*, 1:137, 142, 149. *His drollery*: *Electromagnetic Theory*, 1:135.
8. Heaviside missed at BAAS: *Engineering* 46 (1888): 352; cited in Hunt, "Oliver Heaviside," 52–53.
9. *Pot and kettle*: Heaviside, *Electromagnetic Theory*, 1:203; *Murdering potentials*: letter to FitzGerald, quoted in Rautio, "Twenty-three Years," 1004.
10. *Heaviside not quite in "modern" form*: In most undergrad textbooks, Maxwell's equations are written in terms of $\boldsymbol{E}$ and $\boldsymbol{B}$, but for Heaviside it is $\boldsymbol{E}$ and $\boldsymbol{H}$ that are singled out. Maxwell had distinguished between the magnetic induction or magnetic field, $\boldsymbol{B}$, which is a *flux*, and the magnetic *force*, $\boldsymbol{H}$. When the magnetic field is induced entirely by the magnetic force, then $\boldsymbol{B} = \mu\boldsymbol{H}$, where μ is the coefficient of the magnetic permittivity. This is the definition Heaviside used, so it is straightforward to change from $\boldsymbol{H}$ to $\boldsymbol{B}$ when looking at his equations. He also used Maxwell's definition of the electric displacement $\boldsymbol{D}$, a flux, in terms of the force $\boldsymbol{E}$, namely $\boldsymbol{D} = c\boldsymbol{E}/4\pi$.

Today some authors still use $\boldsymbol{H}$, but using $\boldsymbol{B}$ makes Heaviside's equations more symmetrical. Some authors also use $\boldsymbol{D}$ instead of $\boldsymbol{E}$ in the divergence equation, following Maxwell and Heaviside, but as I mentioned it is numerically proportional to $\boldsymbol{E}$.

11. *Potentials unphysical (for localized energy)*: This is explained in detail in Buchwald, "Oliver Heaviside," 293. *Maxwell's wave equations*: in terms of potential, *Treatise* 2, 434 (art. 784); in terms of magnetic field, "A Note on the Electromagnetic Theory of Light," *Philosophical Transactions of the Royal Society* 158 (1868): 643–57, esp. 655.

    In this 1868 paper, Maxwell's four field equations are not quite the four modern equations, but he deduces from them the wave equation for the magnetic field. This is just what Heaviside was trying to do when he "murdered" the potentials! It's a pity Maxwell didn't take this approach further—but as Heaviside noted (*Electromagnetic Theory*, 1:69), Maxwell didn't do his own theory justice in the *Treatise*: instead he provided a brilliant overview of all the contributions to the study of electromagnetism that had been made so far. He certainly showed how and why he developed his theory from the earlier known work, and how it compared with others' action-at-a-distance models, but he definitely wasn't a self-promoter.

12. These four equations look slightly different in different texts; it depends on how the units of the electrical and magnetic constants are chosen. In particular, along with electric and magnetic constants the speed of light c is often set to 1 (as I've done here), but, following Heaviside, the factor of 4π is often effectively set to 1 by adjusting the units of the constants. (Also, some texts use Heaviside's "div" and "curl" instead of Gibbs's dot and cross.)

13. Heaviside was so entranced by the symmetry between the electric and magnetic fields in the four key Maxwell equations that he added a fictitious magnetic "charge" to the equation $\nabla \cdot \boldsymbol{B} = 0$ (thereby positing that there *are* magnetic monopoles—just as Dirac did nearly half a century later) and a magnetic "current" to the $\nabla \times \boldsymbol{E}$ equation. But these additions don't yet have any known physical basis (aside from artificially generated short-lived quantum monopoles), so they are generally left out of the modern electromagnetic equations; this means that—vector formalism aside—the modern equations are indeed *Maxwell's* equations, as my narrative shows.

14. *Finding* $\nabla \times \boldsymbol{E}$ *from Maxwell's whole vector equation* in his *Treatise on Electricity and Magnetism*, vol. 2 (3rd ed., 1891; reprint, Dover, 1954): For example, on 2:232 (art. 590), Maxwell gives (in words) the definition $\boldsymbol{A} = \int \boldsymbol{E}\, dt$, or

equivalently (equation 29 of his 1865 paper), $\boldsymbol{E} = -d\boldsymbol{A}/dt$. Taking the curl of both sides of this, and remembering that Maxwell defined $\boldsymbol{B} = \nabla \times \boldsymbol{A}$, you have (using the fact that you can interchange the order of derivatives)

$$\nabla \times \boldsymbol{E} = -\frac{d}{dt}(\nabla \times \boldsymbol{A}) = -\frac{d\boldsymbol{B}}{dt}.$$

Alternatively, begin with Maxwell's equation (*Treatise* 2:258 [art. 619]) as given in my narrative,

$$\boldsymbol{E} = \boldsymbol{v} \times \boldsymbol{B} - \frac{d\boldsymbol{A}}{dt} - \nabla\phi,$$

and then take the curl of both sides. Using the identity curl grad = 0, the last term drops out. If there are no moving charges, so the electric field is induced only by a time-varying magnetic field (cf. *Treatise* 2:240–41 [art. 599], 2:433 [art. 783]), then $\boldsymbol{v} = 0$. So again you have $\nabla \times \boldsymbol{E} = -\frac{d}{dt}(\nabla \times \boldsymbol{A}) = -\frac{d\boldsymbol{B}}{dt}$

15. *Maxwell's five vector (quaternion) equations* are in his *Treatise* 2:258–59. (There are also seven definition equations on these pages—just as Heaviside used.) Heaviside specifically said that his form of the equations should still be called Maxwell's equations: *Electromagnetic Theory*, vol. 1, preface (fifth page), and 69. Hertz agreed: Rautio, "Twenty-three Years," 1005.
16. Heaviside, *Electromagnetic Theory*, 1:297.
17. *Gibbs's path to vectors*: he was first inspired by Maxwell, then branched out on his own—independently of Grassmann, whom he discovered several years later. We know this from his letter to Victor Schlegel, published much later by Gibbs's student Lynde Phelps Wheeler, in his book *Josiah Willard Gibbs: The History of a Great Mind* (New Haven, CT: Yale University Press, 1952).
18. *Gibbs's reply to Tait's "monster"*: "On the Role of Quaternions in the Algebra of Vectors," *Nature* 43 (April 2, 1891): 511–13. *Heaviside* (incl. "hermaphrodite monster" quote and citation): *Electromagnetic Theory*, 1:137–38, 301.
19. Peter Guthrie Tait, "The Role of Quaternions in the Algebra of Vectors," *Nature* 43 (April 30, 1891): 608. For a detailed analysis of the vector wars on which I've gratefully drawn, see Michael J. Crowe, *A History of Vector Analysis* (Notre Dame, IN: University of Notre Dame Press, 1967), chap. 6.
20. *Thomson's "war over quaternions"*: Cargill Gilston Knott, *The Life and Scientific Work of P. G. Tait* (London: Cambridge University Press, 1911), 185.
21. Knott, *Life of Tait*, 185; Alexander Macfarlane, "Principles of the Algebra of Vectors," *Proceedings of the American Association for the Advancement of Science* 40 (1891, published 1892): 65–117; Alexander McAulay, "Quaternions

as a Practical Instrument of Physical Research," *Philosophical Magazine*, 5th ser., 33 (June 1892): 477–95; Crowe, *History of Vectors*, 189–97.

22. Martin Rees offers solutions to these problems in *If Science Is to Save Us* (Cambridge: Polity Press, 2022).
23. McAulay quoted in Crowe, *History of Vectors*, 195. *Ida McAulay*: Bruce Scott, "McAulay, Alexander," *Australian Dictionary of Biography*, adb.anu.edu.au.
24. *Tait's review of McAulay*: *Nature* 49 (December 28, 1893): 193–94.
25. Heaviside, "Vectors versus Quaternions," *Nature* (April 6, 1893): quoted in Crowe, *History of Vectors*, 200. *Hydroelectricity*: Scott, "McAulay, Alexander"; Carol Raabus and Leon Compton, "The Engineering Feats of Tasmania's Hydroelectric System," ABC Radio, July 29, 2013.
26. Gibbs, "Quaternions and the Algebra of Vectors," *Nature* 47 (March 16, 1893): 463–64.

CHAPTER 9

1. *Grace Chisholm on Cayley*: I. Grattan-Guinness, "A Mathematical Union: William Henry and Grace Chisholm Young," *Annals of Science* 29, no. 2 (August 1972): 117–18.
2. *History of Newnham*: https://newn.cam.ac.uk/about/history/history-of-newnham/. Women were finally allowed to take full degrees at Oxford in 1920 but not at Cambridge until 1948.
3. Cayley and Tait's letters are in Cargill Gilston Knott, *The Life and Scientific Work of P. G. Tait* (London: Cambridge University Press, 1911), 154–96.
4. At the time of writing, many of these machine-made predictions still need to be verified in the lab, but even so they can point the way for genomics research.
5. *Invariance of the discriminant under translations*: The quadratic equation $ax^2 + bx + c = 0$ has the solution $x = [-b \pm \sqrt{b^2 - 4ac}\,]/2a$. If we transform x to $x' = x + h$, the associated quadratic equation is now $ax'^2 + bx' + c = 0$, whose solution is $x' = [-b \pm \sqrt{b^2 - 4ac}\,]/2a$. The discriminants are the same, $b^2 - 4ac$, but the solutions are not the same: for the translated equation we have $x' = x + h = [-b \pm \sqrt{b^2 - 4ac}\,]/2a$, which implies that $x = [-b - 2ah \pm \sqrt{b^2 - 4ac}]/2a$; but this is not equal to the original solution, $x = [-b \pm \sqrt{b^2 - 4ac}\,]/2a$.
6. *Cayley and Tait quotes*: Michael J. Crowe, *A History of Vector Analysis* (Notre Dame, IN: University of Notre Dame Press, 1967), 212, 214. *Grace Chisholm on Cayley*: Grattan-Guinness, "A Mathematical Union," 117.
7. Crowe, *History of Vectors*, 214.

8. Crowe, *History of Vectors*, 217.

9. He mentioned anti-Semitism in connection with his job-hunting in a letter to Mileva Marić on March 27, 1901: *Collected Papers of Albert Einstein*, vol. 1, ed. John Stachel, David C. Cassidy, and Robert Schulmann (Princeton, NJ: Princeton University Press, 1987; English Supplement translated by Anna Beck), document 94, https://einsteinpapers.press.princeton.edu/vol1-trans/182.

10. In the letter of March 27, 1901, referred to in the previous endnote (https://einsteinpapers.press.princeton.edu/vol1-trans/182), Einstein looks forward to the day when "the two of us together will have brought our work on the relative motion to a victorious conclusion." This has been cited by some scholars as evidence that Einstein and Marić were working together on relativity, although the context, and the words "victorious conclusion," also suggest it might be a metaphor for their marriage plans (being held up by disapproval from relatives, Mileva's struggle to graduate, and Einstein's lack of employment). The only prior (and subsequent) times Einstein mentions relativity in the extant letters to Marić are, as far as I could find, in a letter of September 10, 1899 (*Collected Papers of Albert Einstein*, vol. 1, document 54), where he tells Mileva he's had an idea about how relative motion with respect to the ether affects the velocity of light, adding, "But enough of that!" (because she is studying for her exams): https://einsteinpapers.press.princeton.edu/vol1-trans/155; and again in document 57, September 28, 1899, where there is no mention of "our" theory, and similarly in document 128, December 17, 1901. Evidently, she never responded with comments on the subject, judging from her letters and Einstein's letters to her—rather, she was focused on topics relevant to her exams. If only we knew what went on between them when they were together and didn't need to write letters! Einstein certainly nourished early dreams that they would have a scientific life together—and in document 72 of this volume (August 14, 1900), he tells her that he lacks self-confidence and pleasure in work when he is not with her. Few of Marić's letters from this time survive: those that do are focused on her dreams of getting married, passing her diploma exams, and starting her PhD (and her diploma thesis was on heat and energy, not relativity). As for her "doing Einstein's math for him," their exam results in 1900 suggest that Einstein excelled at math and she failed: document 67 of vol. 1, https://einsteinpapers.press.princeton.edu/vol1-trans/163. Exam results are not everything of course, but it does put to rest claims of Einstein's mathematical

incompetence. In my view we can do more justice to Marić as a female scientific pioneer by examining the prejudice that blighted her career, rather than claiming things for which there is no clear evidence. More information about the relationship, and about the special theory of relativity, is in my short e-book *Young Einstein and the Story of $E = mc^2$* (Sydney: Ligature, 2014), and the references therein. For a very brief, updated overview of the evidence for the claim of Mileva Marić as coauthor, see Ann Finkbeiner, "The Debated Legacy of Einstein's First Wife," *Nature* 567 (2019): 28–29.

11. *Maxwell's idea* was expressed in his entry "Ether" in the 9th edition of *Encyclopaedia Britannica* (1878), 8:568–72, and in more detail in a letter to David Peck Todd, March 19, 1879, a few months before he died. Todd recognized its importance and sent it to Stokes, who communicated it to the Royal Society, which published it in its proceedings: "'On a possible method of detecting the motion of the solar system through the luminiferous ether' by the late Professor J. Clerk Maxwell," *Proceedings of the Royal Society*, January 22, 1880, 108–10. *Michelson studied this letter,* for he worked in Todd's office. See also Robert Shankland, "Michelson and His Interferometer," *Physics Today* 27, no. 4 (1974): 37, DOI: 10.1063/1.3128534; Shankland includes a photo of Michelson's interferometer. *In their paper reporting their results, however*, Michelson and Morley mention the satellite method as a possible future experiment in light of their negative result, but they do not cite Maxwell; presumably they didn't know his letter had been published: Albert A. Michelson and Edward W. Morley, "On the Relative Motion of the Earth and the Luminiferous Ether," *American Journal of Science*, ser. 3, 34, no. 203 (November 1887): 345.

12. *Lorentz electron theory*: Maxwell had assumed charge was continuously distributed—hence the charge density term ρ and current density J in his equations; Lorentz showed that these densities are an approximation or average of the distribution of charges (points), and that Maxwell's equations are singular at these points, but hold everywhere else. *Einstein on Lorentz*: Albert Einstein, *Ideas and Opinions* (1954; New York: Three Rivers Press, 1982), 73–76, and Banesh Hoffman (with the collaboration of Helen Dukas), *Einstein* (Frogmore: Paladin, 1975), 98.

13. H. Poincaré, "Sur la dynamique de l'électron," *Rendiconti del Circolo Matematica di Palermo* 21 (1906): 18–76. He'd already presented a preliminary "Note" on this paper, to the Académie des Sciences, June 5, 1905, and had written on related ideas several years earlier. For the English version of

Einstein's 1905 paper (originally published in *Annalen der Physik*): A. Einstein, "On the Electrodynamics of Moving Bodies," in H. A. Lorentz et al., *The Principle of Relativity* (New York: Dover, 1952), 37–65.

14. As early as 1910, Felix Klein, who had been working on the geometry of Lorentz groups that are fundamental in Einstein's special theory, claimed that one could, "if one really wanted to, replace the term 'theory of invariants with respect to a group of transformations' with the term 'relativity with respect to a group'" (quoted in Yvette Kosmann-Schwarzbach [translated by Bertram E. Schwarzbach], *The Noether Theorems: Invariance and Conservation Laws in the Twentieth Century* [New York: Springer, 2011], 70). In 2022 (in *If Science Is to Save Us* [Cambridge: Polity, 2022], 93), Martin Rees suggested the name "theory of invariance" instead of "relativity" would have avoided "misleading analogies with relativism in human contexts."
15. *Grace Chisholm and higher dimensions*: Her recollection is reproduced in Grattan-Guinness, "A Mathematical Union," 128–29.
16. For a beautiful account of these 4-D efforts and their context (including the Booles, and hyper-square analogy), see Nicholas Mee, *Celestial Tapestry: The Warp and Weft of Art and Mathematics* (Oxford: Oxford University Press, 2020).
17. For a modern analysis of (bi-)quaternions in SR, see Joachim Lambek, "In Praise of Quaternions," *Comptes Rendues Mathematical Reports*, Academy of Sciences, Canada, 35, no. 4 (2013): 121–36; https://www.math.mcgill.ca/barr/lambek/pdffiles/Quater2013.pdf. Hamilton himself discussed biquaternions (which have complex coefficients).
18. *"Lazy dog"*: quoted in Michael White and John Gribbin, *Einstein: A Life in Science* (London: Simon and Schuster, 1993), 39. *Minkowski on quaternions*: Scott Walter, "Breaking in the 4-vectors: The Four-dimensional Movement in Gravitation, 1905–1910," in *The Genesis of General Relativity*, ed. Jürgen Renn (Dordrecht: Springer, 2007), 3:212.
19. Minkowski quoted in Constance Reid, *Hilbert* (Berlin: Springer-Verlag, 1970), 105, 112.
20. *The interval* tells you how to take account of the "distance" between two events taking place at two different places and times:

$$\sqrt{(x_2 - x_1)^2 + (y_2 - y_1)^2 + (z_2 - z_1)^2 - (c(t_2 - t_1))^2}.$$

If you observe two events, one after the other, from the *same* point in space, then the interval tells you the time between them according to your own wristwatch (because $x_2 - x_1$, $y_2 - y_1$, $z_2 - z_1$ are all zero). This is called the "proper" time. Similarly, if you measure two events at the same time, the metric tells you the (proper) distance between them (because $t_2 - t_1$ is now zero). But as the Lorentz transformations show, there is no agreement from a relatively moving observer on these times and distances—both of you only agree that the *interval as a whole is invariant.*

By the way, to turn the "signature" (as it's called) of this quadratic interval measure into ++++ rather than +++− Minkowski made the time imaginary, to fit with the original idea of a quadratic form.

21. H. Minkowski, "Space and Time," 1908, English translation in H. A. Lorentz et al., *Principle of Relativity*, 75–91. Turning red: Reid, *Hilbert*, 92. On the 1907 lecture: Walter, "Breaking in the 4-vectors," 219.
22. *On Minkowski's death*: Reid, *Hilbert*, 115.
23. *Klein and Justus*: Crowe, *History of Vectors*, 92.
24. *Invariant space-time interval/constant c*: Speed is distance/time, so in 3-D space, the speed of light can be defined using Pythagoras's theorem for the distance:

$$c^2 = (x^2 + y^2 + z^2)/t^2;$$

another way of writing this equation is, of course,

$$x^2 + y^2 + z^2 - (ct)^2 = 0.$$

The expression on the left is invariant under Lorentz transformations, which means that when the coordinates (x, y, z, t) and (x', y', z', t') are related via a Lorentz transformation, you still get

$$x^2 + y^2 + z^2 - (ct)^2 = x'^2 + y'^2 + z'^2 - (ct')^2.$$

Which means that $x'^2 + y'^2 + z'^2 - (ct')^2 = 0$, too, and so the speed of light in the (x', y', z', t') frame must also be c.

25. William Thomson, "Elements of a Mathematical Theory of Elasticity," *Philosophical Transactions of the Royal Society of London* 146 (1856): 481–98; Augustin Cauchy, "Sur les equations qui experiment les conditions d'équilibre ou les lois du movement intérieur d'un corps solide, élastique ou non-élastique," *Exercises de Mathématiques* 3 (1828): 160–87.
26. Maxwell, *Treatise on Electricity and Magnetism* (Oxford: Clarendon Press, 1873), 2:278–81.
27. Minkowski had called these particular two-index quantities "vectors of the second kind," and Sommerfeld called them "six-vectors." Today they are simply called tensors—in this case, antisymmetric second-rank or second-order

tensors, where the "rank" or "order" refers to the number of indices on its components. (If these tensors are defined through space, rather than at one point, then technically they are tensor fields.)

CHAPTER 10

1. *Einstein's PhD*: Banesh Hoffmann, *Einstein* (Frogmore: Paladin, 1975), 55. *Grossmann seeing Einstein's greatness*: see my *Young Einstein* and references therein.
2. Albert Einstein, *Ideas and Opinions* (1954; New York: Three Rivers Press, 1982), 289.
3. *Einstein to Sommerfeld*: Judith Goodstein, *Einstein's Italian Mathematicians* (Providence, RI: American Mathematical Society, 2018), 95.
4. *Gauss, surveying, and least squares*: Martin Vermeer and Antti Rasilia, *Map of the World: An Introduction to Mathematical Geodesy* (Milton Park, UK: Taylor and Francis, 2019), 181; Frank Reid, "The Mathematician on the Bank Note: Carl Friedrich Gauss," *Parabola* 36, no. 2 (2000). Although Gauss retired from fieldwork in 1825, he directed the survey until its completion in 1844.
5. Actually, Minkowski and Einstein wrote this metric with the plus and minus signs reversed:

$$ds^2 = -dx^2 - dy^2 - dz^2 + c^2t^2, \text{ or } ds^2 = c^2t^2 - dx^2 - dy^2 - dz^2;$$

the choice of signs is called the "signature," and for our purposes the key thing is that the time differential has the opposite sign from the spatial ones.

6. *Outline of Gauss's argument*: The page numbers here (and elsewhere) refer to the English version of Gauss's 1828 paper, *General Investigations of Curved Surfaces of 1827 and 1825*, by Karl Friedrich Gauss, translated by James Morehead and Adam Hiltebeitel, Project Gutenberg, 2011 (from the 1902 edition, Princeton: Princeton University Library); https://www.gutenberg.org/files/36856/36856-pdf.pdf.

I mentioned that for his 2-D surface Gauss transformed his three x, y, z coordinates to functions of two new variables, which he called p, q—so you can write the coordinate transformations (which I'll make linear) as

$$x = f(p, q), y = g(p, q), z = h(p, q).$$

Then the chain rule gives

$$dx = \frac{\partial f}{\partial p}dp + \frac{\partial f}{\partial q}dq = a\,dp + a'\,dq \text{ in Gauss's notation (p. 7).}$$

(Unfortunately Gauss, like Riemann, used dashes instead of different letters.)

Similarly, $dy = bdp + b'dq$, $dz = cdp + c'dq$. If you square these expressions and add, you get (cf. Gauss pp. 18, 20):

$$dx^2 + dy^2 + dz^2 = (a^2 + b^2 + c^2)dp^2 + 2(aa' + bb' + cc')dpdq + (a'^2 + b'^2 + c'^2)dq^2 = Edp^2 + 2Fdpdq + Gdq^2,$$

where Gauss used E, F, G to simplify the expression.

But here's the fascinating thing in hindsight: from the algebraic definition of scalar products, you can see that E, F, G are what we would now call scalar products of the vectors

$$\boldsymbol{v} = a\boldsymbol{i} + b\boldsymbol{j} + c\boldsymbol{k}, \boldsymbol{v}' = a'\boldsymbol{i} + b'\boldsymbol{j} + c'\boldsymbol{k};$$

in other words,

$$E = \boldsymbol{v} \cdot \boldsymbol{v}, F = \boldsymbol{v} \cdot \boldsymbol{v}', G = \boldsymbol{v}' \cdot \boldsymbol{v}'.$$

These two vectors are the *unit tangent vectors* to the coordinate lines p, q as in fig. 10.4 in the narrative. (To see this, consider the infinitesimal displacement vector, using shorthand bracket notation for vectors:

$$d\boldsymbol{r} = (dx, dy, dz) = (a, b, c)dp + (a', b', c')dq = \boldsymbol{v}dp + \boldsymbol{v}'dq;$$

the *tangent vectors* are found by differentiating this with respect to the two coordinates, just as we differentiate an ordinary function to find the slope of its tangent:

$$\frac{d\boldsymbol{r}}{dp} = \boldsymbol{v}, \frac{d\boldsymbol{r}}{dq} = \boldsymbol{v}'.)$$

The geometric definition of the scalar product of two vectors is $\boldsymbol{a}.\boldsymbol{b} = |\boldsymbol{a}||\boldsymbol{b}| \cos\theta$, and as we saw in chapter 9, $|\boldsymbol{a}| = \sqrt{\boldsymbol{a}\cdot\boldsymbol{a}}$. So the angle between these two tangent vectors is

$$\cos\theta = \frac{\boldsymbol{v} \cdot \boldsymbol{v}'}{(\sqrt{\boldsymbol{v} \cdot \boldsymbol{v})(\boldsymbol{v} \cdot \boldsymbol{v}')}} = \frac{F}{\sqrt{EG}}.$$

Later mathematicians will generalize this to arbitrary metrics and dimensions, where the coefficients in the metric are written as g_{ij}: for the 2-D case here we'd have

$$\cos\theta = \frac{g_{12}}{\sqrt{g_{11}g_{22}}}.$$

To find the curvature of the surface, these formulae are applied to triangles whose sides are bounded by the coordinate lines, as in fig. 10.4, and then, as I explained in the narrative, the sum of the angles gives the nature of the curvature.

By the way, if you're familiar with double integrals, then the expression $\sqrt{EG - F^2}$ in the area integral I mentioned in the narrative is the Jacobian. It's the determinant of the matrix of coefficients of the metric, and in terms of a general metric, as in GR, it is written as $\sqrt{-g}$.

Gauss's definition of curvature in terms of angles: p. 46 (and 44).

Harriot's work: see my *Thomas Harriot: A Life in Science* (New York: Oxford University Press, 2019), 160–61, and also John Stillwell, *Mathematics and Its History* (New York: Springer-Verlag, 1989), 249–50.

7. *Hawking on black hole horizon (boundary)*: He published this result in 1972; he also proved it in S. W. Hawking and G. F. R. Ellis, *The Large Scale Structure of Space-time* (Cambridge: Cambridge University Press, 1973), 335–37. For a simple sketch of Hawking's proof and a useful brief history of curvature, see Greg Galloway, "From the Shape of the Earth to the Shape of Black Holes: Aristotle to Hawking and Beyond," Miami University's Mathematics Department, Arts and Sciences Cooper Lecture, November 2017. For an interesting outline of the history of black holes, see the overview on the Nobel Prize website for the 2020 physics prize. Note that some researchers have suggested that the Event Horizon Telescope's (EHT's) first direct image of a black hole could, in fact, be that of a gravitomagnetic monopole rather than a black hole; they have calculated parameters that would distinguish the two possibilities when future, more accurate EHT observations are made: M. Ghasemi-Noedi et al., "Investigating the Existence of Gravitomagnetic Monopole in M87*," *European Physics Journal C* 81, no. 939 (2021); https://doi.org/10.1140/epjc/s10052-021-09696-3.

8. To take just one example, for an account that includes the role Gaussian curvature is playing in the study of the way materials wrinkle, and the possibilities for using these wrinkles in innovative new ways, see Stephen Ornes, "The New Math of Wrinkling," *Quanta* magazine (September 22, 2022); https://www.quantamagazine.org/the-new-math-of-wrinkling-patterns-20220922/.

9. Lewis Campbell and William Garnett, *The Life of James Clerk Maxwell* (London: Macmillan, 1882), 324–25.

10. *Einstein Privatdozent*: Hoffmann, *Einstein*, 86–87. *Gauss on Riemann*: Raymond Flood and Robin Wilson, *The Great Mathematicians* (London: Arcturus, 2011), 160. *Seventy-seven-year-old Gauss*: Goodstein, *Einstein's Italian Mathematicians*, 31. *Gauss devastated*: Stillwell, *Mathematics and Its History*, 253–54.

11. An excellent, more technical account of Riemann's working—and an English translation of the paper—is in Ruth Farwell and Christopher Knee, "The Missing Link: Riemann's 'Commentatio,' Differential Geometry and Tensor Analysis," *Historia Mathematica* 17 (1990): 223–55.

12. Maxwell, *Treatise on Electricity and Magnetism* (Oxford: Clarendon, 1873), 1:333 (art. 280). Note that Thomson and Tait (in their *T&T'*, 1:515) denote

coefficients of elasticity by pairs of letters, as Riemann did, rather than using indices as Maxwell did.

13. Conductivity is a two-index tensor if the material is anisotropic, as Riemann assumed. For isotropic materials the heat spreads out in all directions, so there's no need to worry about variations with direction, and you can use a scalar representation.

14. Bernhard Riemann, translated into English by William Kingdon Clifford, "On the Hypotheses Which Lie at the Bases of Geometry," *Nature* 8, no. 183 (1873): 14–17, and no. 184, 36–37.

15. For detailed analyses of Riemann's 1854 and 1861 papers, and the work of followers including Christoffel, see Olivier Darrigol, "The Mystery of Riemann's Curvature," *Historia Mathematica* 42 (2015): 47–83. See also Farwell and Knee, "Missing Link." An English translation of Christoffel's 1869 paper is given in chap. 8 of Bas Fagginger Auer's "Christoffel Revisited" (master's thesis, Mathematical Institute, University of Utrecht, 2009).

16. If units are chosen so that $c = 1$, the coefficients are 1, 1, 1, −1. Riemann showed that if the metric has constant coefficients, the coordinates can be scaled so that all the coefficients in the metric are 1 (or −1, as Minkowski later showed).

CHAPTER 11

1. For Ricci's biographical details, including political context, I've drawn throughout this chapter on Judith Goodstein, *Einstein's Italian Mathematicians* (Providence, RI: American Mathematical Society, 2018).

2. Goodstein, *Einstein's Italian Mathematicians*, 2.

3. *Ricci to Antonio Manzoni*, November 24, 1872, quoted in Goodstein, *Einstein's Italian Mathematicians*, 6.

4. Goodstein, *Einstein's Italian Mathematicians*, 16.

5. *Ricci and Chisholm on Klein*: quoted in Goodstein, *Einstein's Italian Mathematicians*, 16, 17. NB: Sophie Kovalevsky's Göttingen doctorate in 1874 was "unofficial," like Chisholm's Cambridge degree.

6. Goodstein (*Einstein's Italian Mathematicians*, 27–30) uses Ricci and Bianca's letters to build a tender picture of their courtship.

7. *Ricci's introduction to his 1884 paper*: quoted in Goodstein, *Einstein's Italian Mathematicians*, 32.

8. W. H. and G. Chisholm Young, *Nature* 58, no. 1492 (June 2, 1898): 99–100.

9. *Indices on the Riemann tensor*: Roughly speaking, since this tensor is made up of second derivatives of the two-index metric components, its four indices

relate to which metric component is being differentiated by which pair of coordinates. It's a little more complicated than this because the Riemann tensor is made of sums of derivatives of the metric components, but this is the general idea.

10. Maxwell's process is "additive"—you add the light from the three filters and project the image onto a screen. It is used today in slides and in TV and digital images. Printed images use the "subtractive" method (discovered after Maxwell paved the way), where the three colors are reflected from the pigment on the paper rather than transmitted through the filters/layers of pixels to a screen. The three primary colors of the subtractive method are the "opposites" of Maxwell's—they are cyan, magenta, and yellow.

11. *On the theft of data in training models for AI*: See, e.g., Nick Vincent and Hanlin Li, "ChatGPT Stole Your Work. So What Are You Going to Do?," *Wired* (January 28, 2023), https://www.wired.com/story/chatgpt-generative-artificial-intelligence-regulation/. For writers fighting back, see, e.g., Vanessa Thorpe, "'ChatGPT Said I Did Not Exist': How Writers and Artists Are Fighting Back against AI," *The Guardian*, March 19, 2023.

12. *Benefits and problems with NLP including LLMs*: Much will have changed by the time this book goes to press, such is the pace of AI development, but here are some recent references. For another example of the benefits, see Samantha Spengler, "For Some Autistic People, ChatGPT Is a Lifeline," *Wired*, May 30, 2023; https://www.wired.com/story/for-some-autistic-people-chatgpt-is-a-lifeline/#. On the problems, in addition to the theft of training data, much has been written about the unreliability of some of ChatGPT's output—so although there are clear benefits, the jury is still out on its role in education: see, e.g., Hayden Horner, "ChatGPT: Brilliance or a Bother for Education," Engineering Institute of Technology's news website, March 13, 2023, https://www.eit.edu.au/chatgpt-brilliance-or-a-bother-for-education/. Similarly, in telehealth (and much else), there are advantages and disadvantages: see, e.g., Som Biswas, "Role of ChatGPT in Public Health," *Annals of Biomedical Engineering* (March 2023), published online at https://www.researchgate.net/profile/Som-Biswas-2/publication/369269117. There are obvious social problems with sophisticated AI, from enabling surveillance to creating deep fakes and fake news. I mentioned the problem of bias in the notes for chap. 4, but see also, e.g., Grace Browne, "AI Is Steeped in Big Tech's 'Digital Colonialism,'" *Wired UK* (May 25, 2023); https://www.wired.co.uk/article/abeba-birhane-ai-datasets. On a similar theme, see the

series of articles "AI Colonialism" by *MIT Technology Review*, https://www.technologyreview.com/supertopic/ai-colonialism-supertopic/, which also includes examples where oppressed peoples are fighting back by using AI in positive ways. Then there are environmental issues, e.g., Maanvi Singh, "As the AI Industry Booms, What Toll Will It Take on the Environment?," *The Guardian* (June 9, 2003). More than ever, informed public debate about science and technology is crucial!

13. There's much, much more to NLP and LLMs than tensor products, of course. For my account (and for more on NLP), I'm particularly indebted to Qiuyuan Huang, Paul Smolensky, Xiaodong He, Li Deng, Dapeng Wu, "Tensor Product Generation Networks for Deep NLP Modeling," *Proceedings of NAACL-HLT 2018* (New Orleans): 1263–73; Lipeng Ahang et al., "A Generalized Language Model in Tensor Space," 33rd Annual Conference of the AAAI (2019); and Matthew Kramer, "Word Embeddings," Medium.com, August 31, 2021.

14. "In principle, a quantum computer with 300 qubits could perform more calculations in an instant than there are atoms in the visible universe": Charles Q. Choi, "How Many Qubits Are Needed for Quantum Supremacy?" *IEEE News*, May 21, 2020; https://spectrum.ieee.org/qubit-supremacy#:~:text=Superposition%20lets%20one%20qubit%20perform,eight%20calculations%3B%20and%20so%20on.

15. *Enrico Betti*: Stokes's theorem in *n*-D: Victor Katz, "The History of Differential Forms from Clairaut to Poincaré," *Historia Mathematica* 8 (1981): 161–88, esp. 175. *Betti as soldier, contributor to journal*: Goodstein, *Einstein's Italian Mathematicians*, 7. *Betti and Ricci's papers*: Goodstein, *Einstein's Italian Mathematicians*, 148.

16. Ricci's letter, quoted in Goodstein, *Einstein's Italian Mathematicians*, 9–10.

17. *Ricci's long fight for promotion*: Goodstein, *Einstein's Italian Mathematicians*, 35–43, 59–61.

18. G. Ricci and T. Levi-Civita, "Méthodes de calcul différential absolu et leurs applications," *Mathematische Annalen* 54 (1900): 125–201; see 128 for effort and reward in learning a new skill (my translation).

19. For instance, if, following fig. 11.1, you form a second-order tensor T via the tensor (or outer) product of two *contravariant* vectors $\boldsymbol{a}$ and $\boldsymbol{b}$, you'll have the transformation rule

$$T^{\mu'\nu'} \equiv a^{\mu'} b^{\nu'} = (A^{\mu'}_{\sigma} a^{\sigma})(A^{\nu'}_{\lambda} a^{\lambda}) = A^{\mu'}_{\sigma} A^{\nu'}_{\lambda} a^{\sigma} a^{\lambda} \equiv A^{\mu'}_{\sigma} A^{\nu'}_{\lambda} T^{\sigma\lambda}.$$

Don't forget that the letters used for the tensors, transformation matrix coefficients, and indices here are arbitrary, like x in algebra. But what marvelous flexibility they offer, since you can move the transformation symbols around so easily, giving the rule for any kind of tensor you like!

20. *Unruh effect*: In 1976 the Canadian physicist William Unruh found, with the help of quantum theory, that the general theory of relativity predicts that temperature is not exactly the coordinate-independent scalar I said it was in fig. 11.1. Rather, an accelerating observer will measure a slightly different space-time temperature than a stationary observer will. This "Unruh effect" hasn't yet been detected—you'd need to be traveling close to the speed of light to detect one degree of temperature change. But in 2022 a University of Adelaide team led by James Quach invented a "quantum thermometer" that might very soon prove Unruh, and general relativity, right.

21. Note, though, that here we are talking about vectors in a single frame—unlike the rotation example above, this summation is not about transformations between frames.

22. *Proving invariance of the scalar product for coordinate transformations in n-D*: It's easiest to see this using the differential form of the matrix coefficients in the transformation equations. For example, the first of the 2-D rotation transformation equations is $x' = x\cos\theta + y\sin\theta$. Using partial derivatives, the differential form of this equation is

$$dx' = \frac{\partial x'}{\partial x}dx + \frac{\partial x'}{\partial y}dy,$$

and you can see that $\frac{\partial x'}{\partial x} = \cos\theta$, and so on for the other derivatives—so that these derivatives are just the components that I labeled $A_\sigma^{\mu'}$ in the narrative. For the scalar (or inner) product of the column and row vector you'd have (using the chain rule to get the last term):

$$u^{\mu'}v_{\mu'} = A_\sigma^{\mu'}A_{\mu'}^{\lambda}\,u^\sigma v_\lambda \equiv \frac{\partial x^{\mu'}}{\partial x^\sigma}\frac{\partial x^\lambda}{\partial x^{\mu'}}u^\sigma v_\lambda = \frac{\partial x^\lambda}{\partial x^\sigma}u^\sigma v_\lambda.$$

The repeated indices mean the right-hand side is

$$\frac{\partial x^\lambda}{\partial x^1}u^1 v_\lambda + \frac{\partial x^\lambda}{\partial x^2}u^2 v_\lambda + \ldots + \frac{\partial x^\lambda}{\partial x^n}u^n v_\lambda.$$

These derivatives are with respect to the *independent* coordinates, so the only derivative that makes sense is $\frac{\partial x^\lambda}{\partial x^\lambda} = 1$. It's analogous to school calculus, where we usually have just one independent variable, say x, and then $\frac{dx}{dx} = 1$. So the only possible value for σ on that right-hand side expression above is λ. Which means we have

$$u^{\mu'} v_{\mu'} = u^{\lambda} v_{\lambda}.$$

The expression is the same in the transformed coordinate system (with the dashes) as it is in the original coordinates. (It doesn't matter what letter I use for the repeated indices, because they are just place-holders telling you to sum. So I can swap μ for λ.)

In other words, the scalar product is invariant under this change of coordinates.

23. *Invariance of ds^2*: We saw earlier that the coordinate transformation matrices for contravariant and covariant tensors are inverses, $A^{\mu'}_{\sigma} \to A^{\sigma}_{\mu'}$ (or using derivative notation for the matrix components, $\frac{\partial x^{\mu'}}{\partial x^{\sigma}} \to \frac{\partial x^{\sigma}}{\partial x^{\mu'}}$). So the inverse pairs will "cancel," and we'll have

$$ds^2 = g_{\mu'\nu'}\, dx^{\mu'} dx^{\nu'} = A^{\sigma}_{\mu'} A^{\lambda}_{\nu'} A^{\mu'}_{\sigma} A^{\nu'}_{\lambda}\, g_{\sigma\lambda}\, dx^{\sigma} dx^{\lambda} = g_{\sigma\lambda}\, dx^{\sigma} dx^{\lambda}.$$

The distance measure ds^2 has the same form and the same value in each frame. (Remember it's the pattern of the indices that matters, not the choice of letters.)

24. *Beltrami on Ricci's tensors*: Goodstein, *Einstein's Italian Mathematicians*, 49.

25. Ricci and Levi-Civita, "Méthodes de calcul différential absolu," 128 (my translation).

CHAPTER 12

1. G. Ricci and T. Levi-Civita, "Méthodes de calcul différential absolu et leurs applications," *Mathematische Annalen* 54 (1900): 125–201.

2. R. H. Dicke ("The Eötvös Experiment," *Scientific American* 205, no. 6 (December 1961): 84–95), suggests it's unclear whether or not Einstein knew about Eötvös's result during his early thinking about gravity, but that Einstein would certainly have heard if the experiment had shown that Galileo's law was wrong. For the 2022 test: Pierre Touboul et al., "MICROSCOPE Mission: Final Results of the Test of the Equivalence Principle," *Physical Review Letters* 129 (2022): 121102.1–121102.8.

3. Urbain LeVerrier was the first to calculate this discrepancy; with updated measurements his method gave about 43 arc seconds.

4. Quoted in Abraham Pais, *Subtle Is the Lord* (Oxford: Oxford University Press, 1982), 178. Some translate "happiest" as "most fortunate."

5. Because the strength of Earth's gravity increases the closer the falling observer gets to the center of the Earth, there are measurable differences between the falling observer and the uniformly accelerating one. Similarly, ocean tides are caused because one side of the earth is closer to the moon and feels its pull more strongly.

6. *Einstein's appointment at Prague*: Banesh Hoffmann, *Einstein* (Frogmore: Paladin, 1975), 94. In 1882, the University of Prague, known as Charles University, had split into a Czech and a German part in the wake of Czech nationalism and ethnic disputes: https://cuni.cz/UKEN-298.html.

7. *Potential form of Newtonian gravity, from Newton's laws* $F = ma = \frac{GmM}{r^2} \Rightarrow a = \frac{GM}{r^2}$. In Cartesian coordinates, designate the components of the gravitational acceleration a by X, Y, Z, and note that the vector $\boldsymbol{a}$ is in the same direction as r, the distance between the two masses. Putting one mass, m, at the origin, then $\boldsymbol{r}$ is the position vector of the second mass, M, which is at the point (x, y, z). The horizontal component of $\boldsymbol{a}$ is found from $\boldsymbol{a} \cdot \boldsymbol{i} = a \cos\theta = \frac{ax}{r} = \frac{GMx}{r^3}$, and similarly for the other components. Differentiating these (using $r = \sqrt{x^2 + y^2 + z^2}$ and the chain rule), and adding, you get $\frac{\partial X}{\partial x} + \frac{\partial Y}{\partial y} + \frac{\partial Z}{\partial z} = 0$. Since acceleration is proportional to the (conservative) force, we can write its components in terms of a potential $V: X = \frac{\partial V}{\partial x}, Y = \frac{\partial V}{\partial y}, Z = \frac{\partial V}{\partial z}$, and so the above equation becomes Laplace's equation, $\frac{\partial^2 V}{\partial x^2} + \frac{\partial^2 V}{\partial y^2} + \frac{\partial^2 V}{\partial z^2} = 0$. If there is a continuous distribution of matter with density ρ, then Poisson's equation holds instead, and the right-hand side is $4\pi G\rho$.

8. *Charge and mass density caveats*: In electromagnetism there is a need to distinguish between charge density and point charges. In the case of gravity, Newtonian and Einsteinian, the distinction is between an average distribution of matter—across the solar system or in a nebula or galaxy—and a point source like a single star or planet. (Newton proved that spherical bodies act as if all their mass is concentrated at a point, the center of the body.) What this means is that the equations are singular at a point—that is, they don't work—but they're fine outside this point source, where they are called "vacuum equations." And they're fine for an average distribution of matter with density ρ. For more, see Peter Gabriel Bergmann, *Introduction to the Theory of Relativity* (New York: Dover, 1976), 175–77.

9. Einstein to Besso, quoted in Hanoch Gutfreund and Jürgen Renn, *The Road to Relativity: The History and Meaning of Einstein's "The Foundation of General Relativity"* (Princeton, NJ: Princeton University Press, 2015), 9. *"Serious mistakes"* quoted in Judith Goodstein, *Einstein's Italian Mathematicians* (Providence, RI: American Mathematical Society, 2018), 102–3.

10. Quoted in Goodstein, *Einstein's Italian Mathematicians*, 104. I'm also indebted to Goodstein for my summary of Abraham and Einstein's relationship.

11. Einstein's plea was recollected by his fellow ETH student and professor Louis Kollros; quoted in N. Straumann, "Einstein's 'Zürich Notebook' and His Journey to General Relativity," *Annals of Physics* (Berlin) 523, no. 6 (2011): 488–500, esp. 490.
12. In his 1916 paper (*The Foundation of the General Theory of Relativity*, 1916, English translation in H. A. Lorentz et al., *The Principle of Relativity* [New York: Dover, 1952], 113), Einstein defined the general principle of relativity this way: "The laws of physics must be of such a nature that they apply to systems of reference in any kind of motion." In other words, these laws must keep the same form for all observers (all frames of reference), and this means they must be expressed in tensor form.
13. Albert Einstein, *Ideas and Opinions* (1954; New York: Three Rivers Press, 1982), 309.
14. *Any coordinate transformations?* There were many confusing aspects that Einstein had to try to sort out. For instance, what about transformations that don't change the location of points, such as from Cartesian to polar coordinates? Einstein realized the problem, which is why he struggled with the idea of general covariance, as we'll see. See John D. Norton, "General Covariance and the Foundations of General Relativity: Eight Decades of Dispute," *Reports on Progress in Physics* 56 (1993): 791–858, esp. 833–34. Further, on a manifold these transformations are found for points, not for the whole space. For an analysis of this subtlety, see Norton, "General Covariance," 804; and John Earman and Clark Glymour, "Lost in the Tensors: Einstein's Struggle with Covariance Principles 1912–1916," *Studies in History and Philosophy of Science* 9 (1978): 4, 251–78, esp. 254.
15. Einstein, *Ideas and Opinions*, 288.
16. I used the word "minimizing," but technically I mean "extremizing" the integral, for the route is longest on time-like geodesics (because of time dilation as opposed to space contraction); but we don't need to worry about this here.
17. *"Caught fire"*: Einstein's recollection, quoted in Straumann, "Einstein's 'Zürich Notebook,'" 490.
18. *Is $F^{\mu\nu}$ a tensor*: yes, because (cf. chap. 11) it transforms like this:

$$F^{\mu'\nu'} = A^{\mu'}_{\sigma} A^{\nu'}_{\lambda} F^{\sigma\lambda},$$

where in Minkowski space-time the transformation matrices A represent the Lorentz transformations (LTs). Under LTs, however, where time and space coordinates are intertwined, vectors whose components are functions of space and time (such as velocity or the electric and magnetic field vectors) don't transform quite so simply as we saw in chapter 11.

19. Different authors prefer one name or the other, but fairness is restored because in differential geometry there's a dual tensor, denoted with a star as in the box, so both names get used.

20. This symmetry is warranted physically by considering an element of the matter and mathematically by considering what happens when you lower the indices: $g_{\mu\nu} T^{\mu}_{\sigma} = T_{\nu\sigma}$, and $g_{\nu\mu} T^{\mu}_{\sigma} = T_{\sigma\nu}$. But the left-hand sides of these equations are the same because $g_{\mu\nu} = g_{\nu\mu}$, which means $T_{\nu\sigma} = T_{\sigma\nu}$.

21. *Local laws*: Einstein's mass-energy conservation law is $T^{\mu\nu}{}_{;\nu} = 0$, but this is a local conservation law. The notion of global gravitational energy conservation is still especially problematic, but even local concepts such as the energy density of a gravitational field are hard to define physically. That's because $T^{\mu\nu}{}_{;\nu} = 0$ is a mathematical analogy; we'll see more in the next chapter.

On Earth, a "local" region must be small enough that there's no measurable curvature of the surface—otherwise the inverse-square law of gravity shows that gravitation varies from place to place, and Galileo's constant gravitational acceleration of 32 feet/sec/sec, which I used in fig. 12.1, no longer applies. In the solar system, a "local" region can be large, as long as the gravitational field from the sun and planets is roughly constant. Further afield, "local" can cover a huge area—perhaps half the distance between two stars. (I owe these estimates to Bertrand Russell's brilliant *The ABC of Relativity*, originally published in 1925, excerpted in *The World Treasury of Physics and Mathematics*, ed. Timothy Ferris [Boston: Little, Brown, 1991], 194–202.)

For detailed discussion on Einstein's struggles with energy conservation and covariance, see Straumann, "Einstein's 'Zürich Notebook'"; Earman and Glymour, "Lost in the Tensors"; and Galina Weinstein, "Why Did Einstein Reject the November Tensor in 1912–1913, Only to Come Back to It in November 1915?," *Studies in History and Philosophy of Modern Physics* 62 (2018): 98–122. The "November tensor" is the Ricci tensor.

For detailed discussion on Einstein's attempt to explain *Entwurf*'s lack of general covariance, and its philosophical significance even today, see John D. Norton, "The Hole Argument," *Stanford Encyclopedia of Philosophy*, online, updated 2019; https://plato.stanford.edu/entries/spacetime-holearg/.

22. *Grossmann on trouble with tensors*: quoted in Earman and Glymour, "Lost in the Tensors," 258–59. *Einstein's "heavy heart"*: quoted in Straumann, "Einstein's 'Zürich Notebook,'" 489.

23. *Einstein to Ehrenfest*, document 173, in *Collected Papers of Albert Einstein*, vol. 8, ed. Robert Schulman, A, J. Knox, Michel Janssen, and Jósef Illy; English

translation by Ann M. Hentschel (Princeton, NJ: Princeton University Press, 1998); available online thanks to the Press and the Einstein Papers Project, https://einsteinpapers.press.princeton.edu/vol8-trans/195. Einstein to Besso, quoted in Goodstein, *Einstein's Italian Mathematicians*, 105–6. See also Earman and Glymour, "Lost in the Tensors," 264ff., for discussion of why Einstein's colleagues rejected *Entwurf*.

24. Quoted in Earman and Glymour, "Lost in the Tensors," 260.

25. Stern quoted in Hanoch Gutfreund, "Otto Stern—with Einstein in Prague and in Zürich," Springer Link, June 20, 2021, open access, https://link.springer.com/chapter/10.1007/978-3-030-63963-1_6?error=cookies_not_supported&code=bb7fb68a-a41c-4c71-ace0-33a64d5f7756.

26. *Mileva's sad letter*: quoted in Roger Highfield and Paul Carter, *The Private Lives of Albert Einstein* (London: Faber and Faber, 1993), 128. Einstein's acrimonious demands on Mileva are painfully outlined in letters of July 1914, e.g., document 22, *Collected Papers of Albert Einstein*, vol. 8, https://einsteinpapers.press.princeton.edu/vol8-trans/60.

27. *Declaration to the Cultural World*: Constance Reid, *Hilbert* (Berlin: Springer-Verlag, 1970), 137–38.

28. Einstein to Levi-Civita, document 60, *Collected Papers of Albert Einstein*, vol. 8, https://einsteinpapers.press.princeton.edu/vol8-trans/99.

29. David E. Rowe, "Einstein Meets Hilbert: At the Crossroads of Physics and Mathematics," *Physics in Perspective* 3 (2001): 379–424, esp. 393–96.

30. As we saw in chapter 11, homogeneous coordinate transformations keep equations such as $\boldsymbol{a} \cdot \boldsymbol{b} = 0$ invariant. Similarly Einstein said, in his 1916 overview (*The Foundation of the General Theory of Relativity*, 1916, English translation in H. A. Lorentz et al., *The Principle of Relativity* [New York: Dover, 1952], 121), that "if a law of nature is expressed by equating all the components of a tensor to zero, it is generally covariant." For discussion, see Norton, "General Covariance," 833–34. Norton notes (834) that covariance of the metric is more restrictive than the usual transformations of tensor analysis.

31. Einstein's November 1915 letters to his family are in *Collected Papers of Albert Einstein*, vol. 8 (surrounded by letters to Hilbert!), e.g., documents 142–43, https://einsteinpapers.press.princeton.edu/vol8-trans/174, document 150, https://einsteinpapers.press.princeton.edu/vol8-trans/177. On November 5, 1915, Marić had indicated her own willingness for Einstein to see more of their boys, document 135, https://einsteinpapers.press.princeton.edu/vol8-trans/169. Einstein never managed a good relationship with his younger

son, Eduard, who was brilliant but highly sensitive, later developing schizophrenia. Einstein paid for his care, but the emotional burden fell on Marić.

32. *Einstein to his friend Ehrenfest*, quoted in Banesh Hoffmann, *Einstein* (Frogmore: Paladin, 1975), 125; *Einstein to Hilbert*, November 18, 1915, *Collected Papers of Albert Einstein*, vol. 8, document 148, https://einsteinpapers.press.princeton.edu/vol8-trans/176; *Einstein to Besso*, November 17, 1915, in *Collected Papers of Albert Einstein*, vol. 8, document 147, https://einsteinpapers.press.princeton.edu/vol8-trans/176. Note that Einstein did not yet have his full field equations, but he did have the correct vacuum equations, which is what he needed to calculate the geodesic path of Mercury; the result deviated from a Newtonian ellipse, and this gave him the discrepancy in the motion of the perihelion. For the derivation see general relativity textbooks such as Ray d'Inverno, *Introducing Einstein's Relativity* (Oxford: Clarendon Press, 1992), 195–98 (including 198 for comparison of observed values of perihelion precession with the values calculated using general relativity).

33. *No explicit equations in November 20 paper*: Leo Corry and his colleagues Jürgen Renn (Max Planck Institute for the History of Science) and John Stachel (Center for Einstein Studies, Boston University) compared the proofs with the published version in "Belated Decision in the Hilbert-Einstein Priority Dispute," *Science* 278 (1997): 1270–73. More references are listed below. *Hilbert November 20 proofs not generally covariant*: see, e.g., Vladimir P. Vizgin, "On the Discovery of the Gravitational Field Equations by Einstein and Hilbert: New Materials," *Physics-Uspekhi* 44, no. 12 (2001): 1289; Gutfreund and Renn, *Road to Relativity*, 33; Jürgen Renn and John Stachel, "Hilbert's Foundation of Physics: From a Theory of Everything to a Constituent of General Relativity," in *The Genesis of General Relativity*, ed. Jürgen Renn (Dordrecht: Springer, 2007), 4:858–59.

34. Following Mie, Hilbert used a very different approach from Einstein—an elegant Lagrangian (variational) approach. But Einstein, too, had already tried this approach, in his 1914 paper, which Hilbert had read. Einstein included his variational method for discussing the conservation laws in his 1916 overview paper as well, using a Hamiltonian rather than a Lagrangian. For discussion on the two approaches see Rowe, "Einstein Meets Hilbert," 414–15.

35. *Einstein-Hilbert priority, and on the different routes of Einstein and Hilbert*: Tilman Sauer, "Einstein Equations and Hilbert Action: What Is Missing on Page 8 of the Proofs for Hilbert's First Communication on the Foundations of Physics?," *Archives for History of Exact Sciences* 59 (2005): 577–90. Vizgin,

"On the Discovery of the Gravitational Field Equations," 1283–98. Leo Corry, Jürgen Renn, and John Stachel, "Belated Decision," 1270–73. F. Winterberg, "On 'Belated Decision in the Hilbert-Einstein Priority Dispute,' Published by L. Corry, J. Renn and J. Stachel," *Z. Naturforsch* 59a (2004): 715–19. John Earman and Clark Glymour, "Einstein and Hilbert: Two Months in the History of General Relativity," *Archives for History of Exact Sciences* 19 (1978): 291–308. Renn and Stachel, "Hilbert's Foundation of Physics," 4:857–973 (for Einstein's account of his derivation, see, e.g., his letters to Sommerfeld (document 153) and Ehrenfest (document 185) in *Collected Papers of Albert Einstein*, vol. 8). David E. Rowe, "Einstein Meets Hilbert," 379–424. Ivan T. Todorov, "Einstein and Hilbert: The Creation of General Relativity," preprint online at arXiv:physics/0504179v1, April 25, 2005. Galina Weinstein, "Did Einstein 'Nostrify' Hilbert's Final Form of the Field Equations?," online at *Physics ArXiv:1412.1816*, December 11, 2014. And more! Note that there is much misinformation online about the priority issue. For instance, V. A. Petrov ("Einstein, Hilbert and Equations of Gravitation," blog online at https://arxiv.org/pdf/gr-qc/0507136.pdf) claims Einstein couldn't have derived the correct advance of Mercury's perihelion when he said he did, because he didn't yet have the final equations; he implies Einstein must have seen Hilbert's work (on the trace term), but the perihelion equations need only the vacuum solution where there is no trace; Petrov also implies that Hilbert derived the Bianchi identities, but this obscures the fact that Hilbert did not adequately understand the role these identities play in conservation of energy (as I'll discuss in the next chapter). Petrov cites Winterberg (above), who cites C. J. Bjerknes (as Petrov does), but Bjerknes is not a credible scholar: see, e.g., John Stachel, "Anti-Einstein Sentiment Surfaces Again," *Physics World* 16, no. 4 (2003): 40.

36. Rowe discusses this well in "Einstein Meets Hilbert," 408; he does note (418) that Hilbert should have changed the submission date on the published paper, although this was standard practice at the time. I should add that today, papers are published with the original submission date *and* revision dates.

37. Renn and Stachel, "Hilbert's Foundation of Physics," shows in detail the ways that Hilbert modified his published paper after reading Einstein's and notes the ways he'd originally tried to present essential Einstein contributions as his own (see, e.g., 920–21).

38. Hilbert quoted in Reid, *Hilbert*, 142. Einstein's poignant reconciliation letter is quoted in, e.g., Vizgin, "On the Discovery of the Gravitational Field Equations," 1289.

39. We saw earlier that in Minkowski space-time the scalar product (and hence the divergence sum) has a sign change for the t-component. But the point here is that these are *definitions*, so just keep your eye on the pattern of the indices in the divergence terms.

40. This is because it is always possible to find a locally inertial ("free-falling") frame at a point, one in which special relativity holds (and the Christoffel symbols used in the covariant derivative are zero). By the rules of tensor analysis, the form of a tensor equation will be invariant in any frame—as long as you use (torsion-free) covariant rather than partial derivatives, to take account of the curvature of space-time. This rule also assumes we are replacing the flat Minkowski metric by the general curved metric.

41. In "nonrelativisitic" units, though, $k = 8\pi G/c^4$, where G is the proportionality constant in Newton's law of gravity; this highlights the fact that Einstein derived his equations by analogy with Newton's and ensured that they reduced to Newton's in weak gravitational fields such as Earth's.

42. These two equations are equivalent because contracting the indices on Einstein's original equation, and noting that $g^{\mu}_{\mu} = 4$ (by definition of $g_{\mu\nu}$ and $g^{\mu\nu}$ as inverses, as we saw in chap. 11), you find that $R^{\mu}_{\mu} = -kT^{\mu}_{\mu} \equiv R = -kT$. (Note that in his original equation Einstein actually had $-k$ as his coefficient, but since this is a constant, I've absorbed the minus sign into my definition of k, to fit with the way the equation is usually written today. Also, in his final equation Einstein wrote the left-hand side [Ricci tensor] out in full, in terms of Christoffel symbols, but he'd already defined this expression as $R_{\mu\nu}$.)

43. The difference between Einstein's earlier equation, $R_{\mu\nu} = kT_{\mu\nu}$, and these two final forms hinges on the scalar T or Hilbert's equivalent scalar R. (These scalars are called "traces.") Whether Einstein or Hilbert realized this first is at the heart of the "priority dispute," because it is only implicit in Hilbert's November 20 Lagrangian formulation. Since Einstein and Hilbert exchanged papers, they certainly influenced each other, but it is likely they each took this final step independently, for they followed different routes. Today, Hilbert's approach is widely used, and he is commemorated in the so-called Einstein-Hilbert action associated with the Lagrangian.

44. *Recent tests of Einstein's theory*: See, e.g., Pierre Touboul et al., "MICROSCOPE Mission: Final Results of the Test of the Equivalence Principle,"

Physical Review Letters 129, no. 21102 (September 14, 2022); Ignazio Ciufolini et al., "An Improved Test of the General Relativistic Effect of Frame-Dragging Using the LARES and LAGEOS Satellites," *European Physical Journal C* 79, article no. 872 (2019); Gemma Conroy, "Albert Einstein Was Right (Again): Astronomers Have Detected Light from Behind a Supermassive Black Hole," ABC News, July 29, 2021, https://www.abc.net.au/news/science/2021-07-29/albert-einstein-astronomers-detect-light-behind-black-hole/100333436; Geraint Lewis, "Astronomers See Ancient Galaxies Flickering in Slow Motion Due to Expanding Space" (a test of Einstein's predictions about time slowing down), *The Conversation*, July 4, 2023; Jet Propulsion Laboratory blog (August 24, 2022), "NASA Scientists Help Probe Dark Energy by Testing Gravity"—they found that Einstein's equations hold firm. (Pavel Kroupa, University of Bonn, disagrees, in "Dark Matter Doesn't Exist," *IAE News*, July 12, 2022. But a new study used general relativity's prediction of gravitational lensing to map dark matter: Robert Lea, "New Dark Matter Map Created with 'Cosmic Fossil' Shows Einstein Was Right (Again)," *Space* [April 18, 2023].) *On precession of black holes*: Brandon Specktor, "One of the Most Extreme Black Hole Collisions in the Universe Just Proved Einstein Right," *LiveScience*, October 13, 2022. *On frame dragging*: See, e.g., Charles Q. Choi, "Spacetime Is Swirling around a Dead Star, Proving Einstein Right Again," *Space.com*, January 31, 2020. And much more!

For a popular overview of the eclipse expeditions (and light bending around a black hole), see my article "A 'Revolution in Science' 100 Years Later," *Cosmos Magazine* 83 (2019): 29–35. Note it was Johann Soldner who had earlier used Newton's theory to predict light bending, getting half the general relativity result. Also note that accusations of bias against the leaders of the expedition arose in 1980, but they have been overturned and the 1919 results confirmed.

For a brief overview of gravitomagnetism, see my piece "The Amazing Concept of Gravito-electromagnetism," *Cosmos Magazine* 84 (September 2019): 61–63. An abridged version is at https://cosmosmagazine.com/science/introducing-the-amazing-concept-of-gravito-electromagnetism/.

My own research has been on the second analogy mentioned in the main article. See, e.g., C. B. G. McIntosh, R. Arianrhod, S. T. Wade, and C. Hoenselaers, "Electric and Magnetic Weyl Tensors: Classification and Analysis," *Classical and Quantum Gravity* 11 (1994): 1555–64; and R. Arianrhod, A. W-C. Lun, C. B. G. McIntosh, and Z. Perjés, "Magnetic Curvatures,"

Classical and Quantum Gravity 11 (1994): 2331–35. Einstein's gravitational constant is added to the equations for some applications, especially dark energy.

45. Einstein, *Ideas and Opinions*, 289–90.
46. Einstein's cover page was missing from English translations but was recently tracked down by Alicia Dickenstein, who gives Einstein's full acknowledgment in "About the Cover: A Hidden Praise of Mathematics," *Bulletin of the American Mathematical Society*, n.s., 46, no. 1 (January 2009): 125–29.
47. Levi-Civita quoted in Goodstein, *Einstein's Italian Mathematicians*, 151, 155.

CHAPTER 13

1. *Klein and Hilbert on Miss Noether*: quoted in Yvette Kosmann-Schwarzbach, translated by Bertram E. Schwarzbach, *The Noether Theorems: Invariance and Conservation Laws in the Twentieth Century* (New York: Springer, 2011), 45, 66.
2. In his 1993 paper "General Covariance and the Foundations of General Relativity," John Norton gave a fascinating account not just of Einstein's struggles but also of the way others responded to or reinterpreted the meaning of his principles of covariance, relativity, and equivalence. He illustrates this with the way textbook accounts evolved over the twentieth century and points out that there is still controversy or confusion. Of course, as I showed at the end of the previous chapter, physical observations so far have confirmed that the equations work brilliantly, no matter their foundations!
3. An English translation of Noether's paper is given in Kosmann-Schwarzbach, *Noether Theorems*, 3–22.
4. In a weak gravitational field like that on Earth, it turns out that the time component of the "momentum" is, indeed, the sum of the particle's rest mass, gravitational potential energy, and kinetic energy. See, e.g., Bernard Schutz, *A First Course in General Relativity* (Cambridge: Cambridge University Press, 1985), 190.
5. To see the essence of Noether's result, which she fully generalized in her 1918 theorems, let's go back to those translations we talked about. Taking all values of a, these translations from x to $x + a$ form a group (like the Lorentz transformations in fig. 9.3). Groups relating to invariance are called "symmetry groups." The "symmetry" here is the invariance expressed by the fact that V is independent of x. If the same symmetry applies in each direction, then

all the components of $\boldsymbol{p}$ are conserved and we have the familiar law of conservation of momentum. What Noether showed was that the symmetry groups for classical mechanics *and* for special relativity are finite, but the symmetry group in general relativity is infinite, because general relativity allows all possible coordinates and therefore all possible point transformations. This meant that the conservation law for the energy-momentum of the gravitational field is indeed different from the usual conservation laws in mechanics and special relativity. For technical details, see Kosmann-Schwarzbach, *Noether Theorems*, 56–64.

6. *On physical laws as divergence*: Peter Gabriel Bergmann, *Introduction to the Theory of Relativity* (New York: Dover, 1976), 194–97. *On divergence in Noether's theorems*: Kosmann-Schwarzbach, *Noether Theorems*, e.g., 6–10.

7. *On defining energy in general relativity*: Robert M. Wald, *General Relativity* (Chicago: University of Chicago Press, 1984), 84, 286, and for Noether's theorem, 457; S. W. Hawking and G. F. R. Ellis, *The Large-Scale Structure of Space-time* (1973; Cambridge: Cambridge University Press, 1991), 61–62, 73–74, 88–96. *On Noether's theorems*: Kosmann-Schwarzbach, *Noether Theorems*; for a simpler overview, see David E. Rowe, "On Emmy Noether's Role in the Relativity Revolution," *Mathematical Intelligencer* 41, no. 2 (2019): 65–72.

8. Einstein had insisted that any coordinates should be allowed for a truly general theory of relativity—his principle of covariance—but we saw that this allows for coordinates that represent the same point. This means that we can get different forms of the metric that actually represent the same spacetime—just as we saw for the equation of the circle in terms of Cartesian and polar coordinates. To mitigate this situation, the four energy-conservation equations (or the contracted Bianchi identities we'll meet in the next section) impose additional constraints on the choice of coordinates (see note 10).

9. In chapter 12, $T^{\mu\nu}$ and $T_{\mu\nu}$ had the same content because the indices were raised with the Minkowski metric. In general relativity, tensors pick up terms from the metric when the indices are raised, but because this happens on both sides of a tensor equation, the essential content of the equation is the same.

10. These four contracted Bianchi identities give additional information about the Ricci tensor, which means that of the ten Einstein equations, only six are independent. This allows the freedom to choose the four coordinates (the frame) arbitrarily, ensuring that every observer deduces the same laws of physics.

11. T. Levi-Civita (1917), translated by S. Antoci and A. Loinger, "On the Analystic Expression That Must Be Given to the Gravitational Tensor in Einstein's Theory," https://arxiv.org/pdf/physics/9906004.pdf. *On Hilbert's convoluted derivation via a variational approach*, see p. 59 of David E. Rowe, "Einstein's Gravitational Field Equations and the Bianchi Identities," *Mathematical Intelligencer* 24, no. 4 (2002): 57–66; Ivan T. Todorov, "Einstein and Hilbert: The Creation of General Relativity," arXiv:physics/0504179v1; David E. Rowe, "Emmy Noether on Energy Conservation in General Relativity," December 4, 2019, preprint online at https://arxiv.org/pdf/1912.03269.pdf, 21n32; Carlo Cattani and Michelangelo De Maria, "Conservation Laws and Gravitational Waves," in *The Attraction of Gravitation: New Studies in the History of General Relativity*, ed. John Earman, Michel Janssen, and John D. Norton (Boston: Birkhäuser, 1993), 67.
12. It would mean that the scalars R and T would be constant all throughout the universe. These scalars reflect curvature and mass-energy, so T should be different in a vacuum than amongst matter.
13. *Bianchi identities*: David E. Rowe, "Einstein's Gravitational Field Equations and the Bianchi Identities," *Mathematical Intelligencer* 24, no. 4 (2002): 57–66; *Struik and Schouten*: Rowe, "Einstein's Gravitational Field Equations," 66; Kosmann-Schwarzbach, *Noether Theorems*, 43.
14. *On Struik*: Here and in the following paragraphs I'm indebted to David E. Rowe, "Interview with Dirk Jan Struik," *Mathematical Intelligencer* 11, no. 1 (1989): 14–26. The interview also discusses how Struik's Marxist ideas led to his suffering under McCarthyism. Struik was evidently remarkable, and, equally remarkably, he lived to be 106 years old (he died in 2000).
15. Struik quoted in Rowe, "Interview with Struik," 19. Einstein quotes in Constance Reid, *Hilbert* (Berlin: Springer-Verlag, 1970), 142.
16. Quoted in Reid, *Hilbert*, 143.
17. *Einstein in support of Noether*: quoted in Kosmann-Schwarzbach, *Noether Theorems*, 72. *Weimar*: Rowe, "Noether's Role in Relativity," 66.
18. *Mother of modern algebra*: Norbert Schappacher and Cordula Tollmien, "Emmy Noether, Hermann Weyl, and the Göttingen Academy: A Marginal Note," *Historia Mathematica* 43 (2016): 194–97.
19. Struik quoted in Rowe, "Interview with Struik," 16.
20. *Struik*: Rowe, "Interview with Struik," 17. *Hodge*: quoted in Judith Goodstein, *Einstein's Italian Mathematicians* (Providence, RI: American Mathematical Society, 2018), 165.

21. *1928 congress and Hilbert's stirring speech*: Reid, *Hilbert*, 188.
22. See, e.g., the Australian Mathematical Society's *Gazette* 49, nos. 1 (Ole Warnaar's President's column) and 2 (Letters, from Aerwm Pulemotov).
23. Einstein and Cartan quotes are from the letters of December 8, 1929, and February 17, 1930, respectively, in *Elie Cartan–Albert Einstein: Letters on Absolute Parallelism 1929–1930*, ed. Robert Debever (Princeton, NJ: Princeton University Press, 1979).
24. *Einstein to Cartan*: in *Elie Cartan-Albert Einstein*, ed. Debever, 203; *Einstein to Mrs. Grossmann*: in Banesh Hoffmann, *Einstein* (Frogmore: Paladin, 1975, 36.

EPILOGUE

1. Particle colliders enabled physicists to build the Standard Model describing matter and its interaction with the various forces, and CERN points out that there have also been social benefits as well as scientific ones: https://home.cern/news/news/cern/society-benefits-investing-particle-physics. Former particle physicist Sabine Hossenfelder, however, is one who is skeptical about the scientific benefits; see, e.g., her blog post at https://backreaction.blogspot.com/2022/04/did-w-boson-just-break-standard-model.html. Similarly, this post on *Quora* suggests negative results in the LHC are vital for science, but in the sense of ruling out popular theories: https://www.quora.com/Was-building-the-Large-Hadron-Collider-worth-it-Did-they-discover-anything-from-it. Also see, e.g., Nick Scott, "CERN's Grand Ambitions: Are Particle Accelerators Worth It?" *Varsity* (January 26, 2021), https://www.varsity.co.uk/science/20486; Tom Hartsfield, "Please, Don't Build Another Large Hadron Collider," *Big Think* (June 6, 2022), https://bigthink.com/hard-science/large-hadron-collider-economics/.
2. In effect, Dirac used the *square* of $E = mc^2$, which yielded positive and negative solutions: the positive one is Einstein's equation for ordinary matter; the negative one, $E = -mc^2$, refers to antimatter. For a brief introduction, see the excerpt from Dirac's 1933 Nobel Prize address in *The World Treasury of Physics, Astronomy and Mathematics*, ed. Timothy Ferris (Boston: Little, Brown, 1993), 80–85. For a modern analysis, see Luciano Maiani and Omar Benhar, *Relativistic Quantum Mechanics* (Boca Raton, FL: CRC Press, 2016), 113–16.
3. See Bertha Swirles, "The Relativistic Interaction of Two Electrons in the Self-Consistent Field Method," *Proceedings of the Royal Society A* 157, no. 892 (December 2, 1936): 680–96.

4. *Tensors in computational math, esp. NLA*: See, e.g., Lek-Heng Lim, "Tensors in Computation," *Acta Numerica* (2021): 555–764, DOI:10.1017/S09622492921000076.

5. Einstein to Heinrich Zangger, document 152, *Collected Papers of Albert Einstein*, https://einsteinpapers.press.princeton.edu/vol8-trans/179. This letter also shows Einstein's bitterness at his ex-wife for apparently holding up his reconciliation with his son Hans Albert.

INDEX

Page numbers in italics refer to figures.